04T

Foundations of Engineering Mechanics

Ginevsky · Vlasov · Karavosov

Springer

Berlin
Heidelberg
New York
Hong Kong
London
Milan
Paris
Tokyo

Engineering ONLINE LIBRARY

springeronline.com

A.S. Ginevsky · Y.V. Vlasov · R.K. Karavosov

Acoustic Control of Turbulent Jets

With 212 Figures

Springer

Series Editors:
V. I. Babitsky
Department of Mechanical Engineering
Loughborough University
Loughborough LE11 3TU, Leicestershire
United Kingdom

J. Wittenburg
Institut für Technische Mechanik
Universität Karlsruhe (TH)
Kaiserstraße 12
76128 Karlsruhe
Germany

Authors:
A. S. Ginevsky
Ye. V. Vlasov
R. K. Karavosov
Central Aerohydrodynamics
Institute (TsAGI)
Radio St. 17
107005 Moscow
Russia

ISBN 3-540-20143-2 Springer-Verlag Berlin Heidelberg New York

Cataloging-in-Publication Data applied for
Bibliographic information published by Die Deutsche Bibliothek.
Die Deutsche Bibliothek lists this publication in the Deutsche Nationalbibliografie;
detailed bibliographic data is available in the Internet at <http://dnb.ddb.de>.

Springer-Verlag is a part of Springer Science+Business Media

springeronline.com

Typesetting: Camera-ready by authors
Cover-design: de'blik, Berlin
Printed on acid-free paper 62 / 3020 hu - 5 4 3 2 1 0

PREFACE

This book is dedicated to the problem of the influence of acoustical oscillations on turbulent jets. We review in the book long-standing experimental studies of the authors, as well as of other researchers, Russian and foreign, on the development of acoustical methods for control of aerodynamic and acoustical parameters of subsonic and supersonic gas jets.

It is worth noting that the sensitivity of the laminar jet flow to the sound action has been known for 140 years. This phenomenon was detected at chamber music concerts when a doctor (LeConte, 1858), who was among the observers, had noticed that the candle flame oscillated in time with the violoncello sound so that "a deaf person could hear the harmony." However, researchers soon after that showed (Tyndall, 1867) that the laminar jet becomes sensitive to sound in the absence of burning. Lord Raleigh (1886) explained this effect on the basis of his studies of the mixing layer hydrodynamic instability.

Vlasov and Ginevsky found in 1967 that the slight acoustical excitation of a subsonic turbulent jet permitted controlling turbulent mixture over wide limits. It turned out that two effects could be realized according to the acoustical effect frequency: the turbulent mixture intensification in the jet for the low-frequency acoustical effects and the turbulent mixture decay for the high-frequency acoustical excitation with the frequency one order greater than the low-frequency exposure. The State Register of the USSR Discoveries registered discovery #212 "The phenomenon of the turbulence acoustical attenuation in subsonic jets" with priority of March 31, 1967 (the authors are Vlasov and Ginevsky).

The cited papers initiated an unusual boom in studies of the problem of turbulent jet and mixing layer control through acoustical oscillations. Numerous papers of Russian and foreign researchers appeared, which investigated both effects. Let us focus on those that appeared earlier. Both effects, jet turbulence intensification and suppression, were confirmed in the studies of Isataev and Tarasov (1971), as well as of Rockwell (1972). Furletov (1969) and Crow and Champagne (1971) studied the mixing process intensification in a jet under its low-frequency acoustical excitation. Petersen, Kaplan and Laufer (1974) stated the jet turbulence suppression under high-frequency acoustical exposure. Later, this effect was studied in detail in the paper of Zaman and Hussain (1980).

The following studies of many authors showed that the effects of changing aerodynamic parameters of turbulent jets under their acoustical excitation are followed by the corresponding changes in the acoustical characteristics: the jet broad-band noise is amplified under low-frequency excitation (Bechert, Pfizenmaier, 1976, 1977; Moore, 1977) and is attenuated under high-frequency excitation (Moore and Kibens, 1980).

There were fundamental changes in the insight into the physics of turbulence in free jets, mixing layers, wakes, and near-wall flows in the last 3–4 decades, due to the discovery of large-scale coherent structures. Coherent structures are large-scale periodic vortex clusters that are generated as a result of mixing layer instabilities. They evolve and interact with each other against the background of small-scale turbulence. These structures have scales commensurable with the mixing layer cross section size and are characterized by a rather large life period. The interest in coherent structure studies was the result of their important role in turbulent mixing, burning and aerodynamic noise generation. The most important aspect of the coherent structure existence in jet flows is the ability to control turbulence by means of direct action on the structures.

Another important achievement in turbulent studies over this period was also the development of adequate numerical methods for mathematic simulation of unsteady 3-D turbulent flows with appropriate account of coherent structures.

In the present book the study of acoustical methods of control for turbulent jet flows is tied up with the study of coherent structures and illustrated by examples of their mathematical simulation. The examples of practical implementation of acoustical control for turbulent jet flows are also considered. The authors' results are published in many papers in Russian and foreign journals, they were also presented at numerous Russian and foreign conferences.

The book is intended for researchers, post-graduate students, senior students of physics specialties and engineers dealing with studies and with practical implementation of acoustical methods of control for aerodynamic and acoustical characteristics of turbulent jet flows.

The book reviews the studies of turbulent jet acoustical excitation performed by the authors, their followers and collaborators.

The book also reflects upon the achievements of Russian and foreign researchers. The vast bibliography of experimental and theoretical research in periodical and primarily acoustical excitation of turbulent jet flows, as well as of research in coherent structures in the initial region of jets, is included.

In Chapter 1 the main laws of subsonic turbulent jet propagation are considered and the important role of coherent structures in turbulent mixing processes is elucidated. The main characteristics of coherent structures [frequency, convection rate, interaction processes (the pairing)], as well as their origin, evolution, and breakdown at the end of the initial region are also considered. The ideas of global mechanism of feedback within the initial region resulting from the pairing of coherent structures are briefly described. The relation between coherent structures and jet flow hydrodynamic instability is considered. The problem of generation of subsonic turbulent jet aerodynamic noise and the role of coherent structures in forming near and far acoustical fields of jets are illuminated. The existing conceptions of position of main noise sources within the jet initial region are presented.

Aerodynamic, acoustical and geometric parameters characterizing the initial conditions of issue and determining jet spread laws and receptivity to different periodical excitations are indicated. The classification of acoustical methods of turbulent jet (passive and active) control is provided.

The second chapter is devoted to the experimental studies of the action of weak acoustical harmonic disturbances on aerodynamic characteristics of turbulent jets. The intensification of mixing (turbulence generation) for low-frequency excitation and attenuation of mixing (turbulence suppression) for high-frequency excitation are demonstrated. The study results on influence of acoustical excitation level, flow regime in the initial boundary layer at the nozzle exit section and initial turbulence on realization of both effects (intensification and attenuation) of mixing are presented.

Here we show that for low-frequency excitation, the increase in its level enhances the effect. After that, the saturation begins and the following increase in its level does not cause the intensification of mixing. For high-frequency excitation the level increase at first enhances the turbulence suppression effect and then the effect is attenuated with the excitation level increase. Here one can see the tendency to affect sign changing, i.e. intensification of mixing instead of its attenuation. Then the results of the experimental study of the circular jet cross-section deformation for low-frequency transverse acoustical excitation are presented – the jet sections become oval and extended in the irradiation direction. The influence of acoustical disturbances on the circular turbulent jet issuing from an orifice with a sharp edge is considered also.

The jet acoustical excitation is accompanied by the generation of weak vibration disturbances. A specific investigation has shown that these vibrations do not in themselves have a pronounced effect on jet aerodynamic characteristics. However, generation of intensive vibration disturbances gives the possibility to realize both effects that are present under acoustical excitation: intensification and attenuation of mixing.

The results of acoustical excitation of high-speed (near-sonic) turbulent jets for very high levels of harmonic excitation are presented. Here also the above-mentioned effect of mixing intensification is evident for low-frequency excitation. The mixing attenuation effect for high frequencies is not achieved. The consequences of acoustical excitation of turbulent jets, i.e. the changes in mode composition of turbulent pulsations and positions of coherent structure pairing are considered. The mechanisms of jet acoustical excitation are analyzed.

In closing, the influence of coflowing stream on the mixing intensification for low-frequency acoustical excitation of jets, as well as acoustical excitation of non-isothermal submerged gas jets, is considered. Different approaches to increased efficiency of turbulent jet acoustical excitation are presented: transverse acoustical excitation of a jet by two emitters of equal frequency in phase and antiphase located on either side of the jet, multi-frequency acoustical excitation of jets at the main frequency and its subharmonics for specially selected phase shift, excitation of jets by sound of highest azimuth modes, and acoustical excitation of jets for deterioration of acting signal harmonicity.

The third chapter contains experimental results of studies of the influence of acoustical disturbance on the generation of the aerodynamic noise of subsonic jets. The influence of issue initial conditions, mode composition of acoustical excitation and coherent structures on the noise of isothermal and nonisothermal turbulent jets in the near and far fields is considered.

The fourth chapter is devoted to the analysis of the action of intensive low-frequency acoustical disturbances on subsonic jets. Among other things, the coalescence of ring vortices in pairs does not exist within the initial region.

The fifth chapter is dedicated to the analysis of self-excitation of subsonic turbulent jet flows. In particular, generation of the flow self-sustained oscillations for normal and skewed impinging of near-sonic turbulent jets on a baffle, as well as in low-speed wind tunnels with the open test section, is considered. The well-known approaches to suppression of the self-sustained oscillations are indicated.

In the sixth chapter the modern methods of subsonic turbulent jet numerical simulation are considered for a jet under periodical excitation. Particular emphasis is placed on the ability of the methods to describe laws emerged in experiments (the mixing intensification and attenuation under low- and high-frequency excitation, the effect of saturation as the excitation level increases under low-frequency excitation, and the action sign changing as the excitation level increases under high-frequency excitation).

The seventh chapter contains the brief description of aerodynamic and acoustical characteristics of supersonic anisobaric turbulent jets, processes of broadband noise generation and its discrete components, as well as of control methods for such jets (active for acoustical excitation and passive for irradiation of jets by their own noise under the jet screening).

The eighth chapter is devoted to the study of approaches to attenuation of turbojet engine jet noise. The jet system for attenuation of noise of aircraft engine near-sonic propulsive jets is considered. The system consists of the main central jet and surrounding small jets with high-frequency noise acting on the main jet. The efficiency of this method of jet noise attenuation, which is as much as 2–3 dB in the far field and 4–5 dB in the near field, is demonstrated through an example of simulation and experiments. Methods of attenuation of turbojet engine excess noise resulting from aeroacoustical interaction leading to excitation of the propulsive jet by low-frequency noise of the combustion chamber, compressor and turbine are considered on the basis of analysis of Russian and foreign experiments. It is shown that a similar jet system provides a way of perceptible attenuation of the broadband noise of supersonic anisobaric jets (by 5–6 dB) and above all of suppression of its discrete component (15–20 dB) or of its shift into the high-frequency area.

The acoustical approaches to control of self-sustained oscillations in wind tunnels with the open test section are presented in the ninth chapter. The self-oscillation suppression by high-frequency excitation of the mixing layer and the self-oscillation generation by low-frequency acoustical excitation, as well as the self-oscillation suppression by anti-noise, are considered.

The final, tenth chapter is dedicated to interaction of the mixing layer and the cavity. Here, as in the fifth chapter, all effects are determined by acoustical feedback. A number of practical applications connected with attenuation of pressure pulsations in cavities with separated flows, in the test sections of near-sonic wind tunnels with perforated walls of the closed test section, and in blind branch pieces are considered.

The limited volume of the book obliged the authors to exclude of some important parts. This is especially true for the problem of acoustical control of turbulent jets with chemical reactions and combustion, of turbulent wakes and their coherent structures. Also near-wall turbulent jets, annular and plane jets are not considered. Some of these problems are covered in the review by Vlasov and Ginevsky (1986).

The authors enjoyed the support and encouragement of Professors G.P. Svishchiov and V.V. Struminsky at different stages of the studies. We are glad to express our gratitude to Professor Munin for his attention to our work.

The studies were performed under financial support of the Russian Fund for Basic Studies (Grants 93-013-17713, 94-01-01135, 96-01-00037, 96-02-19577, 99-01-00131, 00-01-00152) in 1993–2000.

T.S. Petrovskaya, G.A. Vishnevsky and M.V. Karadji provided a great help in the preparation of the electronic version of the book. The authors express their sincere gratitude to them.

In preparation for the English edition, its content was slightly expanded by the presentation of new results obtained by the authors, as well as of other results (cf. Chapters 2, 5, 6, and 7). Moreover, the misprints detected in the Russian edition have been corrected.

The translation into English was performed by Professor A.S. Belotserkovsky.

Contents

CHAPTER 1 SUBSONIC TURBULENT JETS

1.1. Aerodynamic Characteristics of Turbulent Jets. Coherent Structures

Consider subsonic turbulent jets of incompressible fluid/gas. At the present time, the main laws of propagation of such jets are well studied in theory and experiment. This is true for mixing layers, plane, axisymmetric, and 3-D submerged jets, as well as for jets in accompanying flow. The jet partitioning into three sections (Fig. 1.1), initial, transition and main sections [1.1, 1.13], is generally recognized.

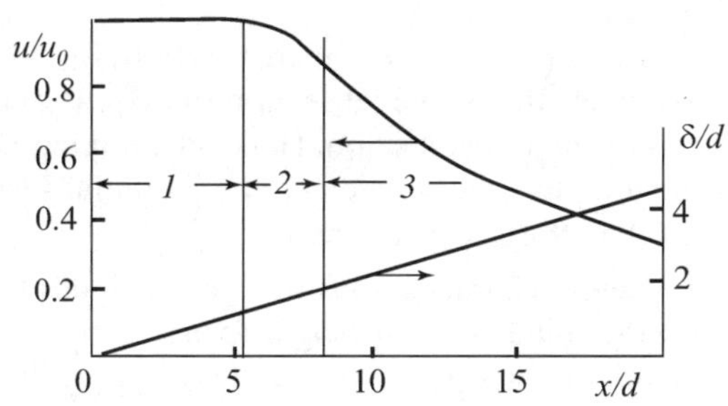

Fig. 1.1. Change of velocity at the jet axis u/u_0 and mixing layer thickness δ/d along a submerged turbulent jet. The initial (*1*), transition (*2*) and main (*3*) regions.

Theoretical studies of hydrodynamic stability of laminar jets show strong instability resulting from the specific shape of the velocity profile with a point of inflection. Experimental studies demonstrate also fast transition to turbulence for rather small values of the Reynolds numbers. Normally, a jet may be considered as turbulent for Reynolds number $\mathrm{Re} = u_0 d / \nu \geq 10^4$. This relation is true in most practical applications. The influence of the Mach number $\mathrm{M} = u_0 / a$ (a is the sound velocity) for subsonic speeds is not very significant. In the absence of the outer accompanying flow ($u_\infty = 0$) we have a submerged jet and $m = u_\infty / u_0 = 0$, for $m < 1$ and $m > 1$ we have a jet in an accompanying flow. In the case of isobaric flow when the longitudinal gradient of pressure or velocity in the accompanying flow is absent $u_\infty = \text{constant}$, the changes of velocity, *I*, temperature, *H*, and admixture concentration, J, in a jet satisfy the following conditions of constant excess impulse, excess heat content and excess admixture content:

$$I = \int_F \rho u(u - u_\infty)dF = \text{constant},$$
$$Q = \int_F \rho u(H - H_\infty)dF = \text{constant}, \tag{1.1}$$
$$J = \int_F \rho u(z - z_\infty)dF = \text{constant}.$$

Here ρ is the gas density, H is the total heat content, and z is the substance mass concentration. Hence the jet propagation along the whole extent is determined by the distribution of the flow main parameters in the jet initial section. The diagram of the initial, transition and main sections of a submerged jet is shown in Fig. 1.1.

At small distances from the nozzle ($x/d \le 1-1.5$) the thin mixing layer in the initial section of a submerged jet is characterized by the unique scale of length, i.e. by the initial thickness of the boundary layer, δ_0, or more definitely, by the initial thickness of momentum loss at the nozzle exit section, $\theta_0 = \int_0^\delta \frac{\rho u}{\rho_0 u_0}(1 - u/u_0)dy$.

The mixing layer thickness progressively increases with distance from the nozzle along the flow and occupies the whole jet section (its size is of order of the nozzle diameter, d) at the end of the initial section. Hence the mixing layer characteristics are determined by the unique length scale and the jet initial section characteristics are determined by two length scales: θ_0 and D.

The notion 'aerodynamic characteristics of a jet' includes the laws of changes in the jet parameters: the initial section length, cross section sizes, velocity, temperature, intensity of velocity and temperature pulsations, turbulence scales, spectra, etc.

Turbulence control, as applied to jet flows, commonly approaches the aerodynamic characteristic change: the stirring intensification or attenuation, which reduce mainly the decrease or increase in the jet range, increase of the relative role of small-scale turbulence, etc.

The well-known phenomenon of intermittence in turbulent jets, i.e. the large-scale sinuousness of their instantaneous boundaries (Fig. 1.2), bears witness to the formation of large-scale periodical vortices in jets. It became evident after the discovery of what is called coherent structures: large-scale periodical vortex clusters (vorticity blobs) developing and interacting with each other against a background of small-scale turbulence. The sizes of these structures are commensurable with the mixing layer cross section sizes and characterized by a rather large life period. Their studies have established fundamental features of turbulent shift flows and clarified the revealed effects. Comprehensive reviews of coherent structure studies are contained in [1.10, 1.11, 1.17, 1.26, 1.34, 1.38, 1.41].

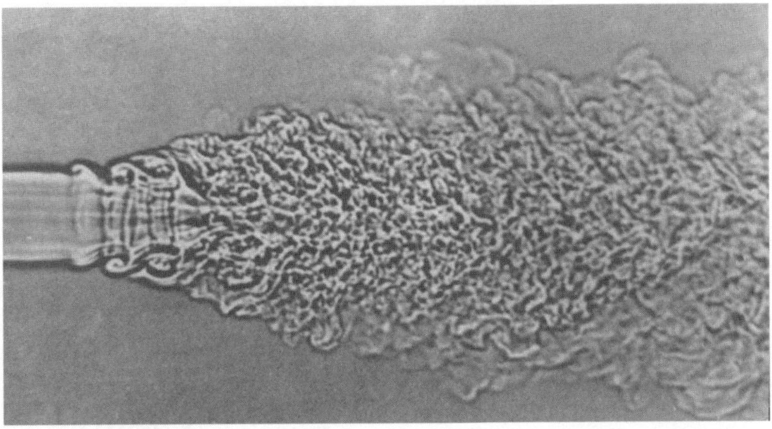

Fig. 1.2. The sinuousness of the instantaneous boundary of a submerged circular turbulent jet.

The availability of relatively long-life vortex clusters interacting with each other introduces deterministic aspects into chaotic, as it was thought before, turbulence and allows modeling of free turbulence laws in a new context.

Most results in the study of coherent structures are obtained by investigating the plane mixing layer and the initial section of a circular jet. There, large 2-D structures were found in the form of overturned waves in visual studies of the mixing layer [1.35]. It was elucidated that the mixing layer growth results from the pairing of adjacent vortices. Each of these pairings leads to drawing unswirled fluid into the mixing layer and to corresponding growth of the mixing layer. However, other experiments show that in some cases (this depends on the issue initial conditions) the vortex pairing is not the only possible mechanism determining the mixing layer growth and ejection. In these cases ejection happens mainly in the process of the development of single vortices and not of their merging in pairs. There were also found 3-D effects in studies of interaction of 2-D vortices in the mixing layer. In addition to transverse periodical vortex structures in the plane mixing layer, longitudinal vortex structures (pair vortices with opposite rotation) are generated. The interaction of longitudinal and transverse vortices leads not only to the production of larger vortices but also to the generation of small-scale turbulence. This phenomenon manifests itself especially dramatically when analyzing development and interaction of large-scale vortices in the initial section of a circular jet. Here thin vortex rings are generated near the nozzle and their pairing leads to the formation of larger annular vortices, which subsequently lose their azimuthal uniformity and become 'star-shaped'.

Further development of these 3-D structures reduces to their complete or partial pairing. After that their decomposition to erratic balls occurs in the end of the initial section or in the transition section of the jet. Fig. 1.3, *a* demonstrates evolution of coherent structures in the circular jet initial section [1.50] and Fig. 1.3, *b* demonstrates coalescence of two annular vortices in the initial section [1.4].

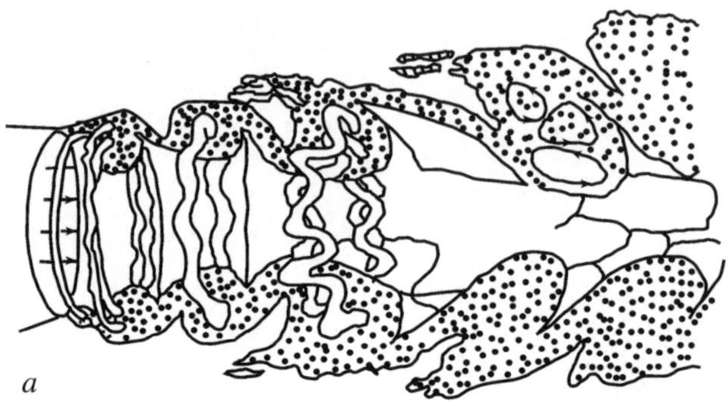

a

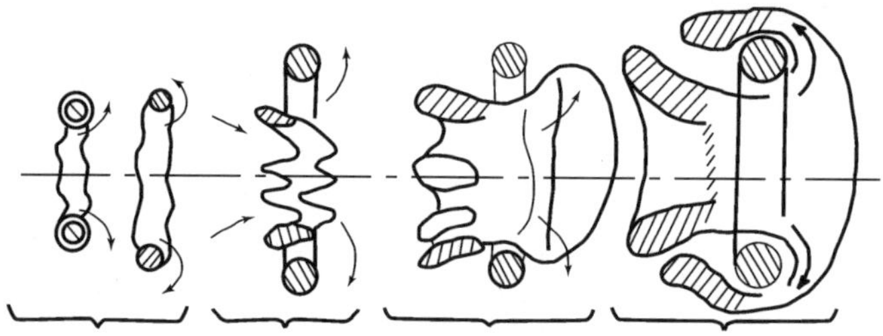

b

Fig. 1.3. Evolution of coherent structures in the initial region of a circular turbulent jet (*a*). Interaction of ring vortices in pairs (*b*).

In order to have quantitative characteristics of coherent structures in jet flows one uses, as a rule, experimentally measured oscillograms of pulsation and pressure, their spectra, autocorrelations, as well as spatial and spatial-time correlations including correlations in narrow frequency bands. The measurements of velocity and pressure pulsation spectra in one-third octave bands on the axis of a circular submerged jet demonstrated (Fig. 1.4, *a*) that, as the distance from the nozzle increases, the characteristic frequency, i.e. the Strouhal number, $\mathrm{St} = fd/u_0$ (f is the frequency corresponding to the spectrum maximum), progressively decreases [1.5]. The number St decreases downstream from 3 to 1 near the nozzle ($x/d < 1$) and to 0.3 – 0.5 at the end of the initial region for $x/d = 1 - 6$. The Strouhal numbers ($\mathrm{St} = fd/u_0$ along the jet axis and $\mathrm{St} = d/u_0T$ within the initial section) are presented in Fig. 1.4 [1.33]. The characteristic frequency of the velocity longitudinal pulsations at the jet axis, f, is determined using the spectrum maximum and the characteristic time period, T, using the measurements of probability density for time intervals between two subsequent zeros of the pressure pulsations with the positive derivative with respect to time. Experimental points in accordance with the data of spectral and correlation measurements [1.4, 1.8] are plotted also. The smooth, and not stepwise, shape of the curve St(x/d) testifies that the points of generation, pairing and disruption of the coherent structures

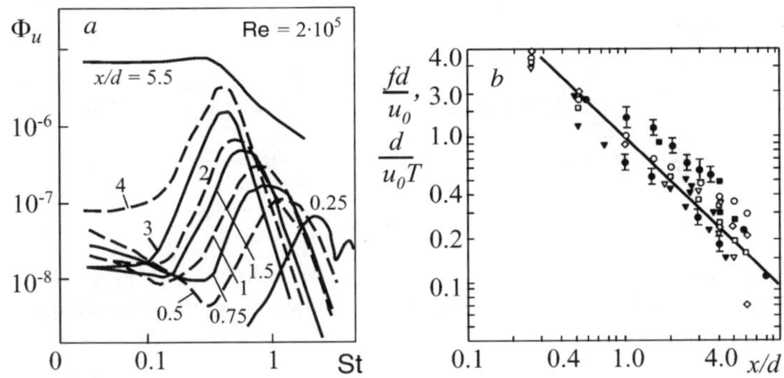

Fig. 1.4. *(a)* Spectra of the velocity pulsations in the potential core of the jet initial region for $x/d = 0.25 - 5.5$. *(b)* The Strouhal numbers St over the initial region of a circular turbulent jet in accordance with the data of spectral and correlation measurements.

are subject to statistical scatter.

The distinct notion of the flow periodical structure in the jet initial section results from measurements of spatial and spatial-time correlations of velocity pulsations in narrow frequency bands. As an example we present [1.14, 1.20] the coefficients of spatial correlation for velocity pulsations in the overall frequency band along the nozzle edge and along the jet axis in its initial section (Fig. 1.5), as well as the coefficients of spatial correlation in narrow frequency bands (Fig. 1.6, *a*).

Two things draw researchers' attention. In Fig. 1.5, the correlation coefficient in the mixing layer near the nozzle passes through zero four times before changing the correlation sign. The distance between two zeros of the coefficient R_{uu} characterizes the longitudinal scale of periodical vortices. In Fig. 1.6, the coefficients R_{uu} in narrow frequency bands (filters with the constant bandwidth f = 10 Hz) are presented. From Fig. 1.6, *a*, it follows that the flow periodicity for St = 0.48 manifests itself to much greater extent than for smaller (St = 0.20) and greater (St = 1.20) frequencies for the initial levels of turbulence ε_0 = 0.5, 5 and 10%. The changes in maximum values of spatial-time correlation of the velocity on the jet axis in narrow frequency bands reinforce this fact.

On the basis of visual studies and measurements of spatial-time correlations of the velocity pulsations in the mixing layer, the average convective velocity of vortex structure transport is determined as $u_c / u_0 = 0.5 - 0.7$.

To what degree do the above relationships depend on the Mach number of the issue? This was investigated in several experimental works, among them [1.44].

The coherence function [1.32], the normalized module of the mutual spectrum of the pressure pulsations in two points, and the corresponding phase of the mutual spectrum (it was used for evaluation of the convection velocity for different frequencies in the longitudinal direction), was determined in near and far acoustical fields.

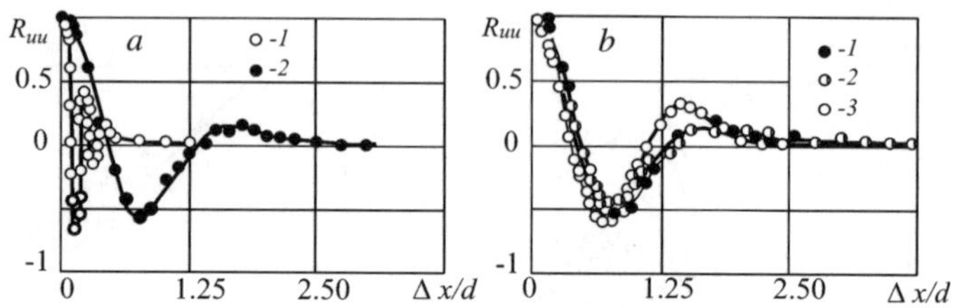

Fig. 1.5. Coefficients of spatial correlation for velocity pulsations in the overall frequency band of longitudinal velocity pulsations along the nozzle edge and along the jet axis. (*a*) $\mathrm{Re} = 2.4 \cdot 10^4$: *1)* $x/d = 0.5,\ y/d = 0.5$, *2)* $x/d = 3,\ y/d = 0$, (*b*) $x/d = 3,\ y/d = 0$: *1)* $\mathrm{Re} = 24 \cdot 10^4$, *2)* $\mathrm{Re} = 1.1 \cdot 10^5$, *3)* $\mathrm{Re} = 3.0 \cdot 10^5$.

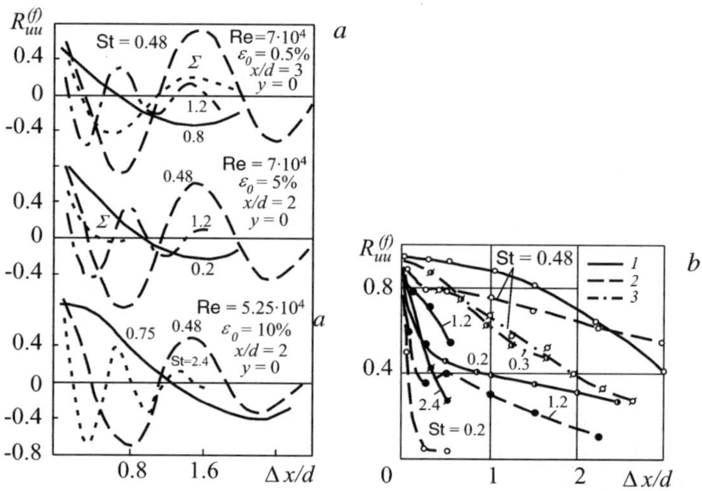

Fig. 1.6. (*a*) The spatial correlation R_{uu} in narrow frequency bands along the jet axis for different values of the initial turbulence $\varepsilon_0 = 0.5\%$, $\varepsilon_0 = 5\%$ and $\varepsilon_0 = 10\%$ (the mark Σ corresponds to the overall frequency band). (*b*) The maximum coefficients of the spatial-time correlation of the velocity pulsations at the jet axis in narrow frequency bands: *1)* $\varepsilon_0 = 0.5\%$, $x/d = 3$; *2)* $\varepsilon_0 = 5\%$, $x/d = 2$; *3)* $\varepsilon_0 = 10\%$, $x/d = 2$.

The coherence function, γ, at the given frequency, *f*, for measurement of the pressure pulsations at two points A and B is determined by the expression

$$\gamma(f) = \left|G_{AB(f)}\right| / \left[G_A^{1/2}(f) G_B^{1/2}(f)\right], \tag{1.2}$$

where $G_A(f)$ and $G_B(f)$ are autospectra of the pressure pulsations at points *A* and *B* respectively and $G_{AB}(f)$ is the complex mutual spectrum at points *A* and *B*.

The velocity of disturbance convection in the longitudinal direction, u_c, is determined from the measured phase, $\varphi(f)$, of the mutual spectrum by means of the relation

$$\varphi = 2\pi f x_{AB} / u = \arctan(\gamma_{im} / \gamma_{rc}) \tag{1.3}$$

where γ_{rc} and γ are the real and imaginary parts of the complex mutual spectrum ($G_{AB} = \gamma_{rc} + i\gamma_{im}$) and x_{AB} is the longitudinal distance between points A and B.

The experiments are performed for the Reynolds numbers $\mathrm{Re} = u_0 d / \nu = (2.6 - 8.3) \cdot 10^5$ and the Mach numbers $\mathrm{M}_0 = u_0 / a = 0.29$ and 0.95. The boundary layer at the nozzle edge is turbulent ($d = 40$ mm).

Fig. 1.7 presents the coherence function, γ, in the near acoustical field of the jet in relation to the Strouhal number $\mathrm{St} = fd/u_0$ when the microphones draw apart in azimuthal and longitudinal directions for two values of the jet issue velocity corresponding to the Mach numbers $\mathrm{M}_0 = 0.29$ and 0.5. A distinguishing feature of the presented results is the conservation of the high degree of coherence at the frequency corresponding to the Strouhal number St = 0.5 for quite considerable distances between the microphones ($\Delta\theta = 180^\circ$ and $x/d = 3.5$). The measurements of the mutual spectrum phase in the jet near field (near the jet boundary) determine the velocity of disturbance convection along the flow, which is equal to $u_c = (0.7 - 0.8)\ u_0$ for frequencies corresponding to St = 0.5.

The measurements of the coherence function in the jet far acoustical field when the microphones draw apart in azimuthal direction and along the rays $\alpha = \mathrm{constant}$ for $R > 1.5$ revealed rather high values at the frequency St = 0.5 (for $\Delta\theta = 60^\circ$, $\gamma = 0.2 - 0.5$ and for $\Delta R / d > 20$, $\gamma > 0{,}6$). Therefore, coherent structures are present not only in the far, but also in the near acoustical field in the whole range of subsonic speeds.

In parallel with the direct approaches to identification of the coherent structures in jets described above, the indirect approaches to evaluation of parameters of the coherent structures have gained acceptance. These approaches reduce to the weak periodical excitation of a jet and measurement of its response to disturbances with different frequencies. With the availability of natural tendency to orderliness the periodical excitation can amplify the latent regular structure above the initial turbulent background and, therefore, make it more distinct [1.8, 1.43]. For this approach to detection of coherent structures the question inevitably arises about their identity with the initial structures occurring in jet flows in the absence of periodical excitation. The answer is ambiguous. The aforementioned indirect approach could be reasonable in the case where weak excitation of the jet does not lead to perceptible change of the averaged flow [1.15]. Nevertheless, even with violation

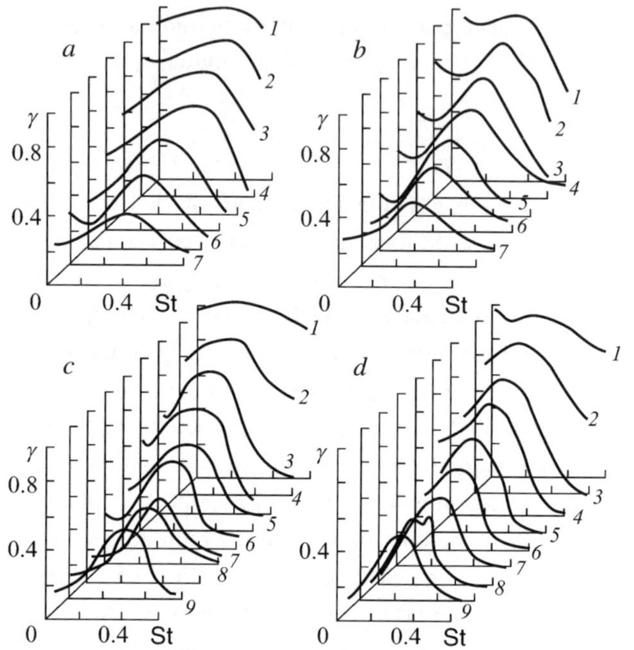

Fig. 1.7. The coherence function in the near acoustical field of the subsonic turbulent jet for $M_0 = 0.29$ and $M_0 = 0.95$. (*a*) $x/d = 2$, $r/d = 1.5$, $M_0 = 0.29$; (*b*) $x/d = 2$, $r/d = 1.5$, $M_0 = 0.95$; *1)* $\Delta x/d = 0.5$, *2)* $\Delta x/d = 1.0$, *3)* $\Delta x/d = 1.5$, *4)* $\Delta x/d = 2.0$, *5)* $\Delta x/d = 2.5$, *6)* $\Delta x/d = 3.0$, *7)* $\Delta x/d = 3.5$; (*c*) $x/d = 4$, $r/d = 2.0$, $M_0 = 0.29$; (*d*) $x/d = 4$, $r/d = 2.0$, $M_0 = 0.95$; *1)* $\Delta\theta = 20^\circ$, *2)* $\Delta\theta = 40^\circ$, *3)* $\Delta\theta = 60^\circ$, *4)* $\Delta\theta = 80^\circ$, *5)* $\Delta\theta = 100^\circ$, *6)* $\Delta\theta = \ 0^\circ$, *7)* $\Delta\theta = 1\ 0^\circ$, *8)* $\Delta\theta = 1\ 0^\circ$, *9)* $\Delta\theta = 180^\circ$.

of the latter condition some integral characteristics of coherent structures, their characteristic frequency and convective transport velocity, differ little from the corresponding characteristics for the unexcited jet.

The response of the initial section of a circular turbulent jet on longitudinal monochromatic acoustical disturbances of different frequencies was first studied in [1.8] for a jet with $\mathrm{Re} = 10^5$ and the initial turbulent boundary layer. Fig. 1.8 demonstrates the obtained amplitude characteristic – the dependency of the mean square value of the velocity pulsation, ε_u, at the point $x/d = 4$, $r/d = 0$ on the intensity of the velocity acoustical pulsations at the nozzle edge, ε_{us}, for different values of the Strouhal number, St_s, calculated from the acting sound frequency, f_s. The main signal depends linearly on the excitation level ε_{us} for all St_s if ε_{us} is small. The harmonics due to nonlinearity slows down the main oscillation rise. The oscillation with St_s = 0,3 is preferential in that it can achieve the maximum possible amplitude because the harmonics for this value of the Strouhal number are generated to the smallest degree.

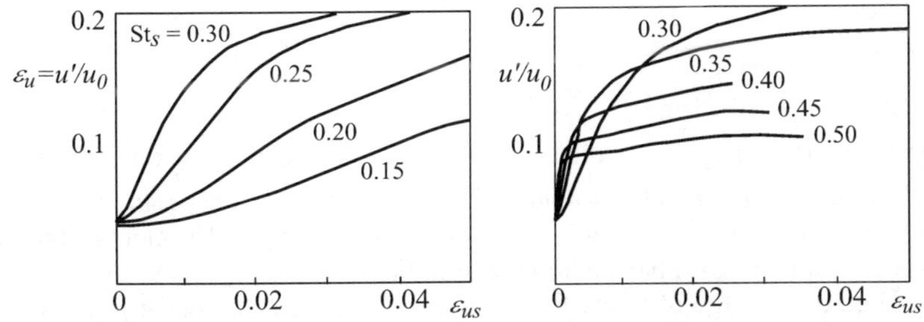

Fig. 1.8. Dependency of the mean square value of the longitudinal velocity pulsation at the point $x/d = 4$, $r/d = 0$ on the intensity of the longitudinal velocity pulsations in the sound wave, ε_{us}, for different values of the Strouhal number, St_s.

The property of the mixing layer of the turbulent jet initial section as the oscillation amplifier manifests itself for the rather uniform initial velocity profile. A severe distortion of the initial velocity profile denudes jets of this property. This fact appears to be connected with attenuation of coherent structures.

Large vortices generating near the end of the initial section have enhanced long-range action and therefore can influence the vortex generation process in the mixing layer near the nozzle, i.e. upstream. As a result the amplitude of the velocity high-frequency pulsations at a fixed point of the mixing layer is modulated by low-frequency pulsations with the characteristic frequency corresponding to the Strouhal number St = 0.3 – 0.5. The frequency analysis of the modulation envelope discloses that considerable energy is contained in the bandwidth corresponding to St = 0.3, which is peculiar to large vortices at the end of the initial section. Therefore, disturbances generated by large vortices are propagated from the area $x/d = (3-5)$ upstream and affect the mixing layer evolution near the nozzle ($x/d < 1$).

All large-scale vortices within the initial section [1.43] influence equally the flow near the nozzle edge because the velocity decrease induced by the vortices is in inverse proportion to the distance x. This is evident from the Biot–Savart law and is compensated by the corresponding linear increase of the vortex circulations. These considerations are the basis for the notion of the global mechanism of feedback resulting from the rapid increase of the vortex circulations during the vortex pairing [1.24]. It was supposed that each vortex pairing along the flow being accompanied by a double decrease of the frequency induces the response at the nozzle edge through the feedback loop, which consists of the subharmonic unstable wave propagating downstream and the acoustical wave propagating upstream.

In addition, the phase difference of these two waves equals necessarily $2N\pi$ where N is an integer [1.16]. Then if x_i corresponds to the i-th pairing position, λ_i is the corresponding wave length of the subharmonic and λ_a is the sound wave length then (this relation is obtained in *Artamonov KI.* Thermo-hydroacoustical stability [in Russian], Moscow, Mashinostroyenie, 1982)

$$\frac{x_i}{\lambda_i}+\frac{x_i}{\lambda_a}=N\,. \tag{1.4}$$

Here $\lambda_i = u_c / f_i; \lambda_a = a / f_i$; u_c is the convective velocity of the coherent vortices downstream and a is the sound speed, i.e. the speed of acoustic waves propagating upstream from the pairing positions. Therefore, the fulfillment of the above phase condition means that an integer number of hydrodynamic and acoustical waves should go in the distance from the nozzle to the i-th pairing position. Relation (1.4) could be presented as follows:

$$\mathrm{St_i}\frac{x_i}{d}=\frac{N}{1+\mathrm{M}_0 u_c/u_0}\frac{u_c}{u_0}\,. \tag{1.5}$$

Let us set here $u_c / u_0 = 0.6$ and $N = 2$. Then we will have St as a function of x/d.

The periodical structures are noted also in the initial section of circular turbulent jets in accompanying flows. Spectral and correlation measurements show [1.42] that here the characteristic frequency increases with an increase of the coflowing parameter, u_∞ / u_0, in accordance with the empirical relation

$$\mathrm{St} = (0.5\ -0.54)/(1-u_\infty/u_0)\,. \tag{1.6}$$

A large body of experiments was also dedicated to the study of coherent structure parameters in the main section of the jet. However these structures do not play any considerable dynamic role in turbulent mixing processes and in periodical exposure to jets.

The review [1.10] presents estimates obtained in different experiments of contributions of large-scale pulsation (coherent structures) kinetic energy in the overall energy for several reference jet flows: the plain mixing layer – 20%, initial section of a circular jet – 50%, main section of a jet – 10%, near wake – 25%, and far wake – 20%.

Coherent structures are generated in jets and mixing layers in the broad range of the Reynolds numbers corresponding to simulations and experiments [1.41].

Comprehensive reviews of the modern state of the art of coherent structure studies for turbulent jet flows are presented in [1.5, 1.10, 1.11, 1.17, 1.26, 1.34, 1.35, 1.38, 1.41, 1.50].

1.2. Coherent Structures and Hydrodynamic Instability

Generation of coherent structures in jet flows is determined by the hydrodynamic instability of these flows. Three types of instability should be identified in initial sections of axisymmetric jets.

A. The Kelvin–Helmholtz 2-D spatial instability is generated for an initial laminar boundary layer at the nozzle edge in the mixing layer ($x/d<1$). Its chacteristics are well described by the linear theory of spatial instability [1.28]. The maximum coefficient of spatial amplification of axisymmetric disturbances corresponds to the Strouhal number $\mathrm{St}_\theta = f_i\theta_0 / u_0 = 0.017$. The same value is valid also for the plane jet mixing. The linear theory results agree very satisfactorily with experiments. In particular, the experiments provide support to the relation $f_i = \mathrm{const} \cdot u_0^{3/2}$, which follows from the above formula when keeping in mind the relation $\theta_0 \sim u_0^{-1/2}$, which is correct in a laminar boundary layer [1.19].

However, a number of experiments give the value $\mathrm{St}_\theta = 0.011 - 0.018$ different from the theoretical ones. This fact is traditionally explained by the influence of the initial conditions (the velocity and pulsation intensity profiles, acoustical disturbances, and the nozzle vibrations). The generation of vortex clusters was discovered along with their subsequent interaction and merging in pairs at the nonlinear stage of the disturbance evolution.

When studying the mixing layer linear stability (the Kelvin–Helmholz instability) theoretically, the velocity profile in the form of the hyperbolic tangent, typical for the formed laminar mixing layer, is given as the initial one. Meanwhile, the Blasius laminar boundary layer characterized by the Tollmien–Schlichting instability is generated at the output section. However, in the theories of the mixing layer linear stability, the velocity profile rebuilding from the Blasius boundary layer to the jet velocity profile is ignored.

For the initial turbulent boundary layer at the nozzle edge and small turbulence scale compared to the wavelength of the maximum augmented mode, one might expect the emergence of the same type of instability as for the initial laminar boundary layer [1.6, 1.12, 1.27]. Then the initial velocity profile will be determined by the nonstationary wave characteristics. Spatially amplifying wave components generated in this case will cause the generation of turbulent annular vortices at the nonlinear phase of evolution, which is observed in experiments.

However, the possibility for calculation of the instability characteristics for the turbulent mixing layer from the initial averaged velocity profile is questioned [1.17] because in this case the velocity pulsations are very large and the averaged velocity profile cannot characterize instability. Nonetheless, thc main mechanisms of evolution and interaction of coherent structures in a jet for the laminar and turbulent boundary layer at the nozzle edge are likely to share many common features. Basically, these mechanisms may be described in the framework of the Euler nonviscous model.

B. The second instability type occurs in the range $x / d = 1 - 6$ and is related to collective interaction of annular vortices in the round jet initial section. In this case low-frequency pulsations generated by the largest vortices prevail. At the end of the initial section the characteristic Strouhal number determined from the nozzle diameter equals to $\mathrm{St} = f_c d / u_0 = 0.3 - 0.5$ and the value scattering is due to influence of the initial conditions at the nozzle edge.

Though interaction of vortex rings within the initial section is the pure nonlinear process, their characteristic frequency was determined in the framework of the stability linear theory in the local parallelism approximation [1.28] where allowances are made for the influence of the finite thickness of the mixing layer near the end of the initial section or, more precisely, the influence of the cross curvature of the mixing layer. It was found that $\mathrm{St} \approx 0.35$ at the end of the initial section in accordance with the known experimental data.

The analysis of linear stability of a weak divergent flow [1.7] qualitatively agrees with the observations and separates the mode with the maximum gain factor for pressure amplitude, which corresponds to the Strouhal number $\mathrm{St} = fd/u_0 \approx 0.4$. As noted in [1.27], these theories implicitly contain nonlinearity because the average velocity profiles measured in experiments and used in calculations already include the result of the Reynolds stress action. These Strouhal numbers correspond to the so-called preferential mode. As it was demonstrated in experiments [1.21], for $d/2\theta_0 > 120$ the Strouhal number of the preferential mode is constant and equal to 0.44.

On the basis of the values St_θ for the initial laminar boundary layer at the nozzle edge and St at the end of the initial section, one can estimate the number of pair vortex coalescences [1.24]:

$$\frac{f_i}{f_c} = \frac{f_i\theta_0}{u_0}\frac{u_0}{f_c d}\frac{d}{\theta_0}. \tag{1.7}$$

Putting $\mathrm{St}_\theta = 0.017$, $\mathrm{St} = 0.3$ and $d/\theta_0 < 150$, we get $f_i/f_c \approx 8 < 2^n$, i.e. there are possible 3–4 pair coalescences of annular vortices within the jet initial section. If 3–4 vortex pair coalescences are realized in the initial section of a circular jet with the initial laminar boundary layer, then for the initial turbulent boundary layer we have only 1–2 pair coalescences.

The collective interaction of a large number of vortices can also be described in the framework of the Euler nonviscous model [1.12]. Calculations performed in the framework of this model show a tendency for vortex pairing and separation of long-wave spectral components.

Owing to the Kelvin–Helmholtz instability mechanism, 2-D waves exponentially increase downstream and roll-up into vortices. According to experimental data, the rolling-up finishes at the downstream point where the main component amplitude peaks with frequency, q. Here the excitation of the subharmonics $f/2$ occurs with the amplitude, which is less than the main amplitude by three orders. The subharmonics rise downstream at the nonlinear stage of instability evolution yields pairing of neighbouring vortices with the pairing point at that section downstream where the subharmonics peak [1.15].

C. The third type of instability is related to generation of azimuthal inhomogeneity of large-scale vortices leading to breakdown of annular vortices into 'balls'. This type of instability is precisely responsible for generation of the above-mentioned 'star-like' structures. Calculations according to the linear theory [1.28] for the first azimuthal mode (n = 1) show that the maximum values of the spatial amplification coefficient for this mode occur at the same frequencies as were determined for axisymmetric disturbances, i.e. for the zero mode (n = 0).

As a first approximation, we can judge about the loss of azimuthal homogeneity of the formed annular vortices, which corresponds to the nonlinear stage of the disturbance evolution in the mixing layer, from the separate vortex ring behavior. Indeed, it is well known that a vortex ring in ideal fluid is unstable and the number of generated azimuthal waves is determined by the vortex core size. Another reason for the possible loss of azimuthal homogeneity comes into force for the interaction of two coaxial annular vortices. Analysis has shown [1.2] that an expanding forward annular vortex to a lesser extent and a contracting back vortex to a much greater extent are sensitive to radial disturbances. The result is the earlier loss of the vortex azimuthal homogeneity (Fig. 1.3, *b*).

Similar forms of instability could be seen also in the initial section of a plane turbulent jet. The flow pattern and instability mechanism in plane and circular jets in the mixing layer near the nozzle are almost identical. For $x/h = 1 - 5$ the flow instability in the initial section of a plane jet is related to the collective interaction of large-scale vortices. At last, violation of the 2-dimensionality of these rectilinear vortices along their span plays the same role as violation of the azimuthal homogeneity of annular vortices in a circular jet [1.15, 1.16].

To judge the axial symmetry violation in a circular turbulent jet (the mode composition of large-scale coherent structures), measurements of 3-D azimuthal correlation of longitudinal velocity pulsations, temperature pulsations [1.37] in the mixing layer, as well as pressure pulsations outside of the jet in its near acoustical field are used. According to the measurements of velocity pulsations azimuthal correlation, $R_{uu}(\psi)$, of the orthogonal Fourier expansion

$$R_{uu}(\psi) = \frac{1}{2}\alpha_0 + \sum_n \alpha_n \cos n\psi \, d\psi, \quad n = 1, 2, ... \tag{1.8}$$

with coefficients

$$\alpha_n = \frac{2}{\pi}\int_0^\pi R_{uu}(\psi)\cos n\psi \, d\psi, \quad n = 0, 1, ...$$

we can evaluate the change of α_n downstream. Here $\alpha_0/2$ corresponds to the zero mode ($n = 0$) and α_n correspond to the higher modes ($n = 1, 2, ...$).

Appropriate experiments and calculations have shown that for $x/d = 1 - 40$ the mode composition of turbulent pulsations substantially changes: for $x/d = 1$ the zero mode prevails and with increase in x/d the relative energy becomes dominant for the 1st mode and increases for modes $n = 2 - 5$. In an analogous way the pulsation correlations for pressure outside the jet calculated from measurement data change for $x/d = 2, 4$ and 6 [1.33]. These results are the natural reflection of deformation of large-scale coherent structures along the jet – from annular vortices in the mixing layer near the nozzle edge to star-like structures at the end of the initial section.

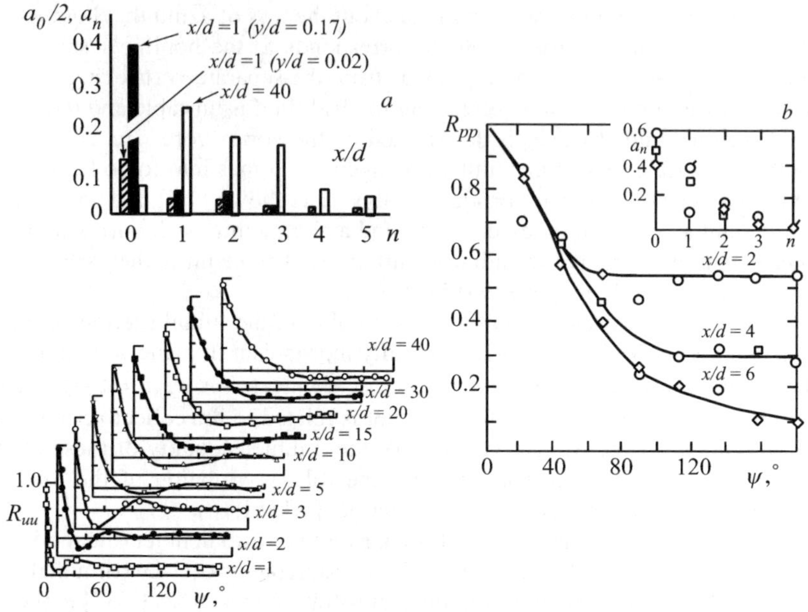

Fig. 1.9. (*a*) The azimuthal correlations of the longitudinal velocity and pressure pulsations in a circular jet for $r/d = 0.02$ and 0.17. (*b*) The mode changes for $x/d = 1$ and 40.

Fig. 1.9 presents the changes in the azimuthal correlations of the velocity and pressure pulsations in circumferential direction for a circular jet at various x/d and the mode composition of the pulsations in a circular jet [1.33, 1.37]. The zero mode corresponds to the axisymmetric vortex ring. The simple interpretation of the first, second, third and fourth modes in the Fourier expansion is demonstrated in Fig. 1.10.

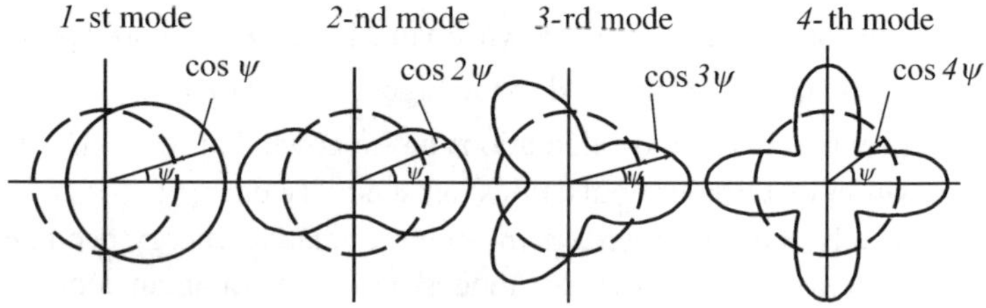

Fig. 1.10. Interpretation of the first – fourth modes in the Fourier expansion.

1.3. Acoustic Characteristics of Subsonic Turbulent Jets

The noise of a free submerged turbulent jet for subsonic issue velocities is produced as a result of large- and small-scale turbulent mixing of gas particles with velocities close to the issue velocities and surrounding gas particles.

The main laws determining the association between the jet acoustical radiation intensity and the flow gas-dynamic and geometrical parameters were established by Lighthill who transformed the Navier-Stokes equation to a nonhomogeneous equation relating the density changes in the surrounding immovable medium to the turbulence characteristics in a jet [1.25]. The analysis of this equation on the basis of the dimensional theory permits obtaining the following relation for the jet sound power:

$$W = k_0 \frac{\rho_0^2 u_0^8 d^2}{\rho_\infty a_\infty^5} . \tag{1.9}$$

Here W is the sound power defined as the acoustical energy emitting by the jet per unit time; u_0 is the issue velocity; ρ_0 and ρ_∞ are densities of the jet gas at the nozzle outlet section and of the environment gas respectively; a_∞ is the sound speed in the environment; and k_0 is an empirical coefficient.

The experimental studies of turbulent jet noise [1.36] have lent support to the validity of 'the eighth power rule' in a broad range of subsonic isothermic jet issue speeds (Fig. 1.11). In this case the value of the coefficient k_0 determined experimentally for isothermic jets with a small degree of turbulence equals $3 \cdot 10^{-5}$ and $k_0 = (1.5 - 2.5) \cdot 10^{-5}$ for turbojet engine jets.

The calculation results [1.30, 1.32] show (Fig. 1.12) that most of the power is radiated by the first ten jet calibers, i.e. x = (0–10) d. The acoustical power radiated by each of the jet unit caliber, i.e. $\Delta x / d = 1$ is practically constant over a length of the whole jet initial section and decreases over the jet main section in accordance with the relation $dW / dx \sim x^{-6}$. About 65% of the jet overall acoustical power is radiated by the initial section.

The jet acoustical power spectrum is demonstrated in Fig. 1.13 in the form of functions $\Delta\varepsilon = \varepsilon_i - \varepsilon_\Sigma$ of the Strouhal number $\mathrm{St} = fd / u_0$. Here ε_i is the sound power level in the one-third octave frequency band and ε_Σ is the overall sound power level.

The maximum in the acoustical power spectrum in the whole range of subsonic issue speeds is observed in the area of the Strouhal numbers where f is the mean frequency of the one-third octave frequency band [1.36].

Fig. 1.14 shows characteristics of sound radiation directivity for a subsonic jet in the far acoustical field (i.e. in the area that is sufficiently far away from the sound source as compared with the source sizes and the radiated sound wavelength). The total sound maximum for isothermic jets is observed at an angle of 30° to the jet axis [1.30].

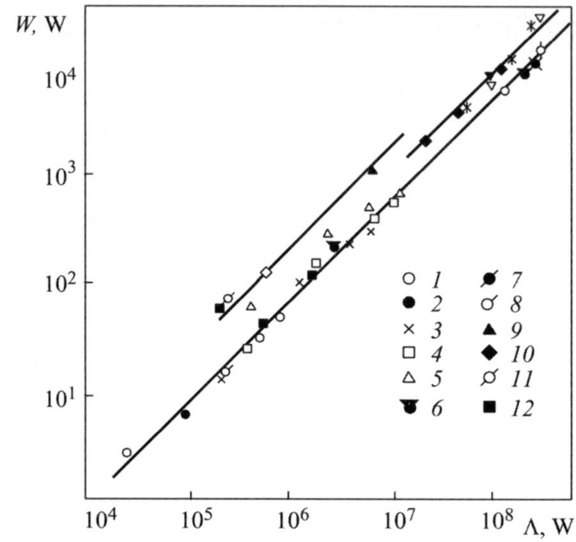

Fig. 1.11. The jet sound power as a function of the Lighthill parameter $\Lambda = \rho_0^2 u_0^8 d^2 / \rho_\infty a_\infty^5$. The nozzle 100 mm, $T_\infty = 288\text{K}$: *1)* $\pi_c = p_{00} / p_\infty = 1.4$; *2)* $\pi_c = 1.6$; *3)* $\pi_c = 1.8$; *4)* $\pi_c = 2.0$; *5)* $\pi_c = 2.2$. Real turbojet engines: Engine # 1 (*6)* the operating mode with the maximum rotation rate; *7)* the nominal rotation rate; *8)* 0.9 of the nominal rotation rate; *9)* 0.8 of the nominal rotation rate); *10)* Engine # 2 ; *11)* Engine # 3 ; *12)* Engine # 4.

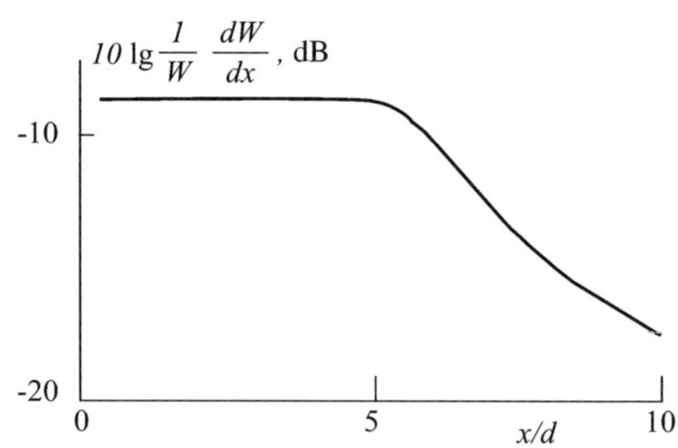

Fig. 1.12. The change in the sound power radiated by the jet unit length along the jet axis.

The temperature rise in anisothermic jets causes the sound speed gradient increase in the jet mixing layer and amplification of the radiation direction departure from the jet axis. The jet sound spatial distribution becomes nonuniform and the acoustical radiation intensity maximum is shifted to the larger angles φ; e.g. the maximum is observed at $\varphi = 40^\circ$ for the jet initial temperature $T_{00} = 800$ K (Fig. 1.15). In Fig. 1.15 characteristics of the jet sound radiation directivity are given in the form of functions $10 \log \Phi$ of the angle φ between the jet axis and

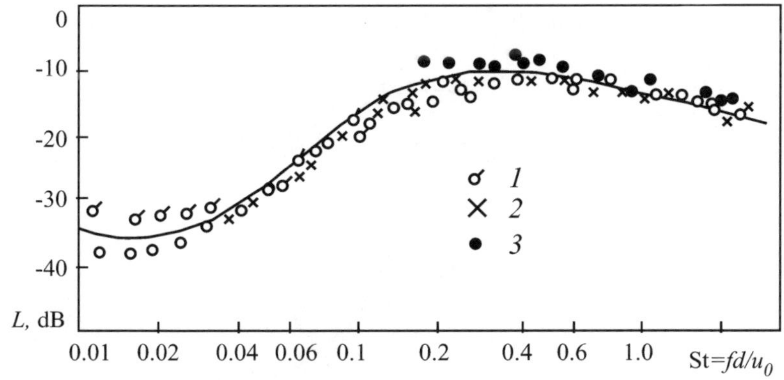

Fig. 1.13. The jet sound power as a function of the Strouhal number: *1)* $d = 30$ mm, $T_{00} = 288$ K; *2)* $d = 100$ mm, $T_{00} = 288$ K; *3)* turbojet engine.

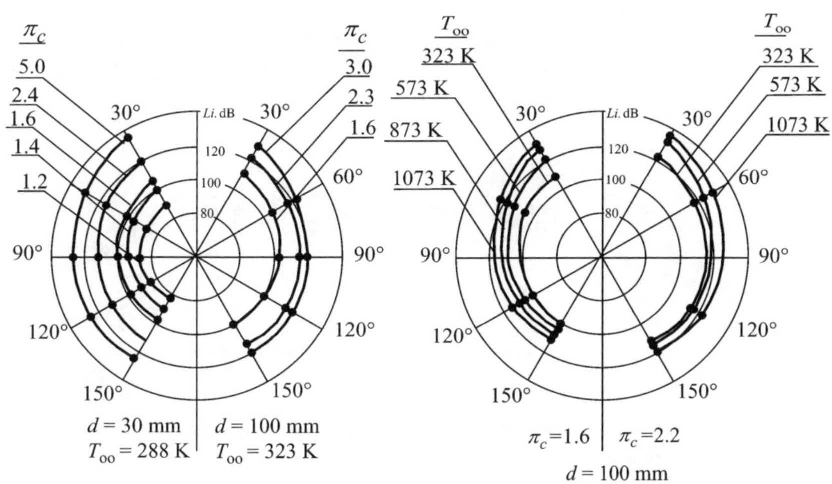

Fig. 1.14. The directivity characteristics of the jet total sound levels ($\pi_c = p_{00} / p_\infty$).

the direction to the sound measurement point. The quantity $10 \log \Phi$ is the directivity factor defined as the difference of the measured sound level and the sound level at the same point from a fictitious source of the same power as the source under study but radiating sound uniformly in all directions.

The jet sound spectra in the far field, measured in the one-third octave frequency bands for different angles φ, are highly distinguished (Fig. 1.16). The sound low-frequency component levels increase with decreasing angle φ [1.30]. Due to the presence of the average velocity gradient at the jet cross sections, the contribution of high-frequency components to the jet noise spectrum

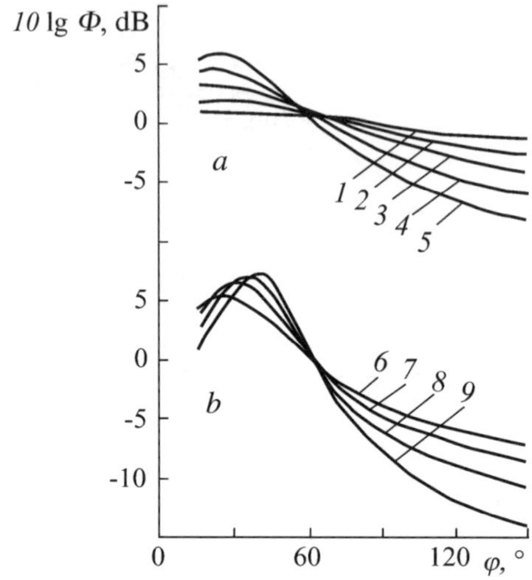

Fig. 1.15. The directivity factor of the total sound of isothermic and anisothermic turbulent jets. *(a)* – Isothermic jets, T_{00} = 288 K: *1)* M_0= 0.2; *2)* M_0 = 0.3; *3)* M_0 = 0.5; *4)* M_0 = 0.7; *5)* M_0 = 1.0. *(b)* Anisothermic jets, M_0 = 0.8: *6)* T_{00} = 288 K; *7)* T_{00} = 600 K; *8)* T_{00} = 800 K; *9)* T_{00} = 1000 K.

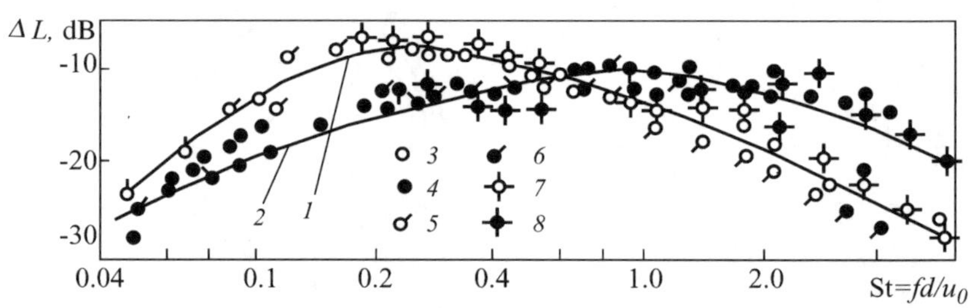

Fig. 1.16. The sound pressure as a function of the Strouhal number for a turbulent jet and a nozzle with diameter d = 100 mm for different values of T_{00} and for a turbojet engine turbulent jet *1)* $20^\circ \le \varphi \le 60^\circ$, *2)* $70^\circ \le \varphi \le 180^\circ$, *3)* T_{00} = 323 K, *4)* T_{00}= 323 K; *5)* T_{00} = 873 K, *6)* T_{00} = 873 K; *7)* turbojet engine, *8)* turbojet engine.

decreases with decreasing φ and the contribution of low-frequency components correspondenty increases. In addition, the maximum becomes more and more distinct in the noise spectra for such a change of the noise spectral composition. The most perceptible change of the noise spectra is observed for small values of the angle φ = 30°– 45°.

The pressure pulsations do not coincide in phase with the velocity pulsations near the border of the turbulent jet mixing area. This is the so-called area of the jet near acoustical field where the relations characteristic for the far acoustical field, i.e. the distance between the sound source and receiver should be large relative to the source sizes and the sound wavelength, are not fulfilled.

The typical distribution of the pressure pulsation levels in the near field of the turbojet engine exhaust [1.48] is presented in Fig. 1.17. The pressure pulsation maximums are observed near the jet border within the initial section.

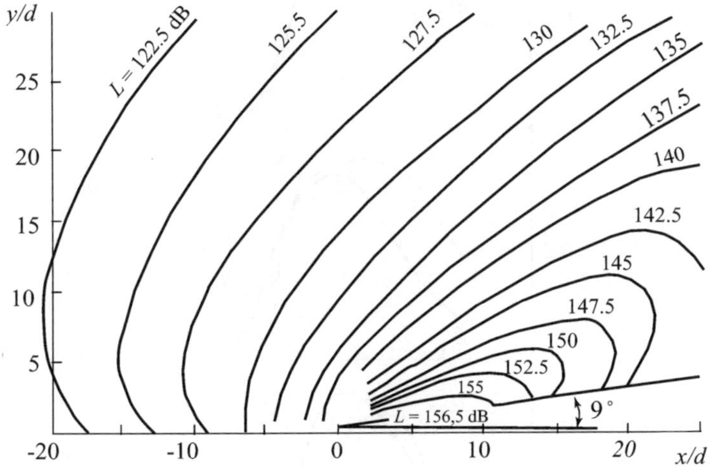

Fig. 1.17. The level curves of the pressure pulsations near the reactive jet of the turbojet engine, $d = 0.5$ m, $u_0 = 600$ m/s.

Variations in the pressure pulsation spectra along the jet border are similar to variations in the turbulent velocity pulsation spectra in the area of most intensive sound generation, i.e. on the line going through the nozzle edge parallel to the jet axis. As the distance from the nozzle edge increases, the frequency of the noise maximum outside the jet on the line parallel to the jet border, as well as the frequency of the maximum f_m in the turbulent velocity pulsation spectrum on the line $y/d = 0.5$, is shifted to low frequencies [1.39] in accordance with the following empirical relation:

$$f_m x / u_0 = 1.35 . \qquad (1.10)$$

At present the important role of the large-scale periodic clusters, i.e. the so-called coherent structures, in turbulent mixing in jets is generally recognized [1.14, 1.41]. These structures evolve against the background of small-scale turbulence; their growth, merging in pairs and subsequent decay within the initial section of subsonic jets determine in many respects the jet aerodynamic characteristics and generation of aerodynamic noise. In particular, it follows from relation (1.10) that near the end of the initial section where $x/d = 4-5$

$$\mathrm{St} = f_m d / u_0 = 0.22 - 0.34 .$$

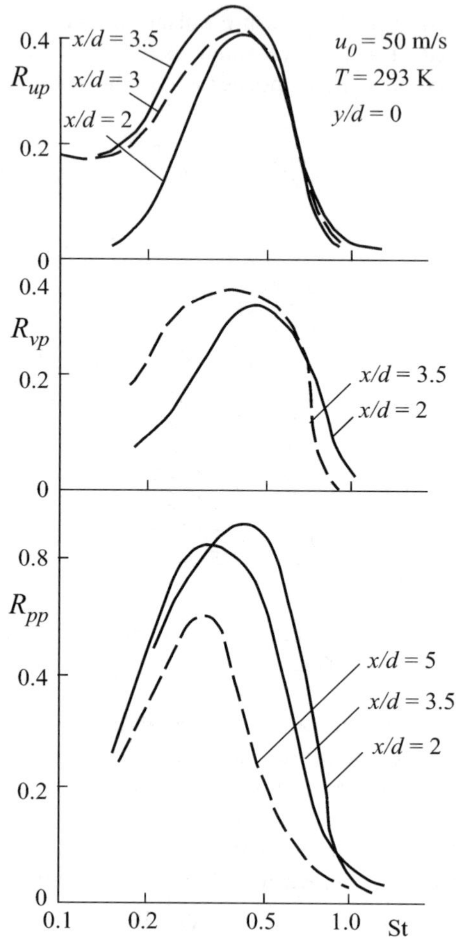

Fig. 1.18. The correlation coefficients for pulsations of the velocity, u', v' and pressure, p', on the jet axis, as well as for pulsations of the pressure in the near acoustical field.

This value of St is close to the corresponding value noted in point 1.1.

Coherent structures can radiate noise during their creation, growth, pairing, and decay at the end of the initial section. Some experimental evidence for the association between coherent structures and jet noise is found.

Fig. 1.18 presents the spatial correlation coefficients, R_{up}, R_{vp} and R_{pp} measured in one-third octave frequency bands for $x/d = 2-5$ at two points: inside the jet and in its near field [1.47]. Inside the jet the longitudinal velocity pulsations, u', the radial velocity pulsations, p', or the pressure pulsations, p', on the axis ($y = 0$) were measured; in the jet near field the pressure pulsations, p', were measured. In all these cases the above correlations have a maximum for

St = 0.4 – 0.5. The availability of considerable correlation between the pressure pulsations in the jet near and far acoustical fields precisely for St = 0.25–0.55 was also found [1.28]. It is shown also that the maximum azimuthal correlation of the pressure pulsations in the jet far acoustical field happens at frequencies which are dominant in the radiated noise spectrum.

Fig. 1.19 presents the directivity factor, $10\log\Phi$, as a function of the angle φ for the total noise and in narrow frequency bands for the Strouhal numbers [1.47] St = 0.15, 0.30, 0.40, 0.46, 1.30, 2.6. It follows that the noise maximum for $\varphi = 30^\circ$ and St = 0.15–0.46 significantly exceeds the similar maximum corresponding to the measured total noise level. This fact testifies that the large-scale coherent structures with the characteristic frequencies corresponding to the Strouhal numbers St = 0.15–0.46 make the greatest contribution in the establishment of the jet noise directivity maximum for small angles φ. A similar conclusion was made in [1.9, 1.29] for subsonic cold and hot jets.

As for the position of main noise sources in the jet, more precisely, in its initial section, there is no consensus on this point. One view [1.23] holds that the jet noise main sources are located at points of annular vortex pairings. This standpoint is allegedly confirmed by the measurements [1.51] demonstrating that jets

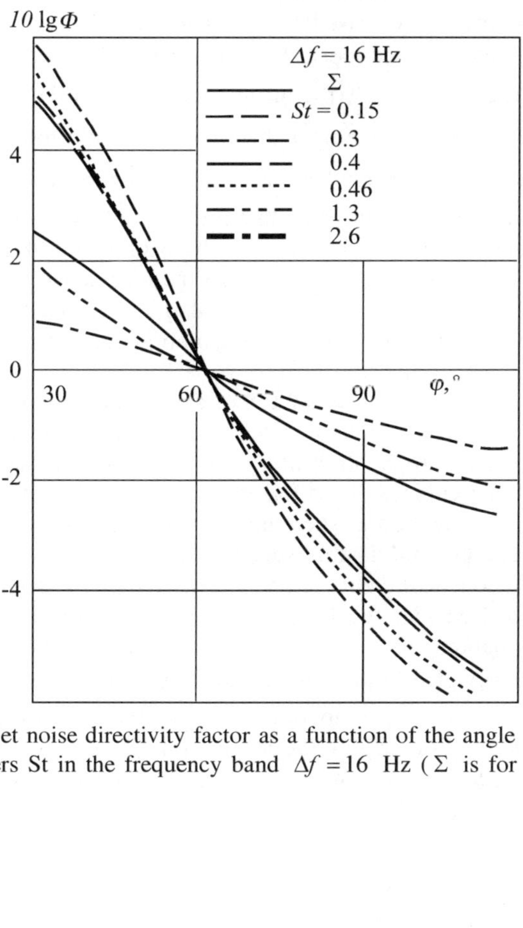

Fig. 1.19. The jet noise directivity factor as a function of the angle φ for different Strouhal numbers St in the frequency band $\Delta f = 16$ Hz (Σ is for the overall frequency band).

with the initial laminar boundary layer at the nozzle outlet section for $M_0 < 0.45$ make more noise than jets with the initial turbulent boundary layer because there is more annular vortex pairings in the first case.

Another point of view [1.17] is that precisely the azimuthal homogeneity decay of toroidal vortices near the end of the initial section is crucial for the generation of most of the jet noise.

1.4. Initial Conditions of Turbulent Jet Issue

Jet instability and its response to different kinds of disturbances noted at the beginning of Chapter 1 make urgent the study and consideration of issue initial conditions: aerodynamic, acoustic and geometric ones.

1.4.1. Aerodynamic and Acoustical Parameters Characterizing the Issue Initial Conditions for Subsonic Submerged and Accompanying Turbulent Jets. Generally, the issue initial conditions are characterized by distributions of the average velocity, temperature, energy, and turbulence scale at the nozzle outlet section. As applied to submerged jets with almost uniform distribution of the above parameters over the section (outside the boundary layer at the nozzle edge), the following parameters are used to characterize the issue initial conditions: $\mathrm{Re} = u_0 d / v$ (the Reynolds number), $M_0 = u_0 / a$ (the Mach number), T_0 / T_∞ (the anisothermic degree), $\varepsilon = u' / u_0$ (the turbulence degree at the centre of the nozzle outlet section), δ_0^* (the displacement thickness), θ_0 (the momentum loss thickness), and $H = \delta^* / \theta_0$ (the boundary layer shape parameter at the nozzle outlet section). The issue initial conditions contain also the flow mode in the boundary layer at the nozzle outlet section (laminar, transition or turbulent modes). In a number of cases it is important to know also the turbulence scale, as well as the presence of the nozzle vibrations – longitudinal and cross ones and their amplitude and spectra. They are characterized by the vibration acceleration that is measured by special-purpose vibration transducers.

Aerodynamic and acoustical characteristics of a jet (this refers equally to experimental plants and real turbojet engines) can change distinctly by the action of acoustical disturbances propagating along the flow via the path of experimental plants or real turbojet engines. Thus the issue initial conditions should be supplemented with the noise level and spectrum in the nozzle outlet section. The presence of discrete components in this spectrum that can noticeably change the jet aerodynamic and acoustical characteristics is particularly essential. For a jet in an accompanying flow, besides the listed parameters, one has to know also the accompanying flow parameters in the plane of the nozzle outlet section, the velocity and turbulence energy profiles, and the coflowing parameter, $m = u_\infty / u_0$. The initial distributions of velocity, temperature and admixture concentrations are also important because they determine the jet invariants – conditions of the constant excess impulse, excess heat content and excess admixture content [1.1, 1.13],

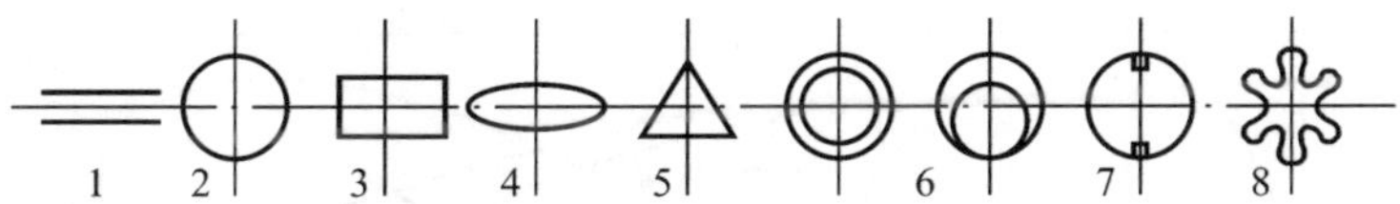

Fig. 1.20. Shapes of a nozzle cross section. 1) plane, 2) circular, 3) rectangular, 4) elliptic, 5) triangular, 6) annular with concentric and eccentric position of the inner and outer nozzles, 7) circular with a generator of longitudinal vortices generating the initial azimuthal heterogeneity of the flow, and 8) lobed.

which are valid in the absence of longitudinal pressure gradient in the accompanying flow.

1.4.2. Geometrical Parameters Determining the Issue Initial Conditions. Geometrical parameters of a unit producing a jet influence essentially its aerodynamic and acoustical characteristics, as well as the configuration of the jet cross section in its initial region. Also, sometimes orifices with the flow separation over the sharp edge are used instead of nozzles to generate plane, circular and elliptic jets.

Traditionally, to compare characteristics of 3-D and circular jets, the equivalent diameter, d_e, corresponding to the nozzle outlet cross section area $F = \pi d_e^2 / 4$ is used as the nozzle characteristic geometrical parameter. To illustrate the influence of the apparatus geometry (a nozzle or an orifice) generating the jet we describe some aerodynamic characteristics of axisymmetric and 3-D turbulent jets.

A. First, we consider two significantly different cases of submerged jets flowing out of a nozzle and orifice with sharp edge (Fig. 1.21). In the first case for the jet shaping with the most possible uniform initial profile of velocity we use a conical nozzle with a smooth contour. The resulting flow compression degree $n \leq 3-10$ (the area ratio for the nozzle inlet and outlet cross sections) enables the flow turbulence degree at the nozzle outlet cross section to be decreased. For the case of the flow issue from an orifice, the separated flow over its sharp edge is realized, the velocity profile at the jet initial section is nonuniform and has its minimum at the center of the jet initial section. Fig. 1.22 presents the average velocities and longitudinal velocity pulsations at the jet axis as functions of the longitudinal coordinate for the case of a jet issue from a nozzle and orifice with 9-fold compression ($D/d_0 = 3$) of the jet [1.49]. For a jet issue from an orifice, the average velocity increases initially along the jet axis, achieves the maximum, and then decreases, similar to a jet flowing out of a nozzle.

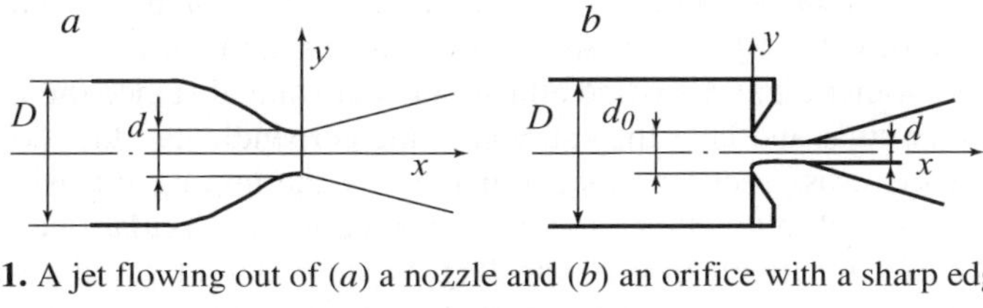

Fig. 1.21. A jet flowing out of (*a*) a nozzle and (*b*) an orifice with a sharp edge.

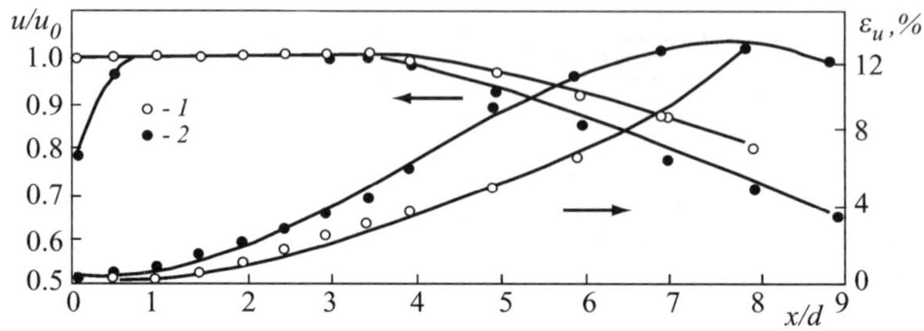

Fig. 1.22. Functions $u(x)$ and $u'(x)$ for jets flowing out of a nozzle and diaphragm (*1)* a nozzle, *2)* an orifice with sharp edges).

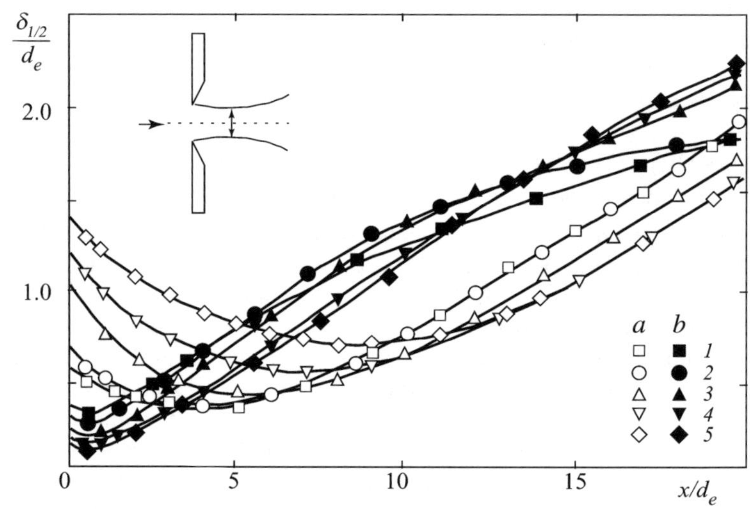

Fig. 1.23. Change along the flow of ordinates of half velocity along the major and minor axes of the elliptic diaphragm with different elongation, λ: *1)* $\lambda = 1.5$, *2)* $\lambda = 2.0$, *3)* $\lambda = 4.0$, *4)* $\lambda = 6.0$, *5)* $\lambda = 8.0$; (*a*) – along the major axis; (*b*) – along the minor axis.

B. The jet issue from an orifice with elliptic cross section. Fig. 1.23 presents the change in the cross section sizes along the flow for two perpendicular sections of a jet with the axis ratio of the elliptic section 3 : 2, 2 : 1, 4 : 1, 6 : 1, and 8 : 1. These dependencies illustrate the phenomenon of reorientation of the jet cross section axes along the flow [1.18]. The cross section isotachs at the initial section near the orifice are homothetic to the orifice ellipse, then at some distance the isotachs become almost circular and later they stretch in the perpendicular direction. This deformation of the cross section is accompanied by the mixing intensification and the increase in the elliptic jet ejection as compared to the circular one. The deformation is caused by interaction of the large-scale coherent structures in the jet mixing layer. Similar effects can be seen also for jets flowing out of elliptic and rectangular nozzles.

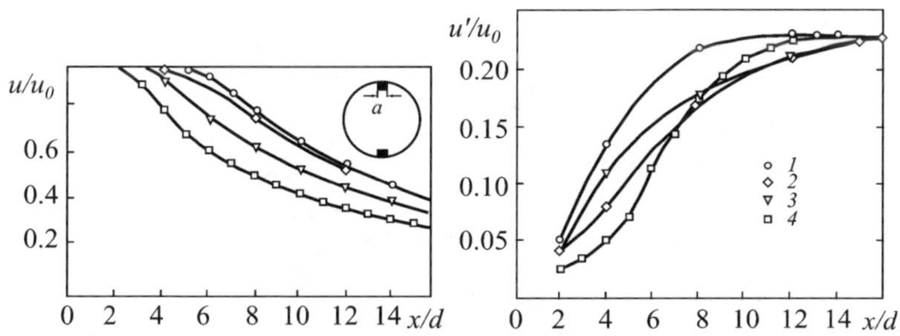

Fig. 1.24. The mixing intensification in a circular jet with square generators of longitudinal vortices (the square side is $a = d/16$); *1)* without generators, *2)* 8 generators, *3)* 4 generators, *4)* 2 generators.

C. The jet issue from a circular nozzle with generators of longitudinal vortices. The mounting of two, four or eight generators of longitudinal vortices (square plates with the side $d/16$) at the outlet section slightly deforms the nozzle cross section and consideraddy changes the jet aerodynamic characteristics [1.3]. Fig. 1.24 presents the change in the average velocity and longitudinal velocity pulsations along the jet axis. The corresponding curves for a circular nozzle without vortex generators are shown for comparison purposes.

This example illustrates the mixing intensification in a jet using vortex generators. In some cases vortex generators at the circular nozzle outlet section initiating the flow azimuthal heterogeneity may cause, according to generator geometrical parameters of the generators, their number and the initial boundary layer thickness, not only the mixing intensification but also the mixing attenuation [1.45]. Fig. 1.25 presents the corresponding dependencies for the average velocity and longitudinal velocity pulsations illustrating this effect. The mixing attenuation for $h/\delta = 1$ seems to be determined by delay in the circular vortex growth in the mixing layer of the jet initial section.

D. The jet issue from a lobed nozzle. To reduce the noise in aviation bypass turbojet engines with the common mixing chamber, lobed nozzles are used. These nozzles provide the mixing intensification and, therefore, the turbulence scale decrease in a jet. By way of example Fig. 1.26 presents the comparison of the average velocity attenuation along a jet axis flowing out of a circular and lobed nozzles [1.22]. Fig. 1.27 shows also the functions $u/u_0 = F_1(x/d_e)$ and $\varepsilon_0 = F_2(x/d_e)$ for circular, rectangular (3 : 1) and lobed nozzles having mixing chamber with length $l/d_e = 1.5$ [1.46].

The above examples illustrate the considerable influence of the nozzle geometry on jet aerodynamic characteristics. These data are important also for estimation of acoustical disturbance influence on turbulent characteristics of jets with different cross sections.

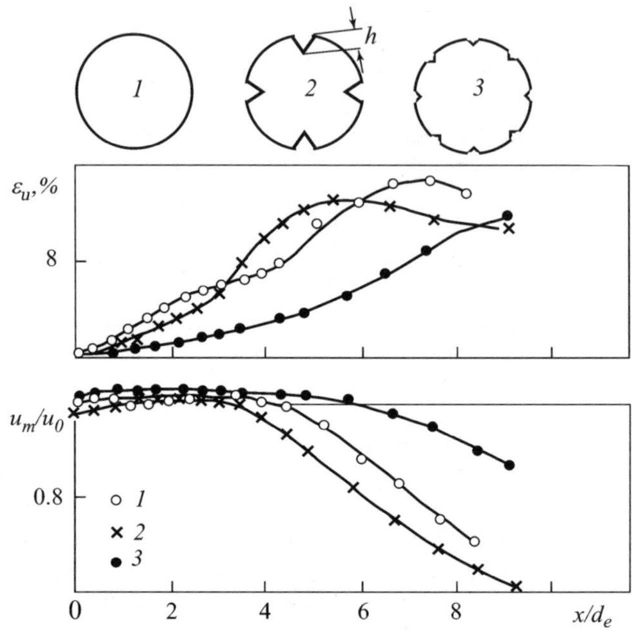

Fig. 1.25. The mixing intensification and attenuation in a circular jet with triangular generators of vortices. The curves $u(x)$ and $u'(x)$. *1)* a nozzle with circular section; *2)* a nozzle with 4 triangular teeth ($h/\delta = 3$); *3)* a nozzle with 8 triangular teeth ($h/\delta = 1$); δ – the boundary layer thickness at the nozzle edge.

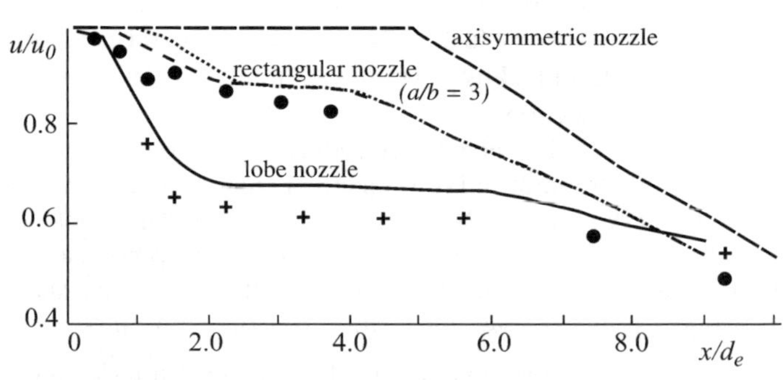

Fig. 1.26. Change in the average velocity along the axis of a submerged jet flowing out of circular, rectangular and lobed nozzles with length $l/d_e = 1.5$; d_e is the nozzle equivalent diameter.

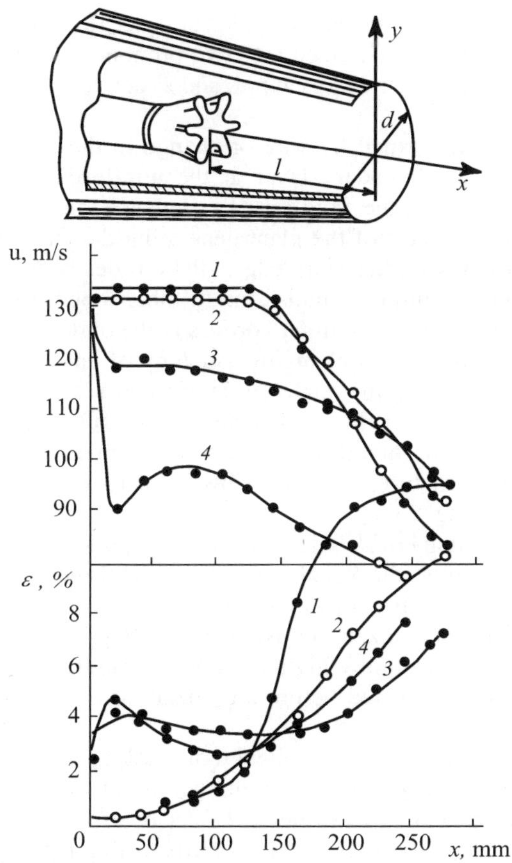

Fig. 1.27. Functions u and $\varepsilon = u'/u_0$ for *1*) circular and lobed nozzles (*2)* 6 lobes, *3)* 12 lobes, *4)* 18 lobes) with the mixing chamber length $l/d_e = 1{,}5$; d_e is the nozzle equivalent diameter.

1.5. Approaches to Turbulent Jet Control

We have already mentioned the important role of coherent structures in the large-scale transition of impulse, heat and mass, as well as in the aerodynamic noise generation in turbulent subsonic jets. The study of generation, interaction and decay of these structures makes possible the considerable refining of the current views of the mechanism of turbulent mixing and noise generation in jets. The dependence of coherent structures on the issue initial conditions and their sensitivity to different periodical disturbances opens up new possibilities for efficient control of aerodynamic, heat and acoustical characteristics of initial sections in turbulent jets, i.e. for the aimed change of these characteristics.

The known approaches to turbulence control in jet flows using action on their coherent structures can be subdivided into two pronounced categories: 1) passive control that is free of the outer energy input, and 2) active control consuming additional energy.

Passive control is performed by way of change in the issue initial conditions (the flow regime in the boundary layer at the nozzle edge, the change in the boundary layer parameters, the flow initial turbulence, and the initial turbulence scale) or change in geometry of the plant generating the jet (configuration of the nozzle or of the orifice with the sharp edges, the nozzles with complex geometry: rectangular, triangular, elliptic, annular, multi-tube, lobed nozzles, and circular nozzles with generators of longitudinal vortices at the outlet section). Passive control provides a way not only of changing topology of large-scale coherent structures, but also of intensifying the relative role of small-scale turbulence in the case of attenuation of large-scale coherent structures. As a rule, passive control causes the mixing intensification, though for some weak actions resulting in attenuation of the jet coherent structures it is possible to achieve the opposite effect – attenuation of the mixing.

Active control is attained by injection of weak periodical (harmonic) disturbances into the apparatus (the nozzle or the orifice) generating the jet. For this purpose acoustical or vibration disturbances, as well as disturbances of the initial boundary layer of a plane nozzle by use of oscillating bands, are common. In a number of cases one can employ high-amplitude periodical pulsations using different kinds of pulsators located before the nozzle or oscillating small wings located behind the nozzle.

With appropriate choice of the frequency of weak acoustical disturbances it is possible to intensify the vortex pairing in the jet initial section or to attenuate the pairing or, alternatively, to cause the early decay of coherent structures. Ultimately, it gives a possibility of generating or attenuating turbulence and strengthening or attenuating the jet noise.

The main advantage of the acoustic control of turbulent jets is its high efficiency: the tone disturbance of very low intensity causes considerable change in aerodynamic, acoustical and other characteristics of a turbulent jet. For example, for acoustical excitement of a jet, the ratio between mean-square velocity pulsations in a sound wave at the nozzle edge, u_s', and the issue velocity, u_0, is $u_s'/u_0 = 0.001 - 0.02$. The essential merit of the acoustical control method for subsonic turbulent jets lies in the fact that its implementation does not demand incorporation of interrupters or moving parts into the flow. Essentially, the acoustical control method for subsonic turbulent jets reduces to control of the jet coherent structures in the jet initial section. When we change the nozzle geometry that causes considerable attenuation of large-scale coherent structures (for example, for a lobed nozzle), the jet periodical excitement does not change the jet characteristics.

The appropriate changes in jet flow geometry that generate the jet self-excitation (for example, self-excitation of a near-subsonic jet for its leaking on a wall, for organization of weak abrupt expansion behind the nozzle, for location of a resonator near the nozzle outlet section, etc.) are intermediate between passive and active control of turbulent jets. In all of these cases the action mechanisms are defined by generation of acoustic oscillations with acoustical feedback.

The mechanisms of turbulent mixing and generation of aerodynamic noise in supersonic anisobaric jets differ considerably from the corresponding mechanisms in subsonic turbulent jets. They are considered in the corresponding chapters of the present book. It is significant that acoustic control methods could be efficient also for supersonic jets. The cases of passive and active control are studied here also. In the latter case the control method using a shield that irradiates the jet with sound waves reflected from the jet gas-dynamic section is the most efficient.

We consider below acoustical approaches to control of aerodynamic and acoustical characteristics of subsonic and supersonic anisobaric turbulent jets. The review of control methods for such jets, as well as of some applications can be found in [1.10, 1.11, 1.14, 1.15, 1.17, 1.40, 1.41].

Notation

x, r – coordinates in the cylindrical coordinate system

d – diameter of the circular nozzle outlet section

$\delta_{1/2}$ – one half-velocity radius in a jet

θ_0 – momentum loss thickness of the boundary layer in the nozzle outlet section

u, v – longitudinal and radial components of the jet averaged velocity

u' – mean-square value of the longitudinal velocity pulsations

$\mathrm{Re} = \dfrac{u_0 d}{\nu}$, $\mathrm{M}_0 = \dfrac{u_0}{a}$, $\mathrm{St} = \dfrac{fd}{u_0}$, $\varepsilon_{u0} = \dfrac{u'}{u_0}$ – Reynolds, Mach, Strouhal numbers, and turbulence level respectively at the nozzle outlet section

R_{uu}, R_{up}, R_{vp}, R_{pp} – coefficients of the spatial correlation for the longitudinal velocity pulsations, for the longitudinal velocity and pressure pulsations, for the transversal velocity and pressure pulsations, and for the pressure pulsations respectively

φ – angle between the jet axis and the line connecting the nozzle outlet section with the point of measuring the pressure pulsations outside the jet

ψ – azimuthal angle

Subscripts 0, 00 and ∞ correspond to the nozzle outlet section, the deceleration parameters and the flow parameters outside the jet.

REFERENCES

1.1. Abramovich GN, Girshovich TA, Krasheninnikov SYu, Secundov AN, Smirnova IP (1984) Theory of turbulent jets (in Russian). Nauka, Moscow

1.2. Acton E (1980) A modelling of large eddies in an axisymmetric jet. J Fluid Mech 98. Pt. 1: 1 – 31
1.3. Bradbury LJS, Khadem AH (1975) The distortion of a jet by tabs. J Fluid Mech 70. Pt. 4: 801 – 813
1.4. Browand FK, Laufer J (1975) The role of large scale structures in the initial development of circular jets. Proc. IV Biennial Symp Turbulence Luquids. Princeton, New Jersey: 333 – 344
1.5. Bruun HH (1979) A time-domain evolution of the large-scale flow structure in a turbulent jet. Proc Roy Soc London: 193 – 218
1.6. Crighton DG (1981) Acoustic as a branch of fluid mechanics. J Fluid Mech 106: 261 – 298
1.7. Crighton DG, Gaster M (1978) Stability of slowly diverging jet flow. J Fluid Mech 77: 397 – 413
1.8. Crow SC, Champagne FH (1971) Orderly structure in jet turbulence. J Fluid Mech 48, # 3: 547 – 591
1.9. Dahan C, Elias G, Maulard J, Perruli M (1977) Coherent structures in the mixing zone of a subsonic hot free jet. J Sound and Vibration 59, # 3: 313 – 333
1.10. Fiedler HE (1987) Coherent structures. Adv Turbul Proc 1st Eur Turbul Conf Lyon: 320 – 336
1.11. Feidler HE, Fernholz HH (1990) On management and control of turbulent shear flow. Progr Aerospace Sci 27, # 4:305 – 387
1.12. Gertsenshtein SYa, Sukhorukov AN (1985) On nonlinear evolution of two-dimensional and three-dimensional waves in mixing layers. Fluid Dynamics 20, #1: 7 – 14
1.13. Ginevsky AS (1969) Theory of turbulent jets and wakes (in Russian). Mashinostroyenie, Moscow
1.14. Ginevsky AS, Vlasov YeV, Kolesnikov AV (1978) Aeroacoustical interactions (in Russian). Mashinostroyenie, Moscow
1.15. Ho CM, Huerre P (1984) Perturbed free shear layers. Ann Rev Fluid Mech 16: 365 – 424
1.16. Ho CM, Nosseir NS (1981) Dynamics of an impinging jet. Part 1. The feedback phenomenon. J Fluid Mech 105: 119 – 142
1.17. Hussain AKMF (1983) Coherent structures – reality and myth. Phys Fluids 26, # 10: 2816 – 2859
1.18. Hussain F, Hussain HS (1989) Elliptic jets. Part 1. Characteristics of unexcited and excited jets. J Fluid Mech 208: 257 – 320
1.19. Isataev SI, Tarasov SB (1972) On velocity pulsation characteristic frequencies and spectra in initial regions of axisymmetric jets (in Russian). In: Applied and theoretical physics 4, Alma-Ata, pp 247 – 252
1.20. Isataev SI, Tarasov SB (1975) On velocity pulsation characteristic scales in the transition area of free axisymmetric jets (in Russian). In: Applied and theoretical physics 7, Alma-Ata, pp 140 – 146
1.21. Kibens V (1981) The limit of initial shear layer influence on jet development. AIAA Pap 1960
1.22. Krasheninnikov SYu, Mironov AK, Vasiliev VI, Zakotenko SN (1998) An analysis of efficiency of mixer/ejector for aircraft jet noise suppression. In: Proc Internat EAA/EEAA Symp "Transport noise and vibration", Tallinn, pp 353 – 356
1.23. Laufer J (1974) On the mechanism of noise generated by turbulence. In: Omaggio a Carlo Ferrari, Torino, pp 451 – 464
1.24. Laufer J, Monkevitz PA (1980) On turbulent jet flow in a new perspective. AIAA Pap 962

1.25. Lighthill MI (1952) On sound generated aerodynamically. Part 1. General theory. Proc Roy Soc, Ser A 211, # 1107: 564 – 587. (1954) 222, # 1148: 1 – 32
1.26. Liu JTC (1989) Coherent structures in transitional and turbulent free shear flows. Ann Rev Fluid Mech 21: 285 – 315
1.27. Mansfeld AD, Rabinovich MI, Sushchik MM (1979) Noise and large-scale coherent structures in turbulence (in Russian). In: Transactions of 2nd All-Union symposium on physics of acoustic-hydrodynamic phenomena and optical acoustics. Moscow, pp 12 – 14
1.28. Michalke A (1971) Instabilität eines kompressiben runden Freisttrahls unter Berücksichtigung des Strahlgrenzschichtdicke. Z Flugwiss 19, # 8-9: 319 – 328
1.29. Morisson GL Emami S, Tatterson GB (1978) An experimental investigation of the coherent structures in an incompressible jet. AIAA Pap 2751
1.30. Munin AG (ed) (1986) Aviation acoustics. Part1 – Noise of subsonic of passenger airplanes in surroundings (in Russian). Mashinostroyenie, Moscow
1.31. Munin AG, Kvitka VE (eds) (1973) Aviation acoustics (in Russian). Mashinostroyenie, Moscow
1.32. Munin AG, Kuznetsov VM, Leontyev EA (1981) Aerodynamic sources of noise (in Russian). Mashinostroyenie, Moscow
1.33. Petersen RA (1979) Influence of wave dispersion on vortex pairing in a jet. J Fluid Mech 89, # 3: 469 – 495
1.34. Rabinovich MI, Sushchik MM (1983) Coherent structures in turbulent flows (in Russian). In: Nonlinear waves. Self-organization, Nauka, Moscow, pp 56 – 85
1.35. Roshko A (1976) Structure of turbulent shear flows: A new look. AIAA J 4, # 10: 1349 – 1357
1.36. Sorkin LI (ed) (1968) Calculations and measurements of characteristics of noise generated in the far noise field by jet airplanes (in Russian). Mashinostroyenie, Moscow
1.37. Sreenivasan KR (1984) The azimuthal correlations of velocity and temperature fluctuations in an axisymmetric jet. Phys Fliuds 27, # 4: 867 – 875
1.38. Sushchik MM (1987) Dynamic of structures in shear flows (in Russian). Nauka, Moscow
1.39. Vlasov YeV (1968) Study of turbulent and acoustical characteristics of subsonic jets (in Russian). Proc TsAGI 1092: 3 – 14
1.40. Vlasov YeV, Ginevsky AS (1980) Problem of aeroacoustical interactions. Soviet Physics Acoustic 26, # 1: 1 – 7
1.41. Vlasov YeV, Ginevsky AS (1986) Coherent structures in turbulent jets and wakes (in Russian). In: Advances in Science and Technology. Series "Mechanics of Fluids and Gases" 20, VINITI, Moscow, pp 3 – 84
1.42. Vlasov YeV, Ginevsky AS, Karavosov RK (1979) Coherent structures in coflowing turbulent jets (in Russian). In: Turbulent jet flows. Part 1. Tallinn, pp 20 – 27
1.43. Vlasov YeV, Ginevsky AS, Karavosov RK (1980) Direct and indirect approaches to experimental detection of coherent structures in turbulent jets (in Russian). In: Mechanics of turbulent flows. Nauka, Moscow, pp 206 – 219
1.44. Vlasov YeV, Ginevsky AS, Karavosov RK, Ukhanova LN (1987) Study of resonance modes for subsonic turbulent jet leaking on a screen (in Russian). In: Problems of turbulent flows. Nauka, Moscow, pp 135 – 147
1.45. Vlasov YeV, Ginevsky AS, Karavosov RK (1989) Influence of flow initial conditions on aerodynamic and acoustical characteristics of turbulent jets (in Russian). Mechanics of nonuniform and turbulent flows. Nauka, Moscow, pp 26 – 34
1.46. Vlasov YeV, Karavosov RK, Lebedeva OV (1991) Acoustic characteristics of a turbofan (in Russian). Aviation Science and Technology 5–6 : 31 – 35

1.47. Vlasov YeV, Ginevsky AS, Pimshtein VG (1992) Control of coherent structures and aero-acoustic characteristics of subsonic and supersonic turbulent jets. Proc DGLR/IAA 14th Aeroacoustics Conf, Aachen: 672-678
1.48. Vlasov YeV, Ginevsky AC, Efimtsov BM, Kuznetsov VM, Munin AG, Samokhin VF, Smoliakov AV, Sobolev AF (1996) Main problems of acoustics (in Russian). Proc. TsAGI 2614
1.49. Vlasov YeV, Ginevsky AS, Karavosov RK, Makarenko TM (1999) Acoustical excitation of turbulent jets issuing from diaphragms (in Russian). In: Modern problems of aero-hydromechanics. Volume 2. Nauka, Moscow, pp 156 – 161
1.50. Yule AJ (1978) Large-scale structure in the mixing layer of round jets. J Fluid Mech 89, P 3: 413 – 432
1.51. Zaman KBMQ (1985) Effect of initial conditions on subsonic jet noise. AIAA J 23, # 9: 1370 – 1373

CHAPTER 2. CONTROL OF AERODYNAMIC CHARACTERISTICS OF SUBSONIC TURBULENT JETS

2.1. Susceptibility of Turbulent Jets to Weak Harmonic Acoustical Disturbances. Effect of Excitation Frequency

Aerodynamic characteristics of turbulent jets and mixing layers can be changed by periodical action on the jet initial section flow. Such action could be realized by generation of periodical change in fluid consumption through the nozzle, by the nozzle vibrations or by excitation of the mixing layer at the nozzle edge using a vibrating band. The listed control approaches are connected with mechanical action on the flow because they require direct action on geometry of a unit generating the jet flow [2.10]. The mechanism of their action on the flow is determined by the jet periodical excitation causing generation of annular periodical vortices at the circular nozzle outlet section. Their interaction with each other essentially changes the flow in the mixing layer of the jet initial region.

Under acoustical irradiation of a turbulent jet the direct interaction of the acoustical field and turbulent pulsations in the jet is practicaly unavailable, as the length of the acoustical waves far exceeds the nozzle reference size (its diameter) or the boundary layer thickness in the initial section of the mixing layer. However, the acoustical waves generate vortex disturbances at flow irregularities [2.6, 2.18], at the nozzle edge in the case of a jet or at the trailing edge dividing two flows over a plate in the case of a mixing layer. These disturbances act on the mixing layer vortex set in the jet initial region much as one can see in the case of mechanical action on a flow. But the acoustical excitation has the important advantage of long-range action, i.e. it does not require the introduction of obstacles or movable units into the flow.

Consider the simplest schemes of single-frequency acoustical excitation of jets. The jet longitudinal (downstream and upstream), transverse and radial irradiation (Fig. 2.1) are recognized. For a jet under longitudinal and radial excitation by sound waves, axisymmetric periodical vortex rings are generated at the nozzle edge and carried away downstream. Skewed vortex rings are generated in jets under transverse acoustical excitation at the circular nozzle cut-off edge [2.42]. They are more intensive at the emitter side and less intensive at the opposite side. This fact, as will be demonstrated later, determines a vague difference in acoustical action effects for longitudinal and transverse excitation.

For a jet under weak longitudinal periodical excitation, sound waves generate at the frequency of the jet instability [2.49] more regular annular vortices (Fig. 2.2, *a*) than for unforced instability initiation (Fig. 2.2, *b*). In both cases these vortices are generated at the nozzle edge. It is worth noting that vortex rings excited by noise are more intensive than in the case of the lack of periodical excitation, they pair and decay at a later time.

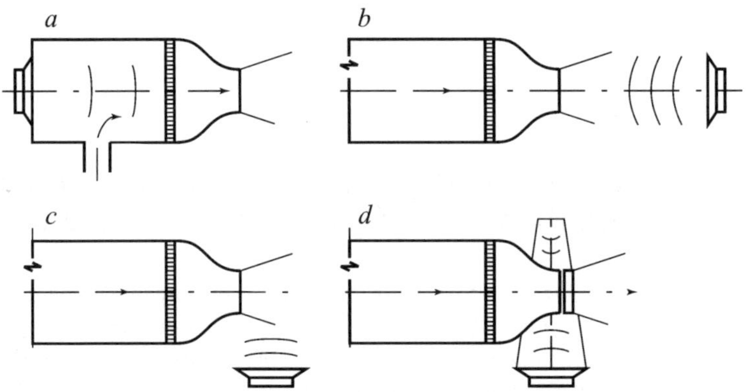

Fig. 2.1. Longitudinal, transverse and radial acoustical excitation of a jet: (*a*) downstream longitudinal irradiation, (*b*) – upstream longitudinal irradiation, (*c*) – transverse irradiation, (*d*) – radial irradiation.

Harmonic (sinusoidal) oscillations characterized by two parameters (the frequency and amplitude), are of the greatest interest for the case of a jet under weak acoustical oscillations. Already first experiments with tone acoustical excitation of low-velocity turbulent jets demonstrated their susceptibility to acoustical oscillations.

Two effects of the opposite character for transverse and longitudinal acoustical irradiation of a turbulent jet depending on the irradiation frequency were first discovered in [2.50, 2.51]. It was found that at frequencies corresponding to the Strouhal numbers $\mathrm{St_s} = 0.2 - 0.6$ (*low-frequency irradiation*) the intensification of mixing occurs, the mixing layer thickness increases, the initial region length decreases, the velocity longitudinal and radial pulsations at the jet axis increase, the periodical vortices within the jet initial region become consolidated, and the jet ejecting ability is strengthened. At frequencies corresponding to Strouhal numbers $\mathrm{St_s} = 1.5 - 5.0$ (*high-frequency irradiation*) the attenuation of mixing occurs, the mixing layer thickness decreases, the initial region length increases, the velocity longitudinal and radial pulsations at the jet axis decrease, the periodical vortices within the jet initial region become pulverized, and the jet ejecting ability is attenuated.

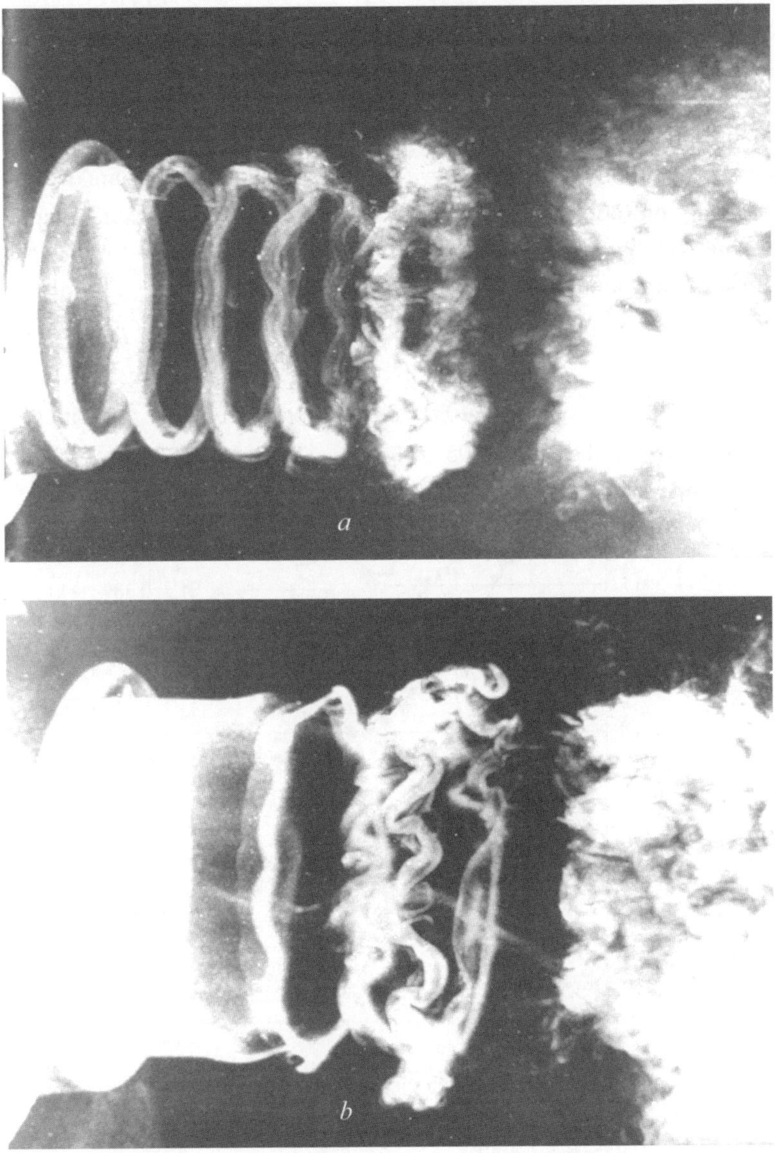

Fig. 2.2. Generation of a series of vortex rings for (*a*) a jet under downstream longitudinal acoustical irradiation and (*b*) in the case of the lack of the jet acoustical excitation.

These effects were first established for low-speed jets (issue speed $u_0 < 100$ m/s, nozzle diameter $d = 10 - 50$ mm, $L = 100 - 120$ dB). The dimensionless parameter that is equal to the ratio between the mean-square value of the velocity pulsations in the sound wave near the nozzle edge without the flow and the issue velocity, i.e. $\varepsilon_{us} = \left(\langle u_s' \rangle\right)^{1/2} / u_0$ for longitudinal irradiation and $\varepsilon_{vs} = \left(\langle v_s' \rangle\right)^{1/2} / u_0$ for transversal irradiation, was used later to estimate the jet acoustical irradiation intensity. In a number of cases $\varepsilon_{us} = u_s' / u_0$ and $\varepsilon_{vs} = v_s' / u_0$ are taken for simplicity with u_s' and v_s' being equal not to the instantaneous values of the velocity pulsations in the sound wave but to their mean-square values. The values of u_s' and v_s' are determined by the values of L using the well-known formula for a plane sound wave $p_s' = \rho a u_s'$ or $p_s' = \rho a v_s'$ where p_s' is the velocity pulsation in the sound wave. Since $L = 20 \log\left(p_s' / p_*\right)$, we have $u_s' / u_0 = p_* 10^{L/20} / \rho a u_0$ where $p_* = 2 \cdot 10^{-5}$ Pa, ρ and a are the density and sound speed in the flow at the nozzle edge respectively. The quantities ε_{us} and ε_{vs} tend to lie within the range 0.05 – 2%.

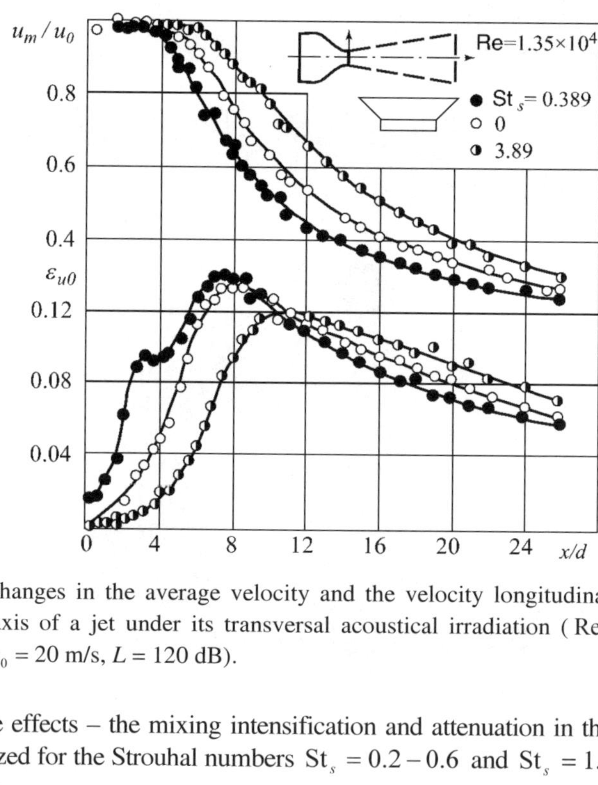

Fig. 2.3. Changes in the average velocity and the velocity longitudinal pulsations along the axis of a jet under its transversal acoustical irradiation ($\mathrm{Re} = 1.36 \cdot 10^4$, $\varepsilon_0 = 1\%$, $u_0 = 20$ m/s, $L = 120$ dB).

Two above effects – the mixing intensification and attenuation in the jet initial region, are realized for the Strouhal numbers $\mathrm{St}_s = 0.2 - 0.6$ and $\mathrm{St}_s = 1.5$ - 5.0 characteristic for coherent structures near the end of the initial region ($x/d = 3 - 5$) and in the near-nozzle part of the mixing layer ($x/d < 1$).

Consider some results of experiments illustrating the above effects for a circular jet. Fig. 2.3 presents the changes in the average velocity and the velocity longitudinal pulsations along the axis of a jet under its transversal acoustical irradiation [2.50] and Fig. 2.4 shows similar results for a jet under its longitudinal and transversal acoustical irradiation [2.14].

Fig. 2.5 presents functions for the average velocity, $u/u_{-} = \varphi_1(\mathrm{St}_s)$, and for the velocity longitudinal pulsations, $u'/u'_{-} = \varphi_2(\mathrm{St}_s)$, for the fixed point $x/d = 8$ at the jet axis that illustrate the ranges of the Strouhal numbers corresponding to the intensification $[u/u_{-} < 1, \; u'/u'_{-} > 1]$ and attenuation $[u/u_{-} > 1, \; u'/u'_{-} < 1]$ of mixing in a jet under its transversal acoustical excitation [2.50, 2.58]. It is worth noting that the nonmonotone saw-tooth nature of the presented functions is determined by the similar nature of the frequency response of the electrodynamic emitter when the maximum level of the sound pressure, *L*, was set at each excitation frequency. Another possible reason for this phenomenon could lie in resonance features of the receiver ejecting the jet. The functions $u/u_{-} = \varphi_1(\mathrm{St}_s)$ and $u'/u'_{-} = \varphi_2(\mathrm{St}_s)$, as a rule, are monotone for a fixed level of the sound pressure $L =$ const. This fact will be illustrated in the next paragraph. Therefore, the main changes in a jet under acoustical excitation happen within the initial and transition regions ($x/d < 8$).

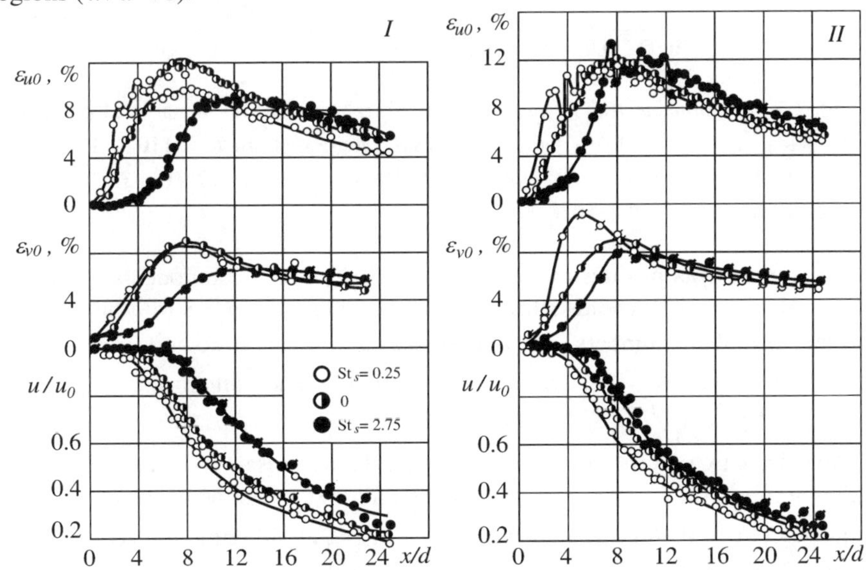

Fig. 2.4. Changes in the average velocity and the velocity longitudinal and radial pulsations along the axis of a turbulent jet under its longitudinal and transversal acoustical irradiation. *I* – longitudinal irradiation, *II* – transversal irradiation. $\mathrm{Re} = 1.4 \cdot 10^4$, $M_0 = 0.059$, $u'_s/u_0 = 0.1\%$, $v'_s/u_0 = 0.4\%$.

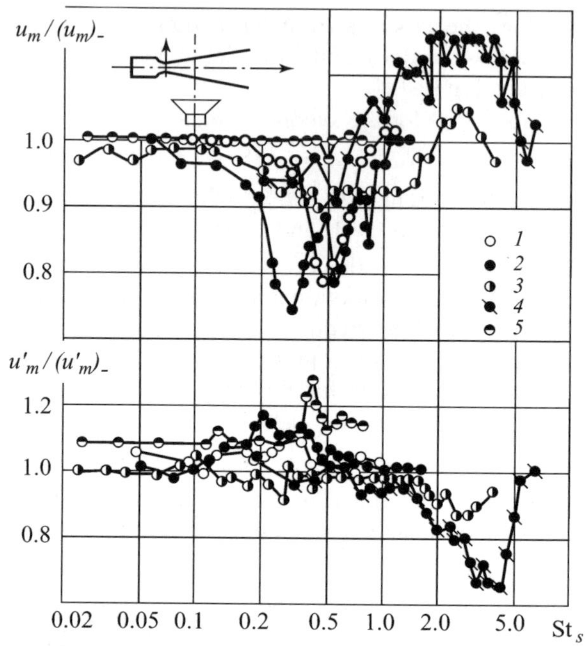

Fig. 2.5. Functions $u_m/(u_m)_- = \varphi_1(\mathrm{St}_s)$ and $u'_m/(u'_m)_- = \varphi_2(\mathrm{St}_s)$ for a fixed point at the jet axis $x/d = 8$ for a jet under transversal irradiation, L = 110 – 120 dB, $u_0 = (13{,}5$ - $114)$ m/s, $d =$ (10 – 55) mm, *1)* $\mathrm{Re} = 2.2\cdot10^5$, *2)* $1.2\cdot10^5$, *3)* $3.1\cdot10^4$, *4)* $2.8\cdot10^4$, *5)* $\mathrm{Re} = 2.4\cdot10$.

Fig. 2.6 presents profiles of the average velocity, three components of the pulsation velocity and Reynolds shift stress at the section $x/d = 8$ of a turbulent jet for the Strouhal numbers of transversal acoustical irradiation $\mathrm{St}_s = 0.39$, $\mathrm{St}_s = 3.89$ and $\mathrm{St}_s = 0$ (without excitation). The other parameters are $\mathrm{Re} = 1.35\cdot10^4$ and $v'_s/u_0 = 0.1\%$ [2.50].

Consider now spectra of the longitudinal velocity pulsations at the jet axis and in the mixing layer ($x/d = 3$) without excitation ($\mathrm{St}_s = 0$), as well as under low-frequency ($\mathrm{St}_s = 0.46$) and high-frequency ($\mathrm{St}_s = 3.4$) longitudinal acoustical excitation. Fig. 2.7 presents the corresponding one third octave spectra [2.58]. One can see that in the near-axis area of the jet initial region the acoustical excitation causes a shift of the spectrum upward (for $\mathrm{St}_s = 0.46$) and downward (for $\mathrm{St}_s = 3.4$). It follows, in particular, that the curves $\mathrm{St}_m(x/d)$ presented in Chapter 1 (Fig. 1.4, *b*) practically coincide for unexcited and excited jets [2.50, 2.54].

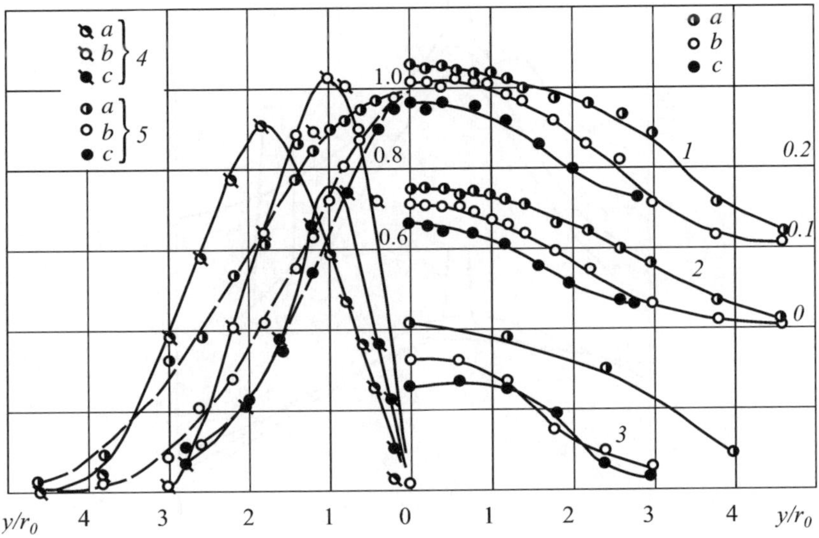

Fig. 2.6. Profiles of the average velocity, three components of the pulsation velocity and Reynolds shift stress at the section $x/d = 8$ of a jet with and without transversal acoustical irradiation ($r_0 = d/2$, Re $= 1.35 \cdot 10^4$). *1)* $\langle u'^2 \rangle^{1/2}/u_m$, *2)* $\langle v'^2 \rangle^{1/2}/u_m$, *3)* $\langle w'^2 \rangle^{1/2}/u_m$, *4)* $100\langle u'v' \rangle/u_m^2$, *5)* u/u_m. (*a*) $\mathrm{St}_s = 0.389$, (*b*) $\mathrm{St}_s = 0$, (*c*) $\mathrm{St}_s = 3.89$.

Intensification of the flow periodicity in the initial region of a jet under its acoustical excitation ($\mathrm{St}_s = 0.5$) is attended (Fig. 2.8) with the change of spatial correlation of the longitudinal velocity pulsations R_{uu} along the jet axis as compared with the case of the lack of excitation [2.25]. As follows from Fig. 2.9, the jet acoustic excitation causes also changes in the Euler time correlation at the jet axis ($x/d = 4$) and in the mixing layer depending on St [2.53].

The consolidation of periodical vortices in the mixing layer of a jet under low-frequency acoustical excitation is accompanied by slight deterioration of the flow isobarity in the near-axis area of its initial region [2.23].

As mentioned above (Section 1.1), the characteristic Strouhal number for the jet high-frequency excitation should be determined not only from the nozzle diameter, but also from characteristic value of the boundary layer at the nozzle edge, e.g. from the momentum loss thickness θ_0. Thus, let us consider data of the corresponding experiment [2.73] for the mixing layer of a circular jet as a function of $\mathrm{St}_d = f_s d/u_0$ and $\mathrm{St}_\theta = f_s \theta_0/u_0$ for $x/d = 4$ where θ_0 is the momentum loss thickness at the initial section of a laminar boundary layer (Fig. 2.10 for a jet under longitudinal acoustical excitation. We can see that the maximum attenuation of the mixing process corresponds to the range $\mathrm{St}_d = 1-3$ and $\mathrm{St}_\theta = 0.012-0.022$.

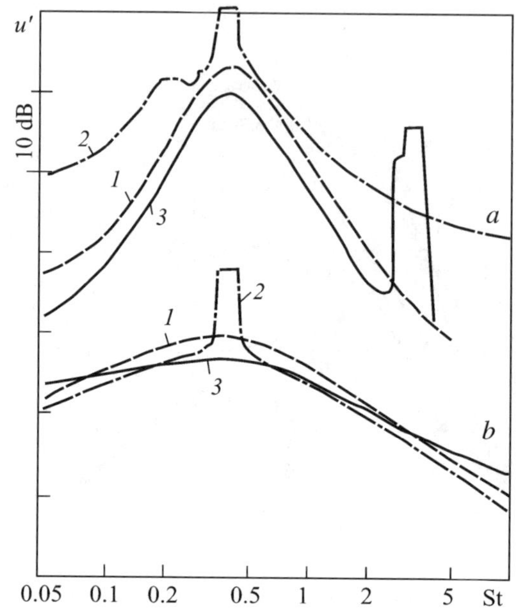

Fig. 2.7. One third octave spectra of the longitudinal velocity pulsations at (*a*) the jet axis and at (*b*) the nozzle edge at the jet section $x/d = 3$ under the jet longitudinal acoustical excitation. *1)* a radiation-free jet, *2)* $\mathrm{St}_s = 0.46$, *3)* $\mathrm{St}_s = 3.4$.

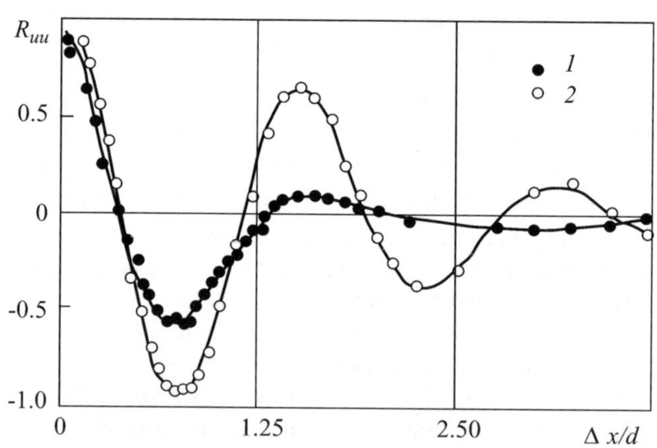

Fig. 2.8. Spatial correlation of the longitudinal velocity pulsations R_{uu} along the jet axis under the jet longitudinal low-frequency acoustical excitation ($x/d = 3$, $y = 0$, $\mathrm{Re} = 2.4 \cdot 10^4$). *1)* a radiation-free jet, *2)* $\mathrm{St}_s = 0.5$.

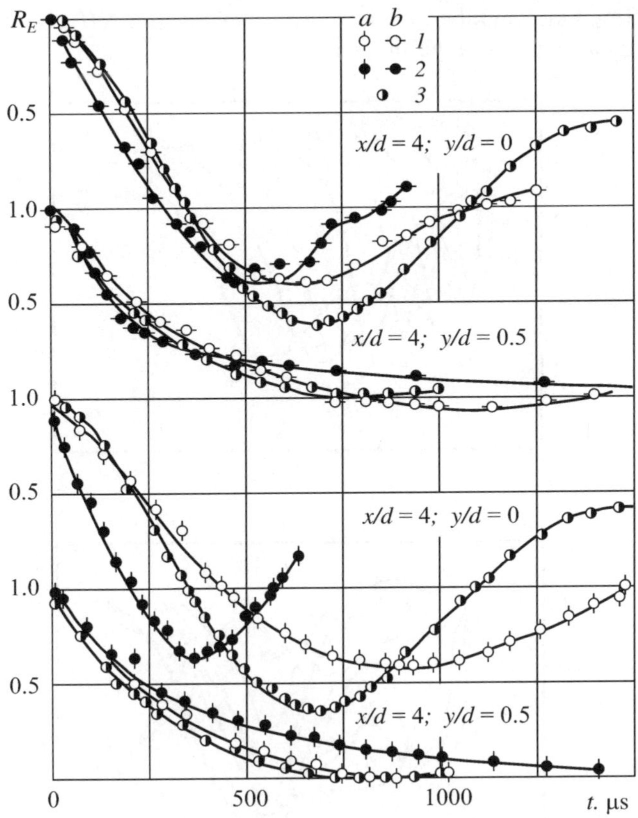

Fig. 2.9. Change in the Euler time correlation of the longitudinal velocity pulsations at the points $x/d = 3$, $y/d = 0$ and $x/d = 3$, $y/d = 0.5$ for a jet under (*a*) transversal and (*b*) longitudinal excitation. *1)* $\mathrm{St}_s = 0.25$, *2)* $\mathrm{St}_s = 2.75$, *3)* a radiation-free jet.

In all these cases, the optimal value of St_θ corresponding to attenuation of the mixing process in the jet equals $\mathrm{St}_\theta = 0.017$. This value is observed at a part of the circular jet mixing layer of the range up to $x = 4d$ or $x = (800-1600)\theta_0$.

In the preface we mentioned that, following the discovery of the phenomenon of the mixing intensification/attenuation in turbulent jets under their low-/high-frequency acoustical excitation [2.50, 2.51], a number of publications supporting these effects have evolved. The earliest of them are [2.7, 2.8, 2.23, 2.41, 2.46, 2.73].

Hence acoustical harmonic excitation of a turbulent jet with moderately small amplitudes is an efficient mean for control of the jet statistical characteristics. The present chapter considers the implementation of such control for different initial conditions of the jet issue (the excitation level, the flow mode in the initial boundary layer, the flow initial turbulence, the mode composition of acoustical disturbances, the flow Mach number, the jet anisothermic degree, effects of departure of the control signal shape from harmonicity, etc.).

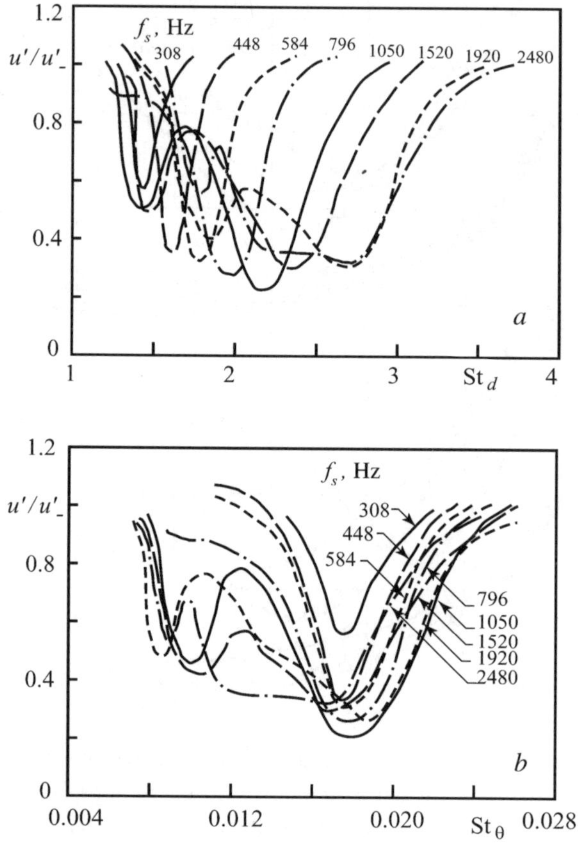

Fig. 2.10. Functions $u'/u'_- = \psi_1(\mathrm{St}_d)$ and $u'/u'_- = \psi_2(\mathrm{St}_\theta)$ for a circular jet under longitudinal acoustical irradiation and fixed frequencies f_s; $x/d = 4$, $u'_s/u_0 = 1\%$.

2.2. Effects of Acoustic Excitation Level

Experimental studies demonstrate that the mixing intensification for a jet under *low-frequency* acoustical excitation increases with the excitation level. However, on gaining a certain level of excitation ($u'_s/u_0 = 1.25\%$), saturation occurs. The subsequent increase of this level scarcely affects the jet characteristics. This conclusion was first justified in [2.47] and later supported in a more comprehensive paper [2.44]. Experiments in the latter work were performed with a jet issuing

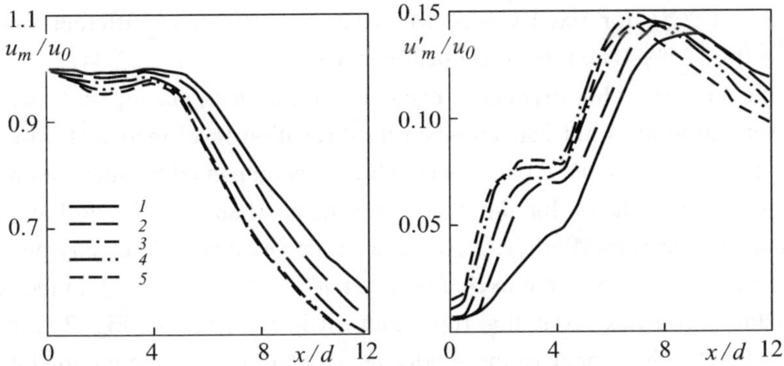

Fig. 2.11. Change of the average velocity / $_0$ and longitudinal velocity pulsations u'/u_0 along the axis of a jet under low-frequency longitudinal acoustical excitation ($M_0 = 0.2$, $St_s = 0.5$). *1)* $u_s'/u_0 = 0$, *2)* $u_s'/u_0 = 0.125\%$, *3)* $u_s'/u_0 = 0.25\%$, *4)* $u_s'/u_0 = 0.875\%$, *5)* $u_s'/u_0 = 1\%$.

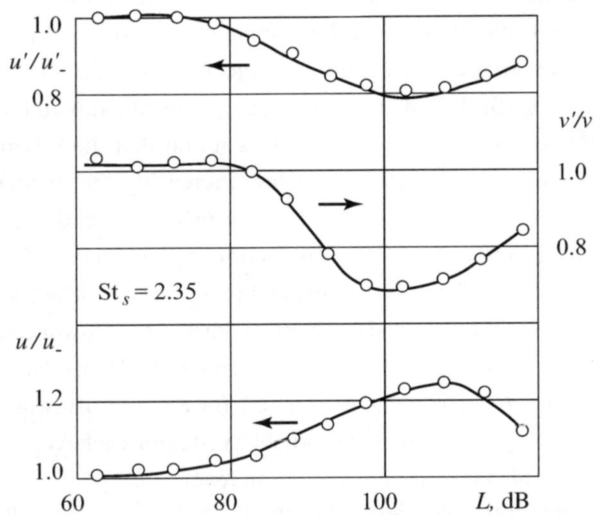

Fig. 2.12. The average velocity, velocity longitudinal and radial pulsations at the axis of a jet under high-frequency acoustical excitation ($St_s = 2.35$) as functions of the excitation level *L*; $u_0 = 10$ m/s.

from a nozzle of diameter $d = 0.088$ m for Mach numbers of the flow $M_0 = 0.2$, 0.3 and 0.54, Strouhal number $St_s = f_s d / u_0 = 0.5$, and the longitudinal acoustical excitation levels $u_s'/u_0 = 0-2\%$. The boundary layer at the nozzle edge was turbulent ($H \approx 1.6$), and the flow initial turbulence at the centre of the nozzle outlet section $\varepsilon_0 = 0.3\%$. Fig. 2.11 presents functions $u_m/u_0 = F_1(x/d)$ and

$u'_m/u_0 = F_2(x/d)$ for fixed values $M_0 = 0.2$, $St_s = 0.5$ and different values of $u'_s/u_0 = 0 - 1.25\%$. The functions illustrate the saturation effect [2.44].

Unlike the case of low-frequency excitation, for *high-frequency* acoustical excitation saturation does not happen when the excitation level increases. Moreover, the excitation level increase over some value is accompanied by attenuation of the effect with the tendency for the action sign to be changed. This follows [2.51] from the dependencies (Fig. 2.12) for the average velocity and velocity pulsations at the jet axis $x/d = 8$ of the excitation level u'_s/u_0 for $St_s = 2.35$ in the case of longitudinal acoustical excitation (upstream). It is evident from Fig. 2.12 that for $u'_s/u_0 = 0.42\%$ the change of the action sign can happen, i.e. the mixing intensification will occur instead of the mixing attenuation.

This phenomenon is described in detail in [2.68, 2.70] for the case of transverse acoustical excitation of a turbulent jet. Fig. 2.13 presents u/u_- and u'/u'_- as functions of the Strouhal number St_s at the point $x/d = 9.5$ for different levels of the sound pressure, L, for two issue speeds u_0 = 10 and 20 m/s. Here the following facts engage our attention. First, these functions are defined for strictly fixed levels of the sound pressure near the nozzle edge and for v'_s/u_0 respectively. Thus the curves $u/u_- = \varphi_1(St_s)$ and $u'/u'_- = \varphi_2(St_s)$ are smooth unlike the similar curves in the previous paragraph. Second, it is found that the mixing attenuation gives way to the mixing intensification when increasing the excitation level for large values of the Strouhal number (for u = 10 m/s, L = 135 dB, $St_s \geq 8$ and for u = 20 m/s, L = 135 dB, $St_s \geq 10$). The action sign change in Fig. 2.13 corresponds to $v'_s/u_0 = 2.76\%$ for u_0 = 10 m/s and to v'_s/u_0 = 1.38% for u_0 = 20 m/s.

The phenomenon of the action sign change is distinctly illustrated by the curves $u/u_- = \varphi(St_s)$ for u_0 = 10 m/s, $St_s = 9$, L = 105 and 135 dB. Here for $x/d > 5$ we have $u/u_- > 1$ for L = 105 dB and $u/u_- < 1$ for L = 135 dB (Fig. 2.14). Therefore, for a jet under high-frequency periodical excitation we have not only the optimal frequency, but also the optimal excitation level.

The above results for a turbulent jet correlate well with the results of a similar study for the mixing layer [2.35]. This paper investigates the influence of the high-frequency excitation amplitude on the turbulence suppression in the mixing layer of a circular jet of diameter d = 0.27 m irradiated through a narrow slot at the nozzle edge in a radial direction. The initial boundary layer was laminar, θ_0 = 0.374 mm, u_0 = 15 m/s, the Strouhal number $St_\theta = 0.006 - 0.025$, amplitude v'_s/u_0 = 0.5%, 2.5%, 3.5% and 4.5%. The characteristic of the turbulence suppression was defined as the ratio u'/u'_- where u' and u'_- are minimum mean-square values of the velocity pulsations with and without excitation respectively (Fig. 2.15) at the section $x/\theta_0 = 200$ where the suppression turned out to be maximum. It was found that for the minimum among the investigated excitation levels, v_s'/u_0, the maximum suppression occurs the Strouhal number $St_\theta = 0.017$ corresponding to the maximum spatial increase of disturbances in complete agreement with the linear theory of stability. As the excitation level increases from 0.5% to 4.5%, the maximum suppression of turbulence in the mixing layer shifts toward higher values of St_θ from $St_\theta = 0.017$ to $St_\theta = 0.022$. For $St_\theta > 0.025$ the values $u'/u'_- > 1$ can be achieved.

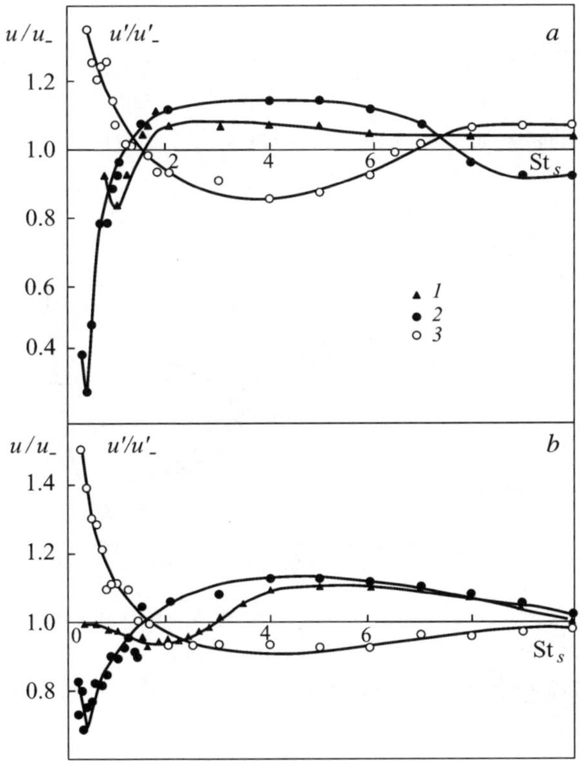

Fig. 2.13. Functions u/u_- *(1, 2)* and u'/u'_- *(3)* of St_s at the point $x/d = 9.5$ of the jet axis for different levels of excitation for (*a*) u_0 = 10 m/s and (*b*) $u_0 = 20$ m/s. *1)* $L = 105$ dB, *2)* $L = 135$ dB, *3)* $L = 135$ dB.

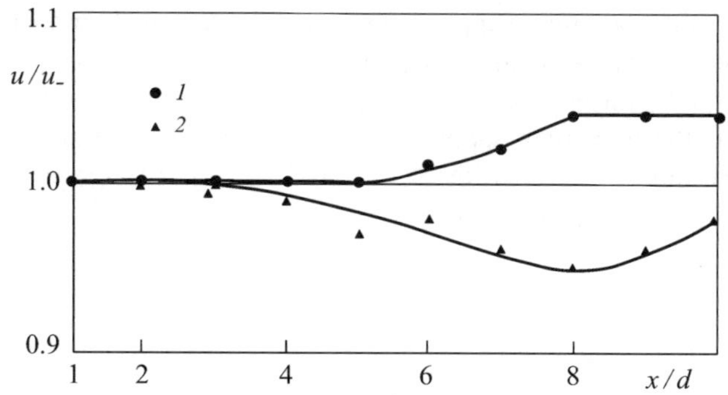

Fig. 2.14. Change of u/u_- along the jet axis for u_0 = 10 m/s, $\mathrm{St}_s = 9$ *1)* L = 105 dB, *2)* L = 135 dB.

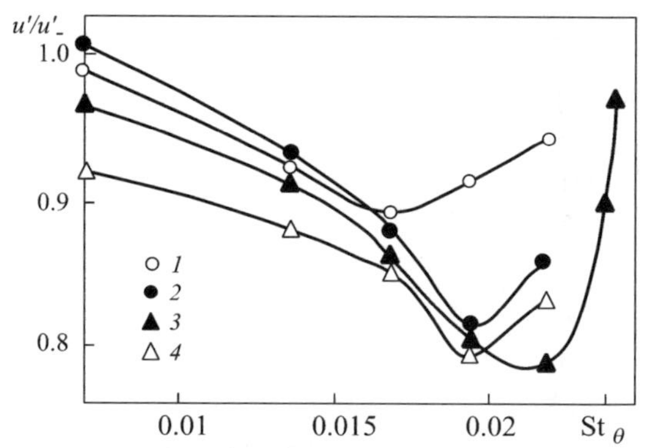

Fig. 2.15. The turbulence suppression degree, u'/u'_-, as a function of the Strouhal number St_θ at the section $x/\theta_0 = 200$ for different levels of acoustical excitation, u'_s/u_0. *1)* $u'_s/u_0 = 0.5\%$, *2)* $u'_s/u_0 = 2\%$, *3)* $u'_s/u_0 = 3\ \%$, *4)* $u'_s/u_0 = 4\ \%$.

2.3. Effect of the Flow Regime in the Boundary Layer at the Nozzle Edge

In the first experiments for acoustical excitation of turbulent jets the flow mode in the initial boundary layer at the nozzle outlet section was not controlled. However, it emerged that the mixing intensification and attenuation effects in a jet under acoustical irradiation depend appreciably on the flow mode in the boundary layer near the nozzle outlet.

The mixing intensification in a jet under its low-frequency excitation is realized regardless of the flow mode in the boundary layer. Some research hypothesized that the mixing attenuation effect in a jet under its low-frequency excitation, which is realized for the initial laminar boundary layer, disappears in the process of turbulization of the boundary layer [2.35, 2.40]. A number of experiments were devoted to clarification of this problem [2.45, 2.58, 2.60, 2.64, 2.66].

The longitudinal acoustical excitation of a turbulent jet for different flow modes in the boundary layer at the nozzle outlet was studied in [2.58]. To change the initial laminar velocity profile, the nozzle cylinder outlet region of diameter $d = 25$ mm and length $x = 1.5d$ may be lengthened till values $x_0 / d = 7.6$ and 20. The boundary layer thickness at the outlet section of these tubes was calculated having regard to the surface cross curvature: the displacement thickness, $\delta_0^* = \int_0^\delta (1-\overline{u})(1-y/r_0)dy$, momentum loss thickness, $\theta_0 = \int_0^\delta \overline{u}(1-\overline{u})(1-y/r_0)dy$, $\overline{u} = u/u_0$, and the shape parameter, $H = \delta_0^*/\theta_0$.

The experimental results are presented in the following table.

Table.

Nozzle	x_0/d	δ_0^*/r_0	θ_0/r	H	$\varepsilon_{\max}$ %
1	1.5	0.0215	0.0083	2.59	3.17
2	7.6	0.0376	0.0160	2.35	11.9
3	20.0	0.0506	0.0334	1.52	15.0

We can conclude from the shape parameter values, *H*, that the cut-off boundary layer was laminar or close to it for nozzles 1 and 2 and turbulent for nozzle 3. This fact agrees with the velocity pulsation measurements in the initial boundary layer at the nozzle outlet (Table data).

The acoustical field was generated by a dynamic loudspeaker mounted on the face wall of the receiver (Fig. 2.16). The Reynolds number was $\mathrm{Re} = u_0 d/\nu = (0.5-1.2)\cdot 10^5$. The successive nozzle lengthening changed naturally the flow mode in the initial boundary layer from laminar to turbulent without use of turbulators. The slight drawback of this method of the boundary layer turbulization is the sensible change in the relative thickness of the initial boundary layer, δ_0/r_0, where $r_0 = d/2$. The sound pressure level at the nozzle outlet section was $L = 120 - 125$ dB.

Presented in Fig. 2.16 are the functions $u/u_{\_} = \varphi_1(\mathrm{St}_s, x_0/d)$ at the point $x/d = 8$ for three variants of the nozzle and show both mixing intensification ($\mathrm{St}_s = 0.2-1.5$) and attenuation ($\mathrm{St}_s = 1.5-6.0$) regardless of the flow mode in the initial boundary layer [2.58]. The realization of the mixing attenuation effect in a jet for large values of St_s was supported in [2.45] for the case of the initial turbulent boundary layer (for $\mathrm{St}_s = 1.5-3.0$, $\mathrm{M}_0 = 0.15$, $L = 130$ dB).

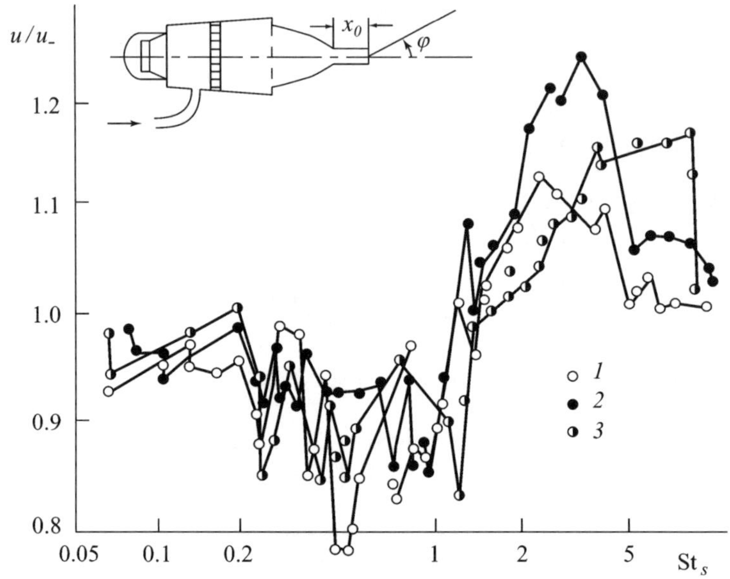

Fig. 2.16. The function $u/u_{-} = \varphi_1(\mathrm{St}_s, x_0/d)$ at the axis point $x/d = 8$ of a jet under longitudinal acoustical excitation. *1)* $x_0/d = 1.5$, *2)* $x_0/d = 7.6$, *3)* $x_0/d = 20.0$.

Chapter 9 describes an approach to the self-oscillations control in wind tunnels with the open working section that is based on low-/high-frequency excitation of a free jet. It is discovered that for high-frequency acoustical excitation ($\mathrm{St}_d = 2-5$) of a jet with diameter d = 1.15 – 2.20 m the mixing attenuation effect is realized regardless of the flow mode in the initial boundary layer of the nozzle outlet section.

Paper [2.66] clarifies conditions for realization of the mixing attenuation effect in the initial region of a turbulent jet for the initial laminar and turbulent boundary layers and for different levels of transversal acoustical excitation. The main parameters of the experiment unit are d = 150 mm, u_0 = 18 m/s, L = 95 – 130 dB, $\mathrm{Re} = 1.89 \cdot 10^5$, and $\mathrm{St}_s = 4.17$. The initial boundary layer at the nozzle cut-off is laminar (the shape parameter H = 2.44); the turbulent boundary layer at the nozzle cut-off (the shape parameter H = 1.54) was obtained using a wire turbulator in the nozzle.

Fig. 2.17 presents functions u/u_{-} and u'/u'_{-} at the jet axis depending on the sound pressure level L at the sections $x/d = 2$ and 6 for the laminar and turbulent boundary layers at the nozzle cut-off. One can see that for the initial laminar boundary layer at both sections, the acoustical irradiation for $\mathrm{St}_s = 4.17$ and $L >$ 115 dB causes the sensible decrease of the velocity pulsations at the jet axis (by 15 – 18%); for the initial turbulent boundary layer similar decrease is achieved only for L = 120 – 130 dB.

For L = 130 dB and the initial turbulent boundary layer we have u'/u'_{-} = 0.82 – 0.88 and u/u_{-} = 1.06. These results lead to the conclusion that the mixing attenuation in a jet under high-frequency acoustical excitation, i.e. the partial turbulence suppression, is realized irrespective of the flow mode in the boundary layer of the nozzle outlet section. However, to realize the effect, one needs an increased level of acoustical excitation for turbulization of the initial boundary layer.

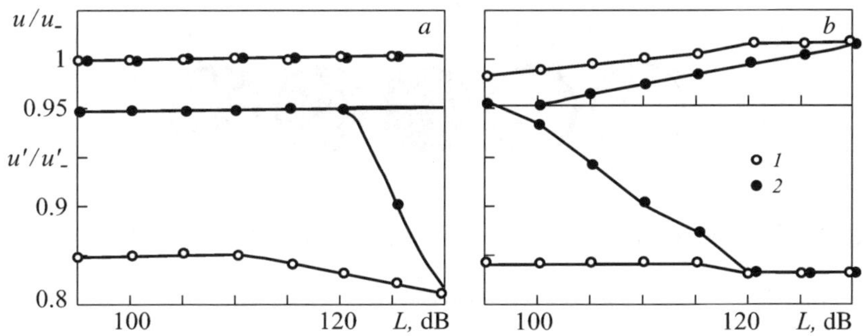

Fig. 2.17. Functions u/u_{-} and u'/u'_{-} at the jet axis depending on the sound pressure level L at the sections $x/d=2$ and 6 for the (*1*) laminar and (*2*) turbulent boundary layers at the nozzle edge. (*a*) $x/d=2$, (*b*) $x/d=6$.

2.4. Influence of Flow Initial Turbulence on the Efficiency of Jet Acoustical Excitation

It is well known [2.13] that the jet initial region is significantly shortened and the turbulent mixing is intensified as the level of the flow initial turbulence increases. Therefore we can reason that the jet acoustical excitation efficiency depends on the initial turbulence level.

Let us consider some results of experimental studies on longitudinal acoustical excitation of turbulent jets for different degrees of the flow initial turbulence [2.45]. The layout of the nozzle, acoustical emitters and turbulizing meshes is depicted in Fig. 2.18. Fig. 2.19 presents the velocity and turbulence intensity profiles at the nozzle outlet section for ε_0 = 0.15, 0.5, 3 and 5%. The experiments are performed for Mach numbers M_0 = 0.05 – 0.35, initial laminar boundary layer ($H \approx 2.4$, $\varepsilon_0 = 0.15\%$) and initial turbulent boundary layer ($H \approx 1.6$, $\varepsilon_0 = 0.15-5\%$). Fig. 2.20 shows functions $u/u_{-} = F_1(\mathrm{St}_s, \varepsilon_0)$ for Mach numbers M_0 = 0.15 and 0.30, $x/d=9$ and excitation level L = 130 dB.

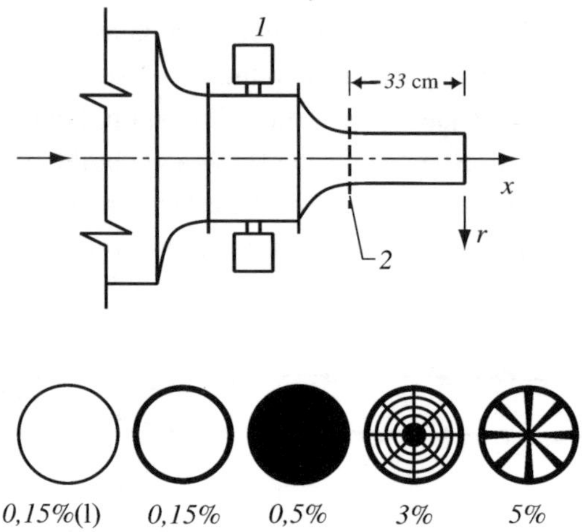

Fig. 2.18. The nozzle ($d = 8.8$ cm), (*1*) acoustical emitters and (*2*) turbulizing meshes.

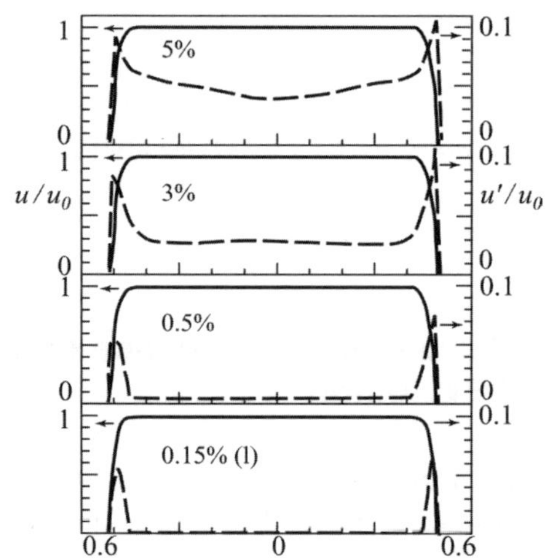

Fig. 2.19. Velocity and turbulence intensity profiles at the nozzle outlet section for various ε_0.

It follows that the turbulence degree increase for low-frequency excitation ($\mathrm{St}_s = 0-1.25$) attenuates the mixing intensification effect with the increase of the Mach number M_0. The mixing intensification attenuation with increasing ε_0 and the Mach number M_0 for high-frequency excitation ($\mathrm{St}_s = 1.5-3.0$) manifests itself to a smaller degree.

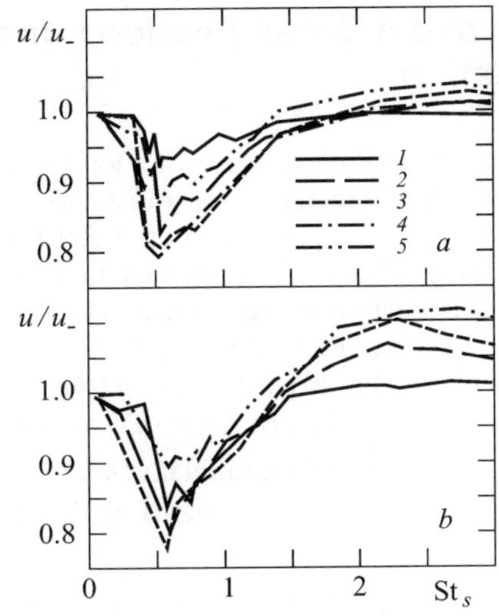

Fig. 2.20. Functions $u/u_{-} = F_1(\mathrm{St}_s, \varepsilon_0)$, (*a*) $\mathrm{M}_0 =$ 0.30 and (*b*) M_0 =0.15, $L =$ 130 dB. *1)* ε_0 $= 5\%$, *2)* ε_0 $= 3\%$, *3)* ε_0 $= 0.5\%$, *4)* ε_0 $= 0.15\%$, *5)* ε_0 $= 0.15\%$ (laminar).

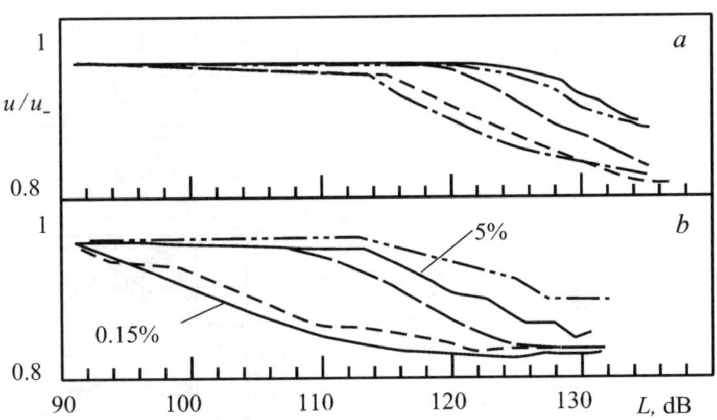

Fig. 2.21. Functions $u/u_{-} = F_2(L)$ for various ε_0, $\mathrm{St}_s = 0.56$ and 0.66. (*a*) $\mathrm{St}_s = 0.66$, $\mathrm{M}_0 = 0.30$; (*b*) $_s =$, $\mathrm{M}_0 = 0.15$.

The joint influence of ε_0 and L on efficiency of the jet low-frequency excitation for $\mathrm{St}_s = 0.33 - 0.66$ is illustrated in Fig. 2.21 by curves $u/u_{-} = F(L, \varepsilon_0)$ for $x/d = 9$.

The presented data illustrate influence of the flow initial turbulence on the efficiency of the jet acoustical excitation. The influence is most pronounced for the jet high-frequency excitation where the elevated initial turbulence considerably attenuates the effect.

2.5. Deformation of Jet Cross Sections under Transversal Acoustical Excitation

In the case of transversal acoustical excitation of a jet, the interaction of sound oscillations with the nozzle edges generates vortex disturbances irregularly along the nozzle edge perimeter. This causes generation of skewed vortex rings and breaks the axis symmetry of the jet flow. The corresponding experimental studies for low-frequency acoustical excitation are performed in [2.37] and for both low- and high-frequency acoustical excitation in [2.63].

The main parameters of the experimental set-up [2.37] are as follows: $d = 135$ mm, $u_0 = 13.5$ m/s, $L = 119$ dB, $\mathrm{St}_s = 0.27$ and 0.45. Fig. 2.22 presents profiles of the average velocity and longitudinal velocity pulsations at the jet sections $x/d = 1$, 3, 5 and 8 for two values of $\mathrm{St}_s = 0.267$ and 0.453 along two

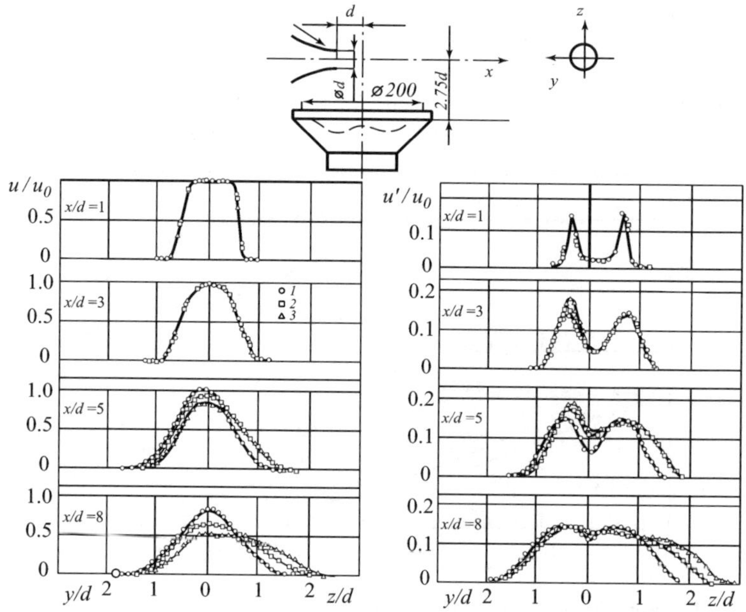

Fig. 2.22. Profiles of the average velocity and longitudinal velocity pulsations along two cross axes y and z for $x/d = 1$, 3, 5 and 8, $\mathrm{Re} = 3.4 \cdot 10^4$: *1)* unexposed jet, *2)* $\mathrm{St}_s = 0.267$, *3)* $\mathrm{St}_s = 0.453$; $L = 119$ dB.

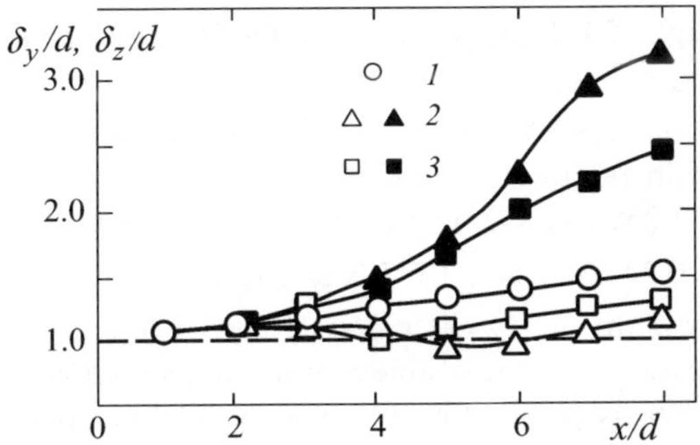

Fig. 2.23. Change of the effective thickness δ_y and δ_z downstream for $\mathrm{Re} = 3.4 \cdot 10^4$. *1)* unexposed jet, *2)* $\mathrm{St}_s = 0.267$, *3)* $\mathrm{St}_s = 0.453$; $L = 119$ dB. Light marks correspond to δ_y / d, black marks to δ_z / d.

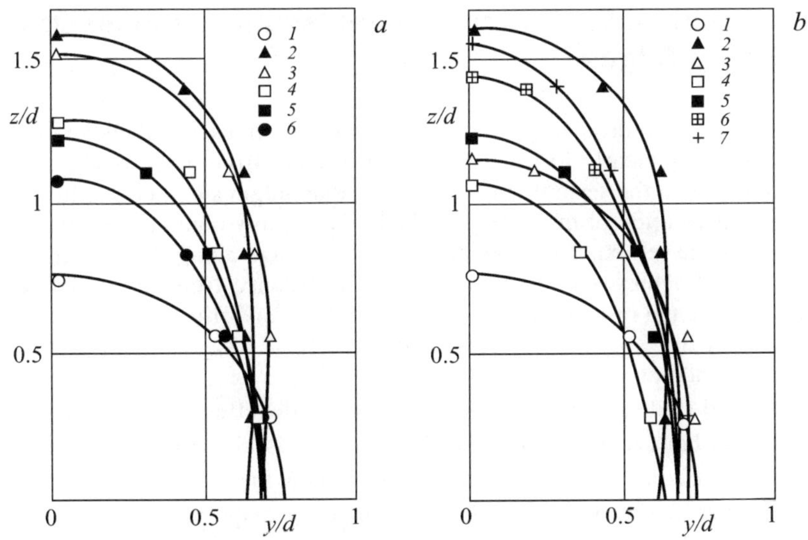

Fig. 2.24. Change of the shape of the jet cross section at $x / d = 8$: (*a*) for $L = 119$ dB: *1)* unexposed jet ($\mathrm{St}_s = 0$), *2)* $\mathrm{St}_s = 0.27$, *3)* $\mathrm{St}_s = 0.29$, *4)* $\mathrm{St}_s = 0.37$, *5)* $\mathrm{St}_s = 0.45$, *6)* $\mathrm{St}_s = 0.80$; (*b*) for $\mathrm{St}_s = 0.27$: *1)* unexposed jet, *2)* $L = 113$ dB, *3)* $L = 119$ dB; for $\mathrm{St}_s = 0.45$: *4)* $L = 113$ dB, *5)* $L = 119$ dB, *6)* $L = 122$ dB, *7)* $L = 125$ dB.

perpendicular axes *y* and *z* where the *z* axis is aligned with the emitter axis. Fig. 2.22 shows also the corresponding velocity profiles for the unexposed jet. It follows that low-frequency transversal irradiation of a jet breaks the initial axial symmetry of the flow. Fig. 2.23 presents the jet effective thickness along the axes *y* and *z* as functions of x / d, i.e. δ_y / d and δ_z / d in the jet region with the length

$x = (0-8)d$ and Fig. 2.24 shows the shape of the jet cross section for $x/d = 8$ and various L and St_s.

Similar studies are performed in [2.63] for the cases of low- and high-frequency excitation. The main parameters of the apparatus are as follows: $d = 20$ mm, $u_0 = 20$ m/s, $\varepsilon_0 = 2.25\%$, $H = 2.09$, i.e. the initial boundary layer is almost laminar, $L = 124$ dB, $\mathrm{St}_s = 0.24$, 0.41 and 3.7. It was found that the cross section deformation exhibits for low-frequency excitation of the jet (the cross section extends in the irradiation direction with the ovalness of the cross section strengthening downstream). For the high-frequency transversal acoustical excitation the jet cross section deformation does not occur, i.e. the flow remains axisymmetric.

Similar deformation of the jet cross-section is observed for low-frequency transversal oscillations of a circular nozzle with large amplitudes [2.52].

2.6. Acoustical Excitation of a Jet Issuing from an Orifice with Sharp Edges

The sensitivity of a turbulent jet to periodical excitation discussed above manifests itself particularly clearly in the case of a jet issuing from a sharp-edged nozzle. This fact is due to the separated flow over the diaphragm sharp edges. In this case the initial distribution of the average velocity is tangibly nonuniform (and has the minimum at the jet axis), the average velocity change along the jet axis is non-monotone and achieves its maximum at some distance from the initial section. Some experimental results for such jets issuing from an orifice with circular, elliptic and triangular sections are contained in [2.22, 2.48], including the case of acoustical excitation.

Theoretical studies of plane and circular jets of ideal fluid issuing from a sharp-edged orifice have demonstrated that these jets on issuing from the orifice grow narrow and at some distance from the diaphragm their cross section does not change further [2.15]. A real turbulent jet behind the diaphragm grows narrow initially and then expands [2.4].

Let us consider the results of experimental studies on excitation of a circular turbulent jet issuing from a sharp-edged orifice [2.65, 2.67]. The measurements were performed at the nozzle with the outlet section of diameter $D = 0.15$ m; the nozzle was partitioned by a plate 5 mm thick with a central hole of diameter $d_0 = 5$ cm with a sharp edge. A narrowing jet issuing from this hole at some distance expanded with the maximum velocity $u_0 = 20$ m/s at the axis in the narrow section.

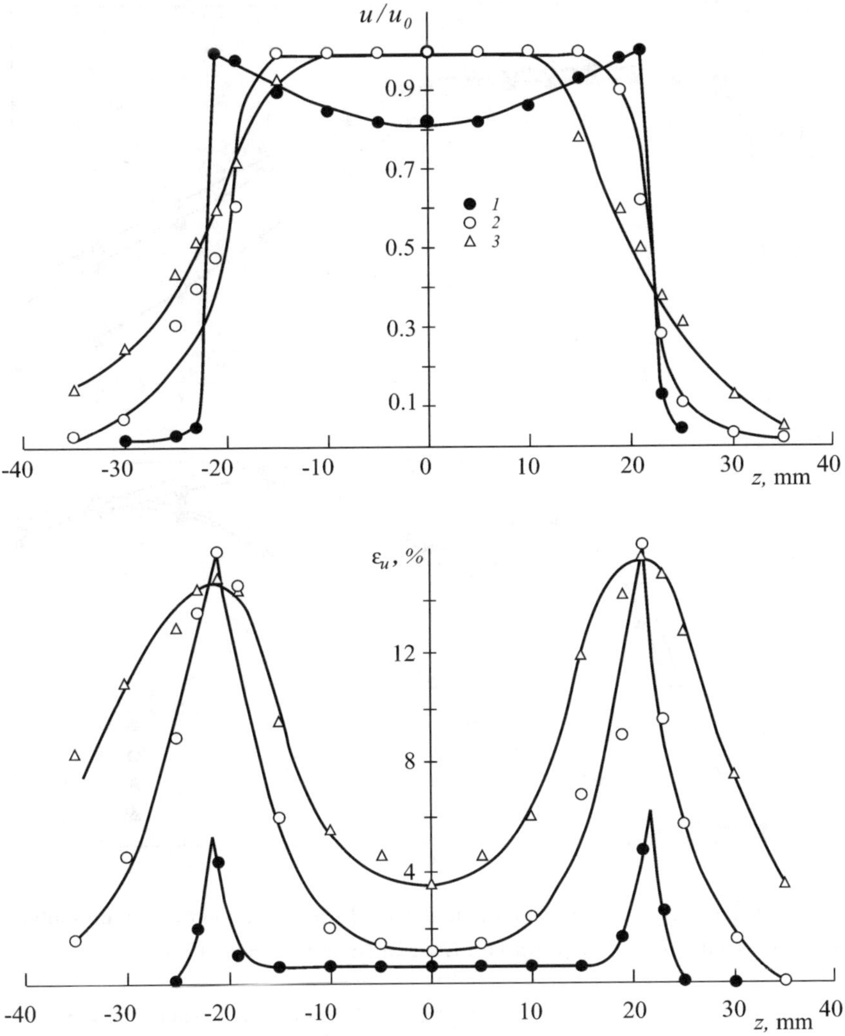

Fig. 2.25. The average velocity and longitudinal velocity pulsation distributions at the initial section of a circular jet issuing from a diaphragm for (*1*) $x/d_0 = 0.12$; (*2*) $x/d_0 = 1$ and (*3*) $x/d_0 = 2$.

Transversal acoustical irradiation was accomplished by an electrodynamic acoustic radiator with the sound pressure level at the diaphragm outlet $L = 110-130$ dB. Fig. 2.25 presents the average velocity and longitudinal velocity pulsation distributions at the initial section 6 mm away from the orifice plane. Fig. 2.26 shows the average velocity and longitudinal velocity pulsations along the jet axis for $L = 120$ dB and $St_s = 0-12.6$. Fig. 2.27 demonstrates the influence of St_s on

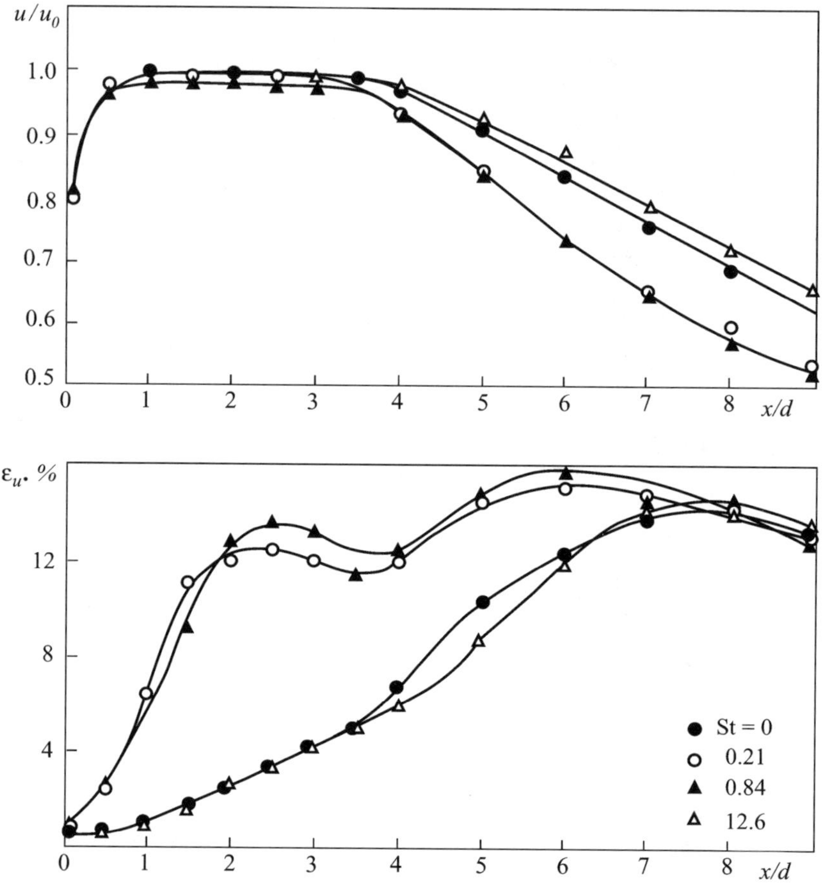

Fig. 2.26. The average velocity and longitudinal velocity pulsation distributions along the axis of a jet issuing from a diaphragm for $L = 120$ dB and $St_s = 0$ - 12.6.

the average and pulsation velocities at the fixed point of the axis ($x/d = 8$) of a jet issuing from a nozzle and a sharp-edged orifice with the same values of d and u_0.

It follows that the values of St_s corresponding to the mixing intensification and attenuation highly differ in both cases. For the case of jet issue from a sharp-edged orifice, the mixing intensification is observed for $St_s = 0 - 4$ and attenuation for $St_s = 4 - 17$. For the case of jet issue from a nozzle these values are 3 – 4 times less due to the small thickness, d, of the mixing layer generating by the separated flow over the orifice sharp edge resulting in the larger initial value of the Strouhal number St_s.

Another important feature of such jets is the lack of deformation of the cross section (that is observed for jets issuing from a nozzle) in the case of low-frequency transversal acoustic irradiation of the jet initial section.

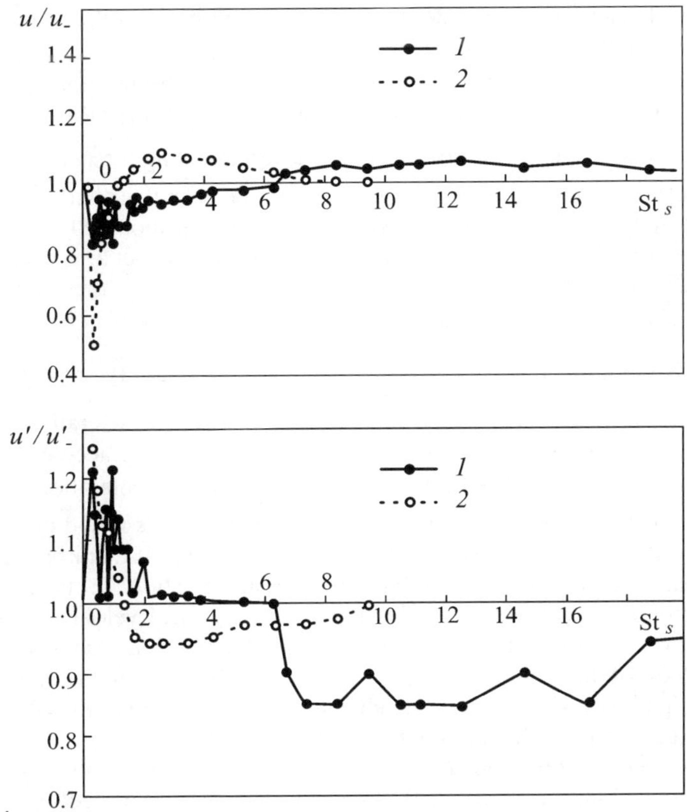

Fig. 2.27. The influence of the transversal irradiation frequency on the average and pulsation velocities at the fixed point of the axis ($x/d = 8$) of a jet issuing from a nozzle and diaphragm. 1) orifice with sharp edge; 2) nozzle.

2.7. Vibration Excitation of a Turbulent Jet

Vibration excitation of a turbulent jet was studied in a set-up described in [2.57, 2.58]. There were two series of experiments. In experiments of the first series a jet was excited by longitudinal acoustical waves with measurements of the nozzle vibrations (longitudinal and transversal) caused by the acoustical irradiation. The nozzle attachment was varied to have a possibility of changing, over wide limits, the nozzle edge vibration acceleration for fixed levels of the acoustical action. In the second series of experiments, without acoustical disturbances, a vibrator generated longitudinal or transversal sinusoidal vibrations of the nozzle (vibration accelerations a_x and a_z). The vibrations varied in amplitude and frequency and were registered by means of vibration transducers. In both cases the jet response on periodical disturbances was defined from measurements of the average velocity at the jet axis at the point $x/d = 8$.

The studies were performed for issue of cold air jets from a nozzle of diameter d = 25 mm with the speed u_0 = 30 m/s and the Reynolds number $\mathrm{Re} = u_0 d/\nu = 0.52 \cdot 10^5$. The nozzles had the variable length (x/d = 1.5, 7.6 and 20.0) to achieve a laminar or transition boundary layer in their outlet sections. The acoustical field was generated by a dynamic loudspeaker mounted on the face wall of the receiver. The sound pressure maximum levels at the jet axis near the nozzle cut-off were as much as L = 115 – 120 dB. The frequency of the sound acting on the jet, f_s, and the vibration frequency, f_v, were monitored by a frequency meter. The nozzle vibration intensity was estimated by the quantity $\varepsilon_v = \left(\left\langle a'^2 \right\rangle\right)^{1/2} / g$, where a' is the vibration acceleration and g is the gravity acceleration.

Considering that the jet acoustical excitation is necessarily accompanied by the nozzle vibrations, the study of influence of the vibration acceleration, ε_v, induced by the sound on the velocity at the jet axis is of interest. It is found that the change of velocity at the jet axis depends chiefly on the acoustical disturbance level and almost unaffected by the nozzle edge vibration acceleration. In the case of a different damping degree of the nozzle attachment, identical velocity changes were achieved for almost equal levels of the acoustical action but for considerably different values of the vibration accelerations. It follows that the mechanism of acoustical action on jet aerodynamic characteristics and the nozzle vibrations generated by the sound irradiation are divorced from each other.

The nozzle edge vibrations in the second series of experiments were generated by a vibrator without acoustical disturbances. The sound emitted by the vibrator was insignificant (L = 70 dB) and could not have any effect on the jet. It was found that the nozzle vibrations with frequencies and amplitudes equal to the corresponding values for the case of sound irradiation did not cause a sensible change in the velocity at the point $x/d = 8$ at the jet axis. However, with considerable increasing vibration acceleration, the nozzle vibrations admit control of the jet parameters for the appropriate choice of frequency f_v and amplitude ε_v. This fact follows from Fig. 2.28 where velocities at the jet axis (x/d = 8) as functions of the Strouhal number $\mathrm{St}_v = f_v d/u_0$ for nozzles 1 – 3 with fixed value of the vibration acceleration $\varepsilon_v = 5$ are presented. Here, similar to the case of acoustical excitation, two effects are realized: mixing intensification and attenuation. However, these effects are weaker than for acoustical excitation.

Fig. 2.29 shows velocities at the point $x/d = 8$ of the jet axis as functions of the vibration intensity, ε_v, with different values of St_v for nozzles 1 – 3. These curves are especially interesting for nozzle 2: here the action sign changes to the opposite one for high-frequency vibrations (St_v = 2.14) and saturation occurs for low-frequency vibrations (St_v = 0.879) as the level of ε_v increases.

One-third octave spectra of the velocity longitudinal and radial pulsations at the jet axis were measured for different distances from the nozzle. The greatest changes were for $x/d = 3 - 5$. Fig. 2.30 presents the spectra of longitudinal, u', and radial, v', velocity pulsations at the axis of a jet under vibration excitation for $x/d = 3$, $\mathrm{St}_v = 0$, 1.39 and 3.98.

In conclusion it should be noted that the described effects for the vibration excitation of a jet correspond to very small mean root values of the vibration amplitudes, $A = \left(\left\langle a'^2 \right\rangle\right)^{1/2} / (2\pi f_v)^2 = 0.006 - 1.4$ μm.

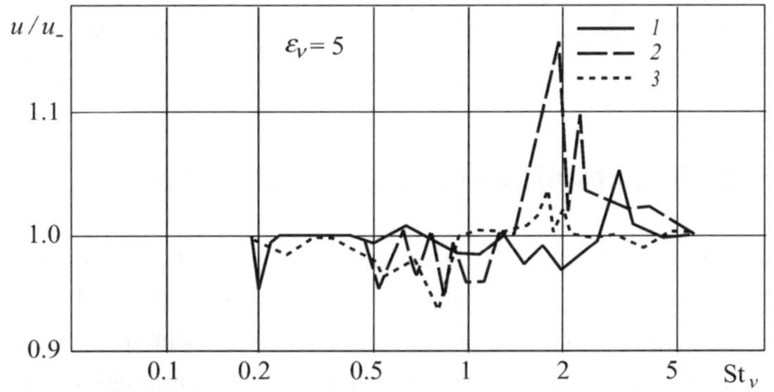

Fig. 2.28. Velocities at the point $x/d = 8$ of the jet axis as functions of the Strouhal number St_v for nozzles 1 – 3 with the fixed value of the vibration acceleration, ε_v.

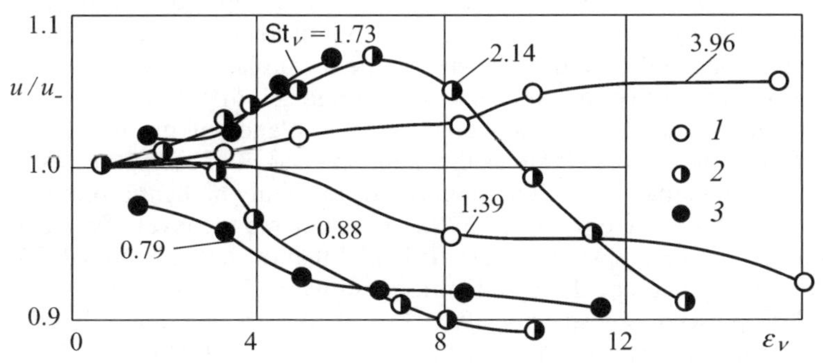

Fig. 2.29. Velocities at the jet axis ($x/d = 8$) as functions of the radial vibration intensity, ε_v, with various St_v for nozzles 1 – 3.

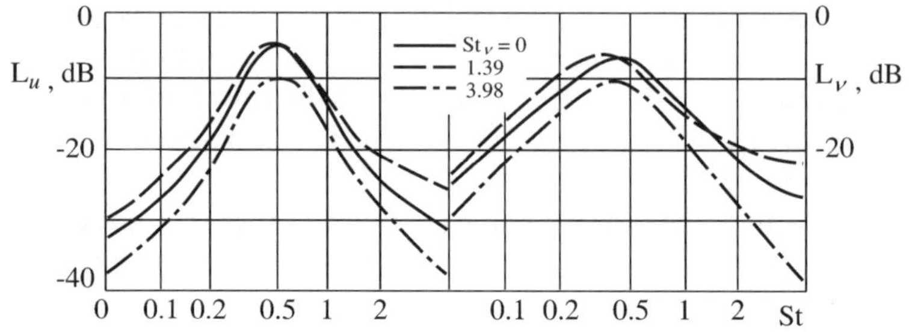

Fig. 2.30. Spectra of longitudinal, u', and radial, v', velocity pulsations at the axis ($x/d = 3$) of a jet under vibration excitation for $St_v = 0$, 1.39 and 3.98.

2.8. Acoustic Excitation of High-Speed Jets

A subsonic jet was irradiated by high-intensity sound (L = 170 dB) with frequency f_s = 14 kHz while the Reynolds number was in the interval $Re = (2.1-19.2)\cdot 10^4$ [2.5]. The experimental set-up scheme is presented in Fig. 2.31. The gas-jet radiator (*1*) being supplied of compressed air from a compressor was placed at one of two focuses of the elliptic concentrator (*2*) truncated in the plane of the other focus (F_2). The air nozzle cut-off (*3*) with the air supply under a pressure of 0.02 – 0.70 atm corresponding to flow Mach number $M_0 = 0.15-0.93$. The nozzle outlet section had diameter d = 4, 5, 6, 9, 12, and 15 mm. The exhausted air from the radiator was completely removed from the concentrator through the grooves (*4*) and did not influence on the jet under study.

The jets under study fitted in the size of the focal plan and were essentially in a sound field uniform over the section (in the range ±1 dB). In the spectrum of the sound signal measured in the concentrator focal plane without the jet, a subharmonic of frequency half as large as the main one was not observed or its level was at least 50 dB less than that of the main frequency. Only higher harmonics, besides the main frequency, were observed in the spectrum and the level of the component with doubled frequency was 15 – 20 dB below that of the main signal. Therefore, the jet was excited practically by a harmonic signal [2.58].

Fig. 2.32 presents the velocity along the jet axis issuing a nozzle of diameter $d = 6$ mm with M_0 = 0.42 (u_0 = 138 m/s) and M_0 = 0.93 (u_0 = 315 m/s) for Strouhal numbers St_s = 0.57 and 0.25 respectively. It follows that the sound action for large subsonic speeds and high amplitudes has the same nature as for rather weak acoustical excitation (p. 2.1). This action for $St_s < 1$ manifests itself as a decrease or even practically complete disappearance of the jet initial region. As follows from the functions $u_0 / u_m = \varphi(x/d)$ presented in Fig. 2.32, the jet opening, i.e. the increase of the angle coefficient of the lines *5 – 7* is faster in the initial region of excited jets.

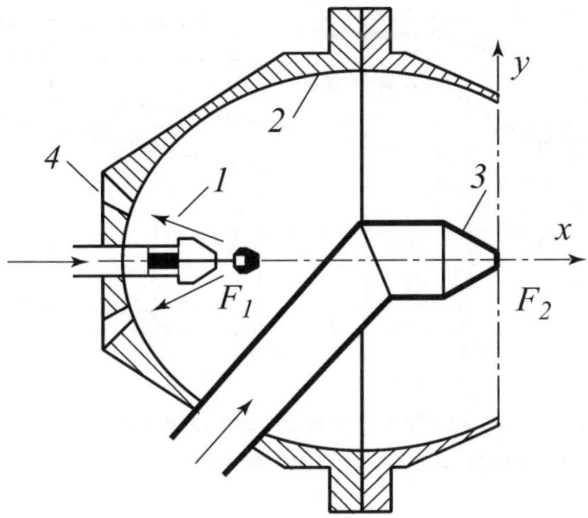

Fig. 2.31. Scheme of the experimental plant: *1)* the sound radiator, *2)* the elliptic concentrator, *3)* the nozzle, *4)* the grooves for removing the radiator exhausted air from the concentrator.

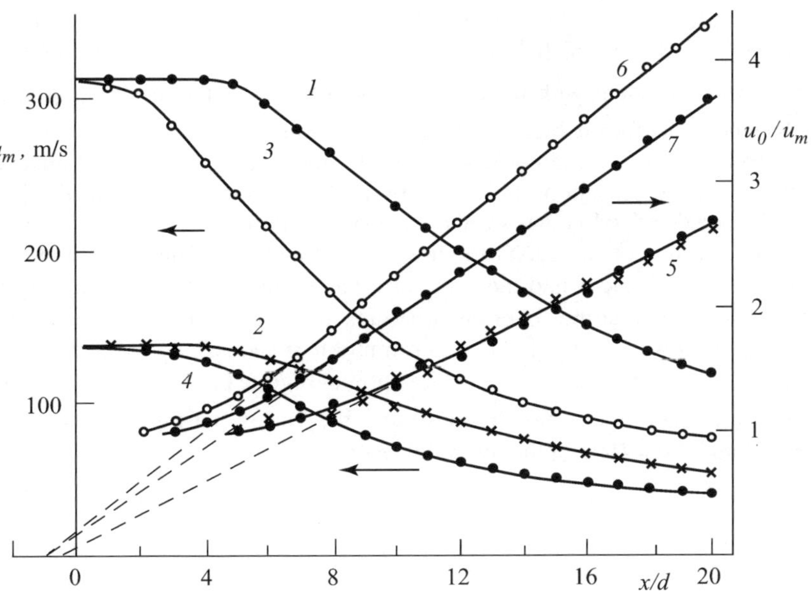

Fig. 2.32. Jet velocities, u_m (curves *1 – 4*), and ratios, u_0/u_m (curves *5 – 7*), as functions of the distance, x/d, along the jet axis. Curves *1, 3, 5, 6* correspond to $M_0 = 0.93$ and curves *2, 4, 7* correspond to $M_0 = 0.42$. Curves *1, 2, 5* are obtained without sound action and Curves *3, 4, 6, 7* are obtained with sound action.

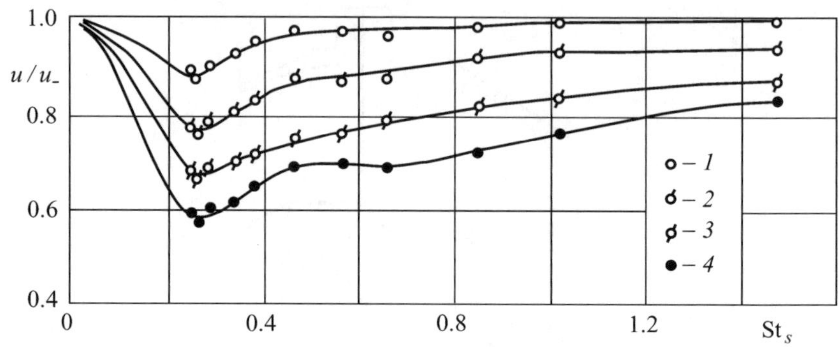

Fig. 2.33. The ratio u/u_{-} as a function of the Strouhal number for different distances from the nozzle of diameter d = 6 mm. Curves *1)* $x/d = 3$, *2)* $x/d = 5$, *3)* $x/d = 7$, *4)* $x/d = 10$.

Fig. 2.33 presents the velocity ratios at the jet axis, u/u_{-} as functions of the Strouhal number for fixed distances from the nozzle cutoff, $x/d = 3$, 5, 7 and 10 for a jet issuing from a nozzle of diameter d = 6 mm. Here u_{-} is the speed at the jet axis without acoustical action. It follows that the mixing intensification has its maximum for $\mathrm{St}_s = 0.25$. It is significant that the ratio u/u_{-} is less than 1 for $\mathrm{St}_s > 1$. It is consistent with the conclusions in Sect. 2.2 that the mixing attenuation does not exist for the high-frequency excitation.

The following experimental result is very interesting. The spectral analysis of turbulent velocity pulsations measured by a hot-wire anemometer at the jet axis and at the nozzle edge line has shown that aside from the frequency of the exciting signal, the spectra contain subharmonics of half frequency. Their amplitudes at the distances $x/d = 3$ and 5 had the same order as pulsations of the main frequency. By this it is meant that the coherent structures are paired with each other in the jet initial region. It was shown that the consumption of acoustical energy does not exceed 2% of the jet kinetic energy for the optimal regime of excitation ($\mathrm{St}_s = 0.25$).

Thus, this experiment has demonstrated the vortex pairing for L = 170 dB, St_s =0.25 and $\mathrm{M}_0 = 0.93$ in the jet initial region.

2.9. Changes in the Mode Composition of Turbulent Pulsations for a Jet under Acoustical Excitation. Localization of Pairing and Decay of Coherent Structures for a Jet under Acoustical Excitation. Mechanisms of the Jet Acoustical Excitation

Let us consider the change in the mode composition of turbulent pulsations in the initial region of a jet under acoustical excitation. Fig. 2.34 presents the azimuthal correlation of the longitudinal velocity pulsations, $R_{uu}(\psi)$, at the inner boundary of the mixing layer ($x/d = 3$, $r/d = 0.24$) of a jet under longitudinal acoustical irradiation with dimensionless frequencies St = 0.32 and 3.70, as well as of a jet without excitation (St = 0). The mode expansion for the above Strouhal numbers is presented also. One can see that the low-frequency action causes noticeable amplification of the zero mode and attenuation of the other modes. Conversely, the high-frequency action causes a slight decrease of the zero mode contribution and the corresponding increase of contribution of the first and higher modes [2.60].

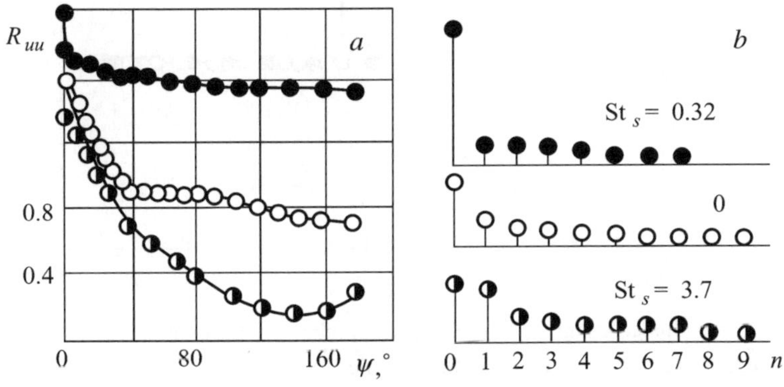

Fig. 2.34. (*a*) Change in the azimuthal correlation, $R_{uu}(\psi)$, in the circumferential direction. (*b*)The mode expansion for three Strouhal numbers $St_s = 0$, 0.32 and 3.70.

A qualitatively similar picture is observed also for measurements in the mixing layer but here the changes in contribution of different modes for a jet under acoustical excitation are considerably less than for the core. This fact is probably connected with the influence of the velocity small-scale pulsations. Hence the increase in the zero mode contribution for the jet low-frequency longitudinal acoustical excitation corresponds to stabilization of annular vortex structures in the mixing layer. The high-frequency excitation causes attenuation of annular structures and amplification of the first mode.

The points of generation of large vortices in the mixing layer and of their subsequent pairing for a jet without periodical excitation are subject to statistical scattering. As a consequence the characteristic frequencies and other parameters (the jet thickness, the average velocity) smoothly change along the flow. The jet acoustical excitation localizes the points of generation, pairing and decay of large vortices. This fact is illustrated by the experimental functions $\mathrm{St}(x/d)$ presented in Fig. 2.35 for high-frequency ($\mathrm{St} = 3.54$) acoustical excitation of a jet [2.26]. Here the Strouhal number reduces by half after each pairing and the pairing point is neatly fixed in the space.

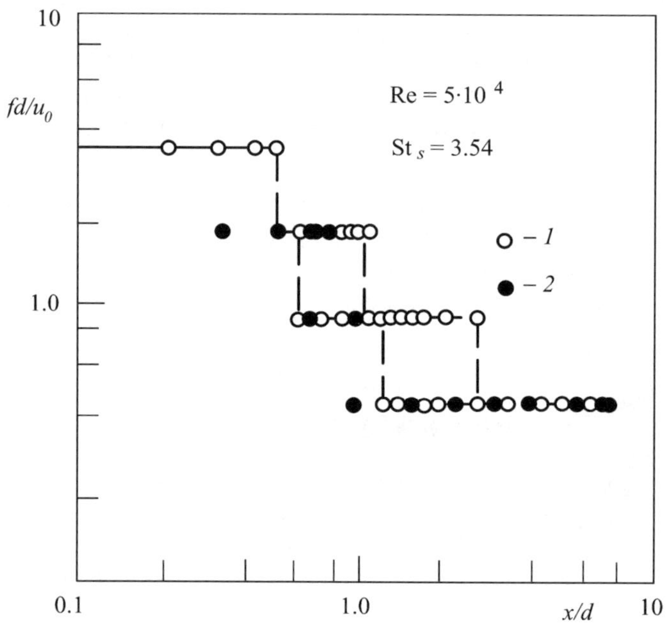

Fig. 2.35. Graduated change of the Strouhal number (*1*) along the jet axis and (*2*) the nozzle edge line of a circular jet under acoustical excitation.

Very thorough experiments with turbulent jets under high-frequency radial acoustical excitation also bear witness to the localization of points of coherent structure pairing [2.28]. The jet main parameters are $d = 50.8$ mm, $u_0 = 30.5$ m/s, $\varepsilon_0 < 0.1\%$, $\mathrm{St}_\theta = 0.0174$. Fig. 2.36 presents changes in the impulse loss thickness along the flow at the length $x/d = 0.1 - 0.8$ with and without acoustical excitation. The nonmonotone nature of the impulse loss thickness $\theta(x/d)$, as opposed to its smooth raise along the flow without acoustical excitement, is noteworthy.

A similar picture is exhibited also for lines of equal values of velocity pulsations and the Reynolds shear stress. Fig. 2.37 presents lines of equal values of the

Reynolds shear stress in a turbulent jet [2.19] without acoustical excitation, as well as for low-frequency excitation when $St_s = 0.3$ and $St_s = 0.85$.

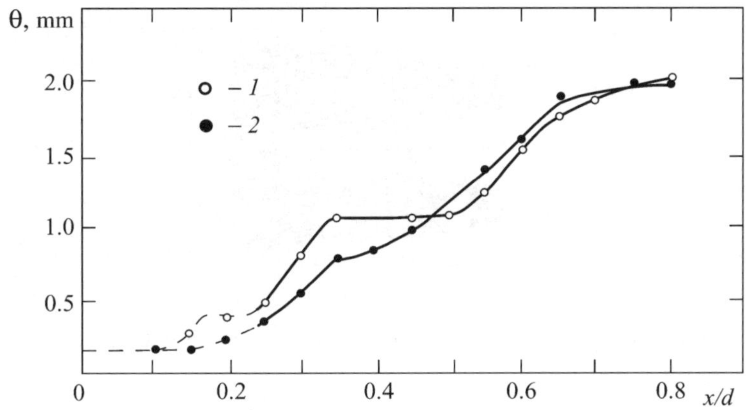

Fig. 2.36. Momentum loss thickness in a jet (*1*) with and (*2*) without acoustical excitation.

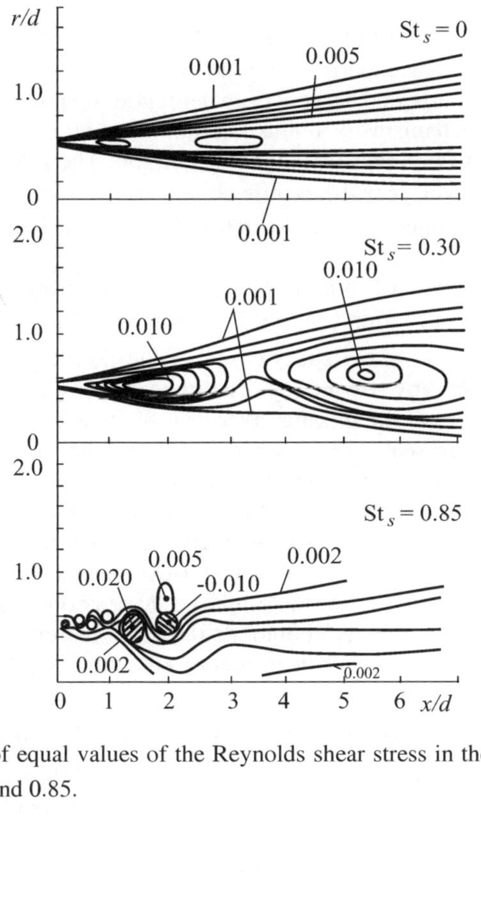

Fig. 2.37. Lines of equal values of the Reynolds shear stress in the jet initial region for $St_s = 0$, 0.30 and 0.85.

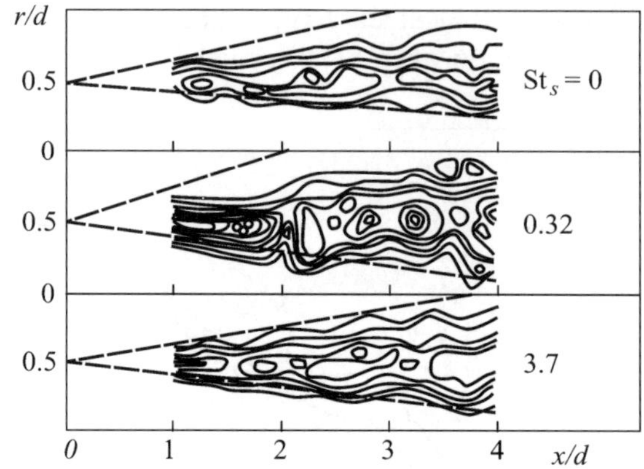

Fig. 2.38. Lines of equal values of the Reynolds shear stress in the jet initial region for $\mathrm{St}_s = 0$, 0.32 and 3.70.

Similar results were obtained in [2.60] for lines of equal values $\langle u'v' \rangle$ in the mixing layer without excitation ($\mathrm{St}_s = 0$), for low-frequency (St_s =0.32) and high-frequency (St_s =3.7) excitation (Fig. 2.38). It should be noted [2.18] that the full control over the flow in experiments with periodical excitement is possible, at most, for a few first vortex coalescences, whereupon the phasing is not retained.

Consider now mechanisms of intensification/attenuation of turbulent mixing in jets under low-/high-frequency acoustical excitation. The most sensitive region of a jet is the mixing layer near the nozzle edge ($x/d = 0 - 0.5$); just here the periodical excitation generates vortices, which define ultimately attenuation or amplification of mixing.

To explain phenomena of mixing intensification in turbulent jet flows for low-frequency periodical excitation (St_s = 0.2 – 0.6), the influence of excitation on the thin mixing layer in the immediate vicinity of the nozzle is considered. The experiments show that low-frequency disturbances modulate the initial high-frequency disturbances in the mixing layer near the nozzle causing ultimately earlier generation of large-scale coherent structures and their subsequent more intensive (with respect to an unexcited jet) rise [2.54].

We now turn to the consideration of possible mechanisms for realization of the mixing attenuation effect under high-frequency periodical excitation. As applied to the initial laminar boundary layer, it is known that the maximum influence of periodical disturbances on the unstable mixing layer is observed in the frequency range corresponding to the maximum disturbance increments near the nozzle edge [2.73]. According to the linear theory of stability [2.33], the increment maximum is achieved at $\mathrm{St}_\theta = 0.017$.

The increase of subharmonic disturbances is retarded for the mixing layer excitation at the frequency corresponding to the above value of St_θ, because of the increase of the ratio between amplitudes of the main harmonic and of the subharmonic disturbances. The latter causes constriction of the area of their resonance interaction and, therefore, deceleration of the instability growth.

The mechanism of the pulsation suppression at the jet axis for high-frequency disturbances is determined by the fast increase of the laminar shear layer instability near the nozzle. Hence, the vortex rolling up and subsequent decay occurs in the shorter region rather than without excitation. This phenomenon slows down or attenuates the subsequent vortex pairings [2.73].

This assertion is questioned in [2.40] based on the fact that then the pulsation levels at the jet axis shall take an intermediate value between two extreme values corresponding to the initial laminar and turbulent boundary layers without excitation. The experiments [2.66] show that this is not the case.

A somewhat different explanation of the above phenomenon pertaining equally to both laminar and turbulent initial boundary layers is given in [2.27]. The explanation, in essence, is this: the high-frequency excitation if the mixing layer near the nozzle ($x/d<1.5$) generates consecutive arranged annular vortices with the distances determined by the excitation frequency and the outflow speed. This suppresses the increase of low-frequency disturbances (with the basic contribution in the turbulence energy) in the mentioned area of the mixing layer under high-frequency sound. Experiments with low initial turbulence levels confirm this conclusion: the low-frequency disturbance increments corresponding to Strouhal numbers $\mathrm{St}_s = 1.3 - 2.5$ decrease for $x/d<1$. Therefore, the small-scale vorticity modulation generated by sound along the shear layer causes the suppression of increase in low-frequency disturbances and stabilization of the shear flow.

The specialized experiments described in [2.74] have demonstrated that the turbulence suppression mechanism for high-frequency excitation of shear flows could not be explained either by interaction of the Tollmien – Schlichting waves in the nozzle boundary layer and the Kelvin – Helmholtz waves in the mixing layer nor by turbulization of the initial boundary layer under acoustical excitation.

The mechanisms of acoustical wave influence on nonlinear evolution of 3-D disturbances in submerged jets are studied in [2.9]. The authors have revealed the "rigid" instability of jet flows and mixing layers with respect to 3-D finite-amplitude disturbances of the "difference" resonance type. A number of phenomena related to aero-acoustical stabilizing and destabilizing influence of acoustical waves on stability and long-range action of jets is explained. The theoretical analysis is performed on the basis of the Navier – Stokes nonstationary 3-D equations without any additional assumptions for calculation of both laminar and turbulent flows.

It is interesting that the calculation of a plane turbulent jet on the basis of the 2-parameter turbulence model [2.3] has demonstrated slight increase of the jet long-range action with decreasing the turbulence initial scale.

2.10. Coflowing Stream Effect on Turbulent Mixing Intensification in Jets under Low-Frequency Acoustical Excitation

As demonstrated in [2.56], the presence of a coflowing stream for values of the accompaniment parameter $m = u_\infty / u_0 < 0.5$ does not change strongly the coherent structure parameters. As a result, the jet response on acoustical excitation turns out to be almost the same as that for a submerged jet. This fact is supported by specialized experiments [2.1].

The effect of longitudinal low-frequency acoustical excitation on the averaged and pulsation flows in subsonic ($M_0 = 0.58$) submerged jets and jets in an accompanying flow (m = 0 – 0.45) is studied. The experimental set-up layout is presented in Fig. 2.39. The ratio between diameters of the outer and inner contours at their outlet section is $D/d = 5$. This value may be considered as an acceptable one because the measurements of the jet parameters are performed in the region of length $x = 10d$.

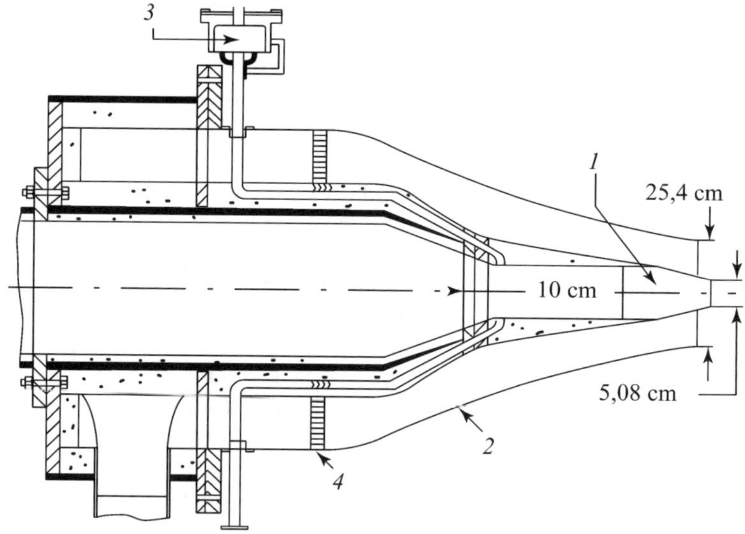

Fig. 2.39. The experimental set-up: *1)* the main nozzle, *2)* the nozzle for generation of the accompanying flow, *3)* the acoustical radiator, *4)* the honeycomb.

Fig. 2.40 presents the changes in the average velocity and longitudinal velocity pulsations along the jet axis ($M_0 = 0.58$) for m = 0, 0.23, and 0.43, Strouhal number St_s =0.5 and sound pressure level L = 141 dB. For comparison Fig. 2.41 presents the similar curves for a jet in an accompanying flow without acoustical excitation.

These results show that, as in the case of submerged jets, the low-frequency excitation of jets in an accompanying flow is an efficient mean for the mixing intensification.

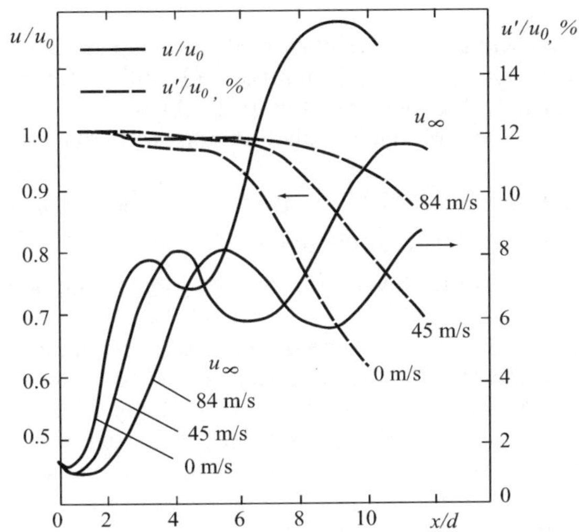

Fig. 2.40. The effect of the accompanying flow on the average velocity and longitudinal velocity pulsations under low-frequency acoustical excitation ($St_s = 0.5$, $L =$ 141 dB).

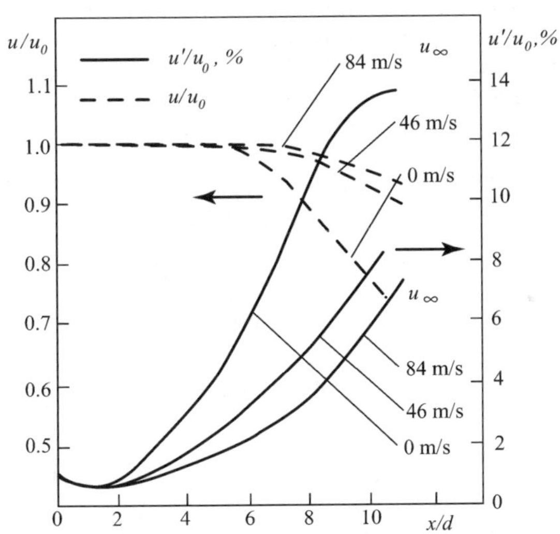

Fig. 2.41. The similar effect without acoustical excitation (M_0 =0.58, u_0 = 195 m/s) (cf. Fig. 2.40).

2.11. Tone Acoustical Excitation of Anisothermic Gas Jets

The results in the present chapter are borrowed from papers [2.30, 2.31] where the analysis of the influence of the Mach number and outflow temperature are based on consideration of the dependency of the local Mach number at a fixed point at the jet axis ($x/d = 9$), and on the Strouhal number with and without longitudinal acoustical excitation.

Fig. 2.42 presents the ratios M/M_{-} as functions of the Strouhal number St_s for cold jets ($T_{00} = 296$ and 292 K) at the point $x/d = 9$ for $M_0 = 0.30$ and 0.80, as well as the sound pressure levels for the Reynolds numbers Re = $3.4 \cdot 10^5$ and $10.6 \cdot 10^5$.

One can judge about influence of the outflow gas temperature on the acoustical action efficiency on the basis of the dependencies M/M_{-} on the Strouhal number St_s for $M_0 = 0.8$ and $T_{00} = 368 - 809$ K (Fig. 2.43). The results demonstrate the effect attenuation down to its disappearance as the temperature increases. In any case the larger levels of excitation are used for larger values of temperature to get the mixing intensification. The preferential value of the Strouhal number corresponding to the maximum of excitation depends slightly on the outflow initial conditions.

The influence of the flow mode in the initial boundary layer of anisothermic jets on efficiency of their low-frequency excitation is studied in [2.31]. Fig. 2.44 presents the dependencies M/M on St_s for $x/d = 9$, $M_0 = 0.8$, $T_{00} = 287$ and 669 K for the nonturbulized initial boundary layer and Fig. 2.45 presents similar dependencies for the turbulized initial boundary layer. It follows that the outflow temperature rise in jets with the thin initial laminar boundary layer causes considerable attenuation of the mixing intensification under acoustical excitation and turbulization of the initial boundary layer eliminates this property.

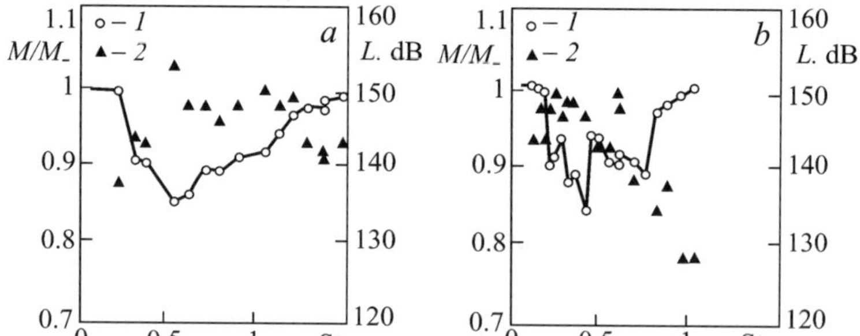

Fig. 2.42. The dependencies M/M_{-} on the Strouhal number St_s for isothermic jets at the point $x/d = 9$ for $M_0 = 0.30$ and 0.80. (*a*) $T_{00} = 296$ K, Re = $3.4 \cdot 10^5$; (*b*) $T_{00} = 292$ K, Re = $10.6 \cdot 10^5$. *1)* the relative Mach number, *2)* the sound pressure level.

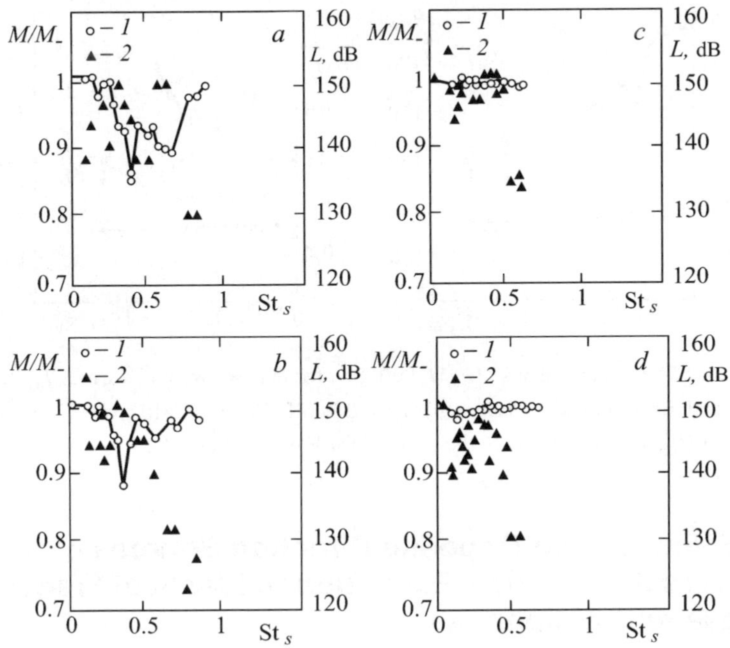

Fig. 2.43. The effect of the outflow gas temperature. The dependencies M / M_{-} on the Strouhal number for $M_0 = 0.8$ and various T_{00}. (*a*) $T_{00} = 388$ K, Re = $7.9 \cdot 10^5$; (*b*) $T_{00} = 488$ K, Re = $5.6 \cdot 10^5$; (*c*) $T_{00} = 672$ K, Re = $3.8 \cdot 10^5$; (*d*) $T_{00} = 809$ K, Re = $3.1 \cdot 10^5$. *1)* the relative Mach number, *2)* the sound pressure level.

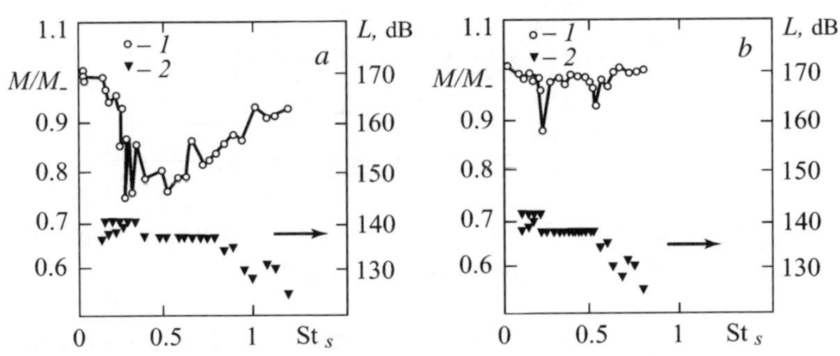

Fig. 2.44. The dependencies / on St_s for $M_0 = 0.8$, various T_{00} for the nonturbulized initial boundary layer. (*a*) $T_{00} = 287$ K, Re = $11.0 \cdot 10^5$; (*b*) $T_{00} = 669$ K, Re = $3.9 \cdot 10^5$. The notations are the same as for Figs. 2.42 and 2.43.

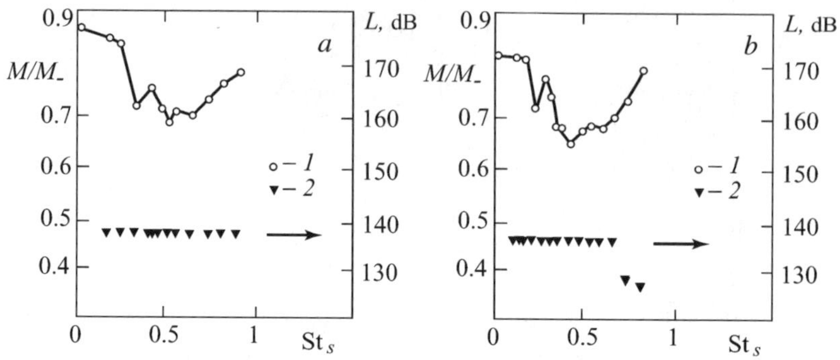

Fig. 2.45. The dependencies M/M_- on St_s for $\mathrm{M}_0 = 0.8$, and various T_{00} for the turbulized initial boundary layer. (*a*) $T_{00} = 279$ K, Re = $11.2 \cdot 10^5$; (*b*) $T_{00} =$ 53 K, Re = $3.9 \cdot 10^5$. The notations are the same as for Figs. 2.42 and 2.43.

2.12. Effect of Two Opposite Direction Sources of Transversal Acoustical Excitation of Identical Frequency in Phase and Antiphase

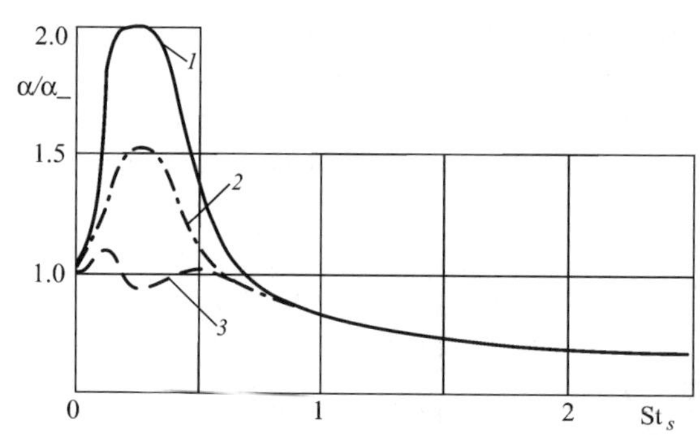

Fig. 2.46. The dependencies of α/α_- on the Strouhal number for $\varepsilon_0 = 0.6\%$ and different phase shifts: *1)* $\psi = 0$, *2)* one speaker, *3)* $\psi = 180°$

Consider a special case of single-frequency transversal acoustical excitation of a turbulent jet by two radiators placed on each side of the jet and emitting sound in phase and antiphase [2.11, 2.55, 2.62]. The main parameters are as follows: Re = $2.8 \cdot 10^4$, $\varepsilon_0 = 0.6\%$ and 2.5%, $L = 105 - 120$ dB, $\mathrm{St}_s = 0 - 2.5$. Fig. 2.46 presents the dependencies of α/α_- on the Strouhal numbers St_s. Curves *1* and *3* correspond to two-sided acoustical irradiation for the phase shift ψ = 0 and $\psi = 180°$. Curve 2 describes the change of α/α_- for the jet under its one-sided transversal acoustical irradiation. Fig. 2.47 demonstrates the corresponding dependencies of the velocity on St_s at the jet axis point $x/d = 8$.

The dependencies of α/α_- on the Strouhal numbers are obtained on the basis of visual studies using the direct shadow method for $\varepsilon_0 = 0.6\%$. The corresponding dependencies of α/α_- on St_s are determined from measurements of velocity at the jet axis for $\varepsilon_0 = 0.6\%$ and 2.5% (Fig. 2.47). Both dependencies could serve as a measure for the efficiency of acoustical action.

As follows from Fig. 2.46, the maximum efficiency of low-frequency action is achieved for $\mathrm{St}_s = 0.25$ and zero phase shift (curve *1*), whereas the minimum efficiency is for $\psi = 180°$ (curve *3*).

Intermediate curve *2* corresponds to the jet transversal irradiation by one radiator. An appreciable mixing attenuation under high-frequency excitation is achieved only for low levels of initial turbulence. The effect does not depend on the phase ($\varepsilon_0 = 0.6\%$ in Fig. 2.46); the effect is weak for the elevated value $\varepsilon_0 = 2.5\%$.

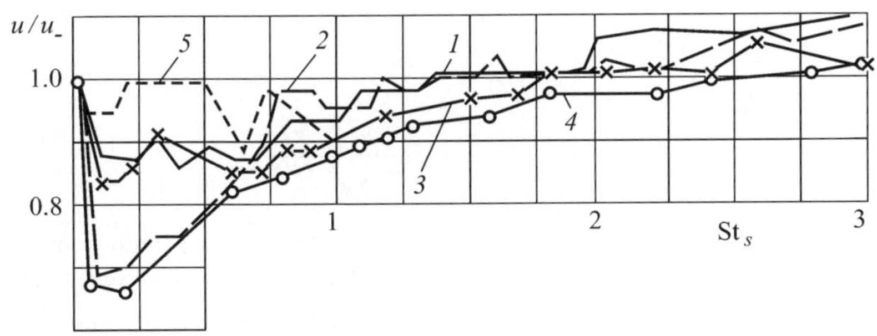

Fig. 2.47. The dependencies of α/α_- on the Strouhal number at the jet axis $x/d = 8$ for different phase shifts. *1)* one loudspeaker ($\varepsilon_0 = 0.6\%$), *2)* two loudspeakers ($\varepsilon_0 = 0.6\%$, $\psi = 0$), *3)* one loudspeaker ($\varepsilon_0 = 2.5\%$), *4)* two loudspeakers ($\varepsilon_0 = 2.5\%$, $\psi = 0$), *5)* two loudspeakers ($\varepsilon_0 = 2.5\%$, $\psi = 180°$).

2.13. Effect of Acoustical Disturbances of Higher Azimuthal Modes

So far we have considered jets under acoustical excitation by plane waves. New possibilities for the control of jets are connected with acoustical excitation by noise of higher azimuthal modes (by helical waves). Some results are described in [2.59]. The experimental apparatus comprised a receiver with a honeycomb and a lattice. The air arrives at the receiver from a compressor and issues through a nozzle of the outlet diameter $d = 40$ mm generating a jet. The sound from four speakers was delivered through cylinder tubes to the nozzle outlet at the section that is offset by 30 mm upstream from the nozzle cut-off plane. The axes of the tubes were perpendicular to the nozzle axis and the tube step in the circumferential direction is 90°. The tube outlet holes were closed flush with the inner surface of the nozzle by a fine-mesh lattice. Signals from different loudspeakers for excitation at the single frequency could be supplied in phase or with phase shift, $\Delta\psi$. The phase shift for two opposite loudspeakers could be equal to $\psi = 0$ or 180° and for all four loudspeakers $\psi = 0$ or 90°. The loudspeaker power was 20 and 150 W. The jet issue speed u_0 = 30 – 60 m/s, Re = (1 – 2) $\cdot 10^5$, the boundary layer at the nozzle cut-off was turbulent.

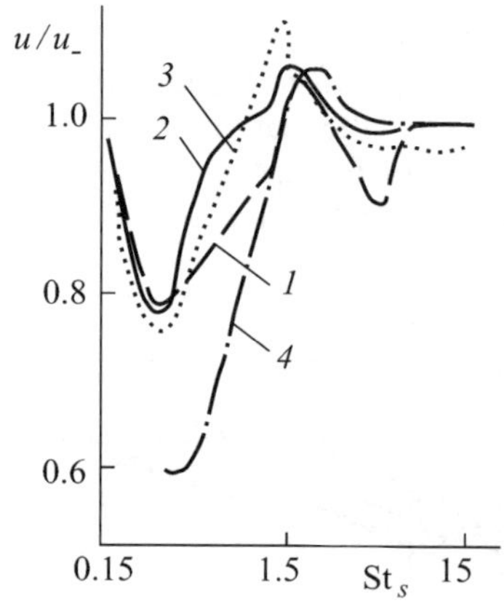

Fig. 2.48. The velocity changes at the jet axis for $x/d = 8$ depending on the Strouhal number, St_s. *1)* jet irradiation by one loudspeaker, *2)* by two loudspeakers in antiphase, *3)* by two loudspeakers in phase, *4)* by four loudspeakers with the phase shift $\psi = 90°$.

Fig. 2.48 presents the dependencies u/u_- at the jet axis for $x/d = 8$ on the Strouhal number, St_s , in the cases of the jet irradiation by one, two and four loudspeakers. Here curve *1* corresponds to the case of one loudspeaker, *2* to the case of two loudspeakers in antiphase ($\psi = 180°$), *3* to the case of two loudspeakers in phase ($\psi = 0$), 4 to the case of four loudspeakers with the phase shift $\psi = 90°$. The following things are particularly interesting. First, the mixing intensification for low-frequency acoustical excitation with $\psi = 90°$ is achieved for the sound pressure level that is less than the corresponding level for a jet under plane waves by 15 – 20 dB. There is 40%-decrease of the velocity at the jet axis point $x/d = 8$ for $St_s = 0.3$. Second, the mixing attenuation effect for high-frequency excitation is weak. The average velocity increase for $St_s = 1.65$ is negligible, probably due to turbulizing influence of fine-mesh lattices on the nozzle walls. Third, the weak distinction in dependencies of u/u_- on St_s in Fig. 2.48 between the case of irradiation by two loudspeakers with the phase shift $\psi = 0$ and 90° casts some doubt. Fig. 2.49 presents changes in the average velocity and in the longitudinal velocity pulsations along the jet axis for the unexcited jet, as well as for the jet irradiated by four loudspeakers ($St_s = 0.3$ and 1.65) with the phase shift $\Delta\psi = 90°$.

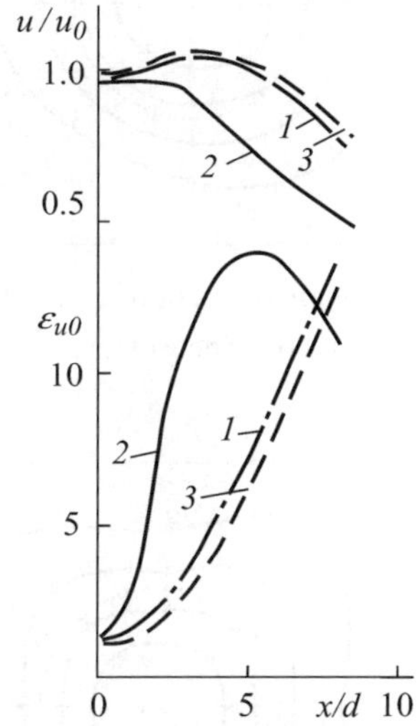

Fig. 2.49. Changes in the average velocity and in the longitudinal velocity pulsations along the jet axis. *1)* the unexcited jet; *2)* irradiation by four loudspeakers, $\psi = 90°$, $St_s = 0.3$; *3)* $\psi = 90°$, $St_s = 1.65$.

It is very interesting to look at the results of studies of the resonance interaction between two spiral waves with opposite signs (+*n* and –*n*). The influence of these waves on the jet causes changes in the large-scale coherent structures and, as a result, distortions of the jet cross-section shape [2.32]. The latter paper contains also the bibliography of the corresponding studies. Let us briefly consider the final results. The experiments were performed for low-frequency acoustical irradiation of a jet ($St_s = 0.66$). It was found that the jet cross-section shape changed from the circular one at the initial section to the oval one (Fig. 2.50 for *n* = +1, –1) at $x/d = 4$. For *n* = +2, –2 the isotachs at the section $x/d = 4$ had almost square shape (Fig. 2.51).

In the case of the jet irradiation by two spiral waves *n* = +1, –1, the isotachs take the oval form with the inclination angle depending on the phase shift between the waves ($\psi = 0$, 90, 180, and 270°) at the section $x/d = 4$ (Fig. 2.52). Finally, in the case of the combined acoustical excitation of jets by plane (*n* = 0) and azimuthal (*n* = 2) waves, the isotachs are also distorted (Fig. 2.53 for $x/d = 3$).

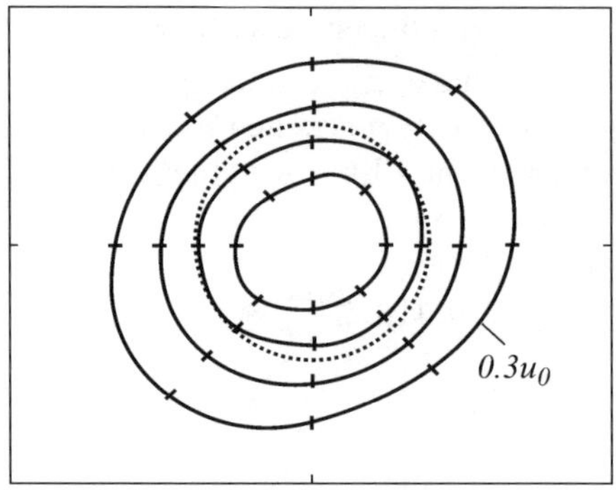

Fig. 2.50. The isotachs at the jet cross-section $x/d = 4$ for acoustical excitation by spiral waves $n = +1, -1$ for $\mathrm{St}_s = 0.65$. The isotachs are from $0.3u_0$ to $0.9u_0$ with the step $0.2u_0$, $\mathrm{Re} = 5{\cdot}10^4$.

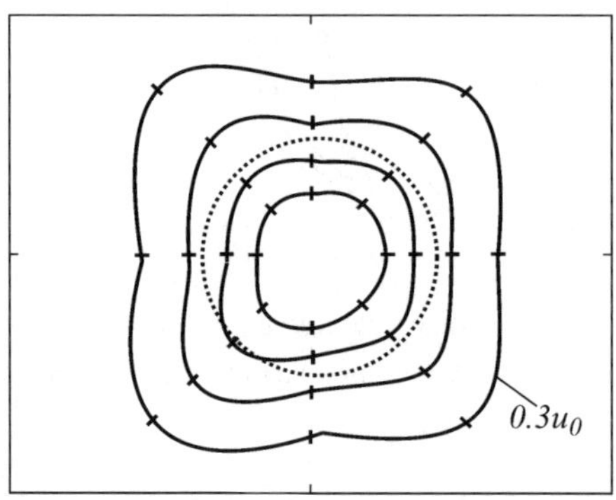

Fig. 2.51. The isotachs at the jet cross-section $x/d = 4$ for acoustical excitation by spiral waves $n = +2, -2$ for $\mathrm{St}_s = 0.65$. The isotachs are from u_0 to u_0 with the step u_0.

2.14. Two-Frequency Acoustical Excitation of Jets. Subharmonic Resonance

The quest for increase in the efficiency of acoustical excitation of turbulent jets produced attempts of using several acoustical signals with different frequencies. When the frequencies are chosen arbitrarily, such excitation is equivalent, in essence, to the change in the spectral composition of acoustical disturbances acting on the mixing layer at the nozzle outlet section with deterioration in the acoustical signal harmonicity. An attempt of such two-frequency excitation of a jet [2.61] has demonstrated that the mixing at low frequencies is able to intensify slightly.

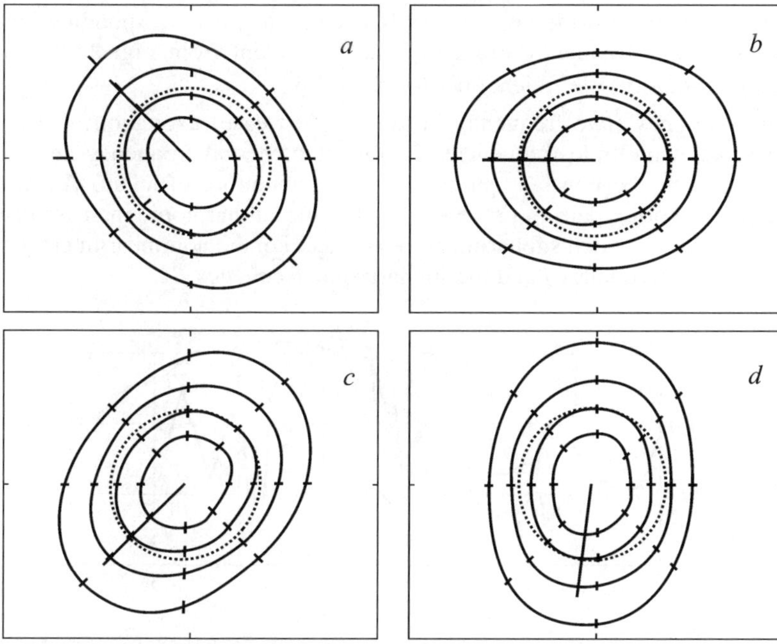

Fig. 2.52. The effect of the phase shift of two spiral waves $n = +1, -1$ on the isotach inclination for $\mathrm{St}_s = 0.65$. The isotachs are from u_0 to u_0 with the step u_0.

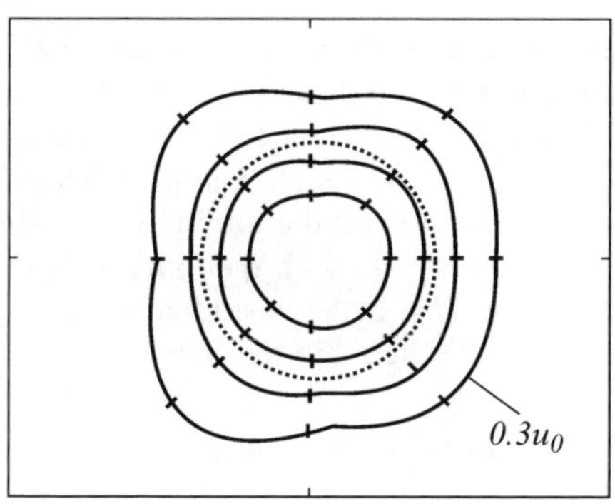

Fig. 2.53. The isotachs at the jet cross-section $x/d = 3$ for the combine acoustical excitation by plane waves ($n = 0$) and by spiral waves ($n = +2$) for $\mathrm{St}_s = 0.65$.

Fig. 2.54 presents values of u/u_- at the axis point $x/d = 8$ (curve *1*) as a function of the Strouhal number St_1 of the longitudinal single-frequency acoustical irradiation ($\mathrm{St}_1 = 0.2 - 3.5$) with the level $L = 110 - 120$ dB. Curves *2* and *3* correspond to the jet longitudinal irradiation by acoustical signals with two constant frequencies $\mathrm{St}_2 = 0.38$ and $\mathrm{St}_3 = 1.93$. In the case of two-frequency irradiation the sound pressure level at the nozzle cut-off is equal to the corresponding level for the case of single-frequency irradiation. As is evident from Fig. 2.54, the two-frequency irradiation is efficient only for St < 1.5.

Multi-frequency (two-frequency in the simplest case) excitation of a jet or a mixing layer at multiple frequencies (at the fundamental frequency and its subharmonics) with the constant phase shift is much more efficient. As noted in Chapter 1, the vortex pairing process related to the subharmonic increase is a consequence of the so-called subharmonic resonance, i.e. the nonlinear interaction between waves of frequency f and the subharmonic frequency $f/2$.

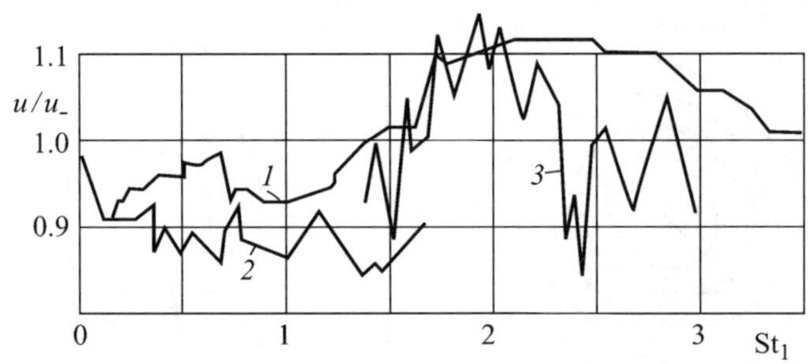

Fig. 2.54. Values of $u/u_{-} = \varphi_1(\mathrm{St}_1)$ at the axis point $x/d = 8$ for longitudinal single-frequency acoustical irradiation, $u/u_{-} = \varphi_1$ $(\mathrm{St}_1, \mathrm{St}_2 = \mathrm{const})$ and $u/u_{-} = \varphi_1$ $(\mathrm{St}_1, \mathrm{St}_3 = \mathrm{const})$ for two-frequency acoustical irradiation.

Therefore, the control of the subharmonic resonance effect could be used for the vortex pairing control and, as a consequence, for the turbulent mixing control [2.21, 2.36, 2.39] by the appropriate choice of the control parameters – Strouhal number, frequency ratio (1/2, 1/4, 1/8), signal amplitudes, and phase shift.

Consider examples of such action on the mixing layer. The two-frequency excitation of the mixing layer past a plate with the laminar boundary layer (the shape parameter at the trailing edge, $H = 2.59$) is studied in [2.21]. The mixing layer is irradiated at two frequencies – at the main frequency f and its subharmonic $f/2$:

$$u_s' = a_f \cos 2\pi ft + a_{f/2} \cos\left\{2\pi\left(\frac{f}{2}\right)t + \psi\right\}.$$

The natural frequency of instability is the Strouhal number $\mathrm{St}_\theta = f\theta_0 / u_e = 0.012$, the excitation levels at the main frequency f and its subharmonic $f/2$ are $u_f' / u_e = 0.01$ and $u_{f/2}' / u_e = 0.01$ for the variable phase shift $\psi = 0 - 180°$.

Fig. 2.55 presents the dependencies of u_f' / u_e and $u_{f/2}' / u_e$ on the initial phase shift at the sections $x/\theta_0 = 30$, 80 and 140 demonstrating the strong dependence of the subharmonic intensity on ψ. For $x/\theta_0 = 80$ and 140 the critical value of $u'_{f/2}$ is at $\psi = 72°$.

Similar results illustrating the important role of the phase shift in two- and three-frequency acoustical excitation of the jet mixing layer at the main frequency and its subharmonics in the ranges $x/d = 0-1.5$ and $x/d = 0.3-0.9$ could be found in [2.38, 2.43]. As is shown there, such excitation gives a possibility for controlling the mixing layer thickness over a wide range.

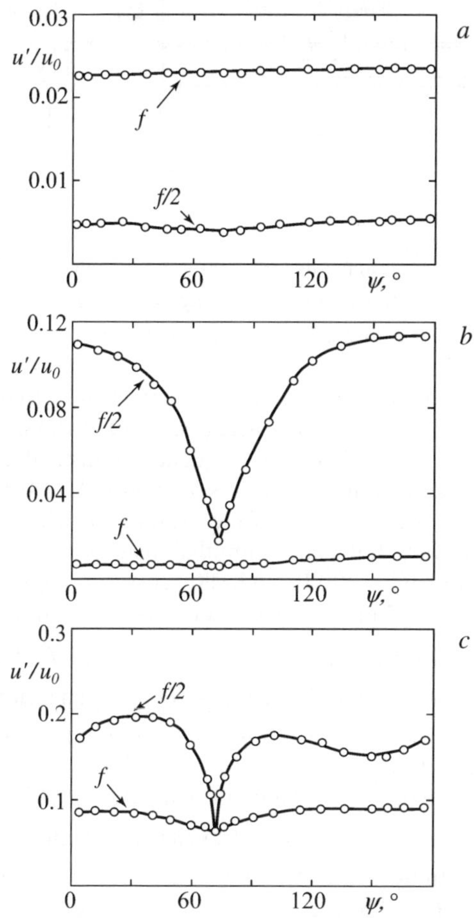

Fig. 2.55. The dependencies of u'_f/u_e and $u'_{f/2}/u_e$ on the initial phase shift, ψ, for two-frequency acoustical excitation at different sections of the plane mixing layer, $x/\theta_0 = 30$, 80 and 140. (*a*) $x/\theta_0 = 30$, (*b*) $x/\theta_0 = 80$, (*c*) $x/\theta_0 = 140$.

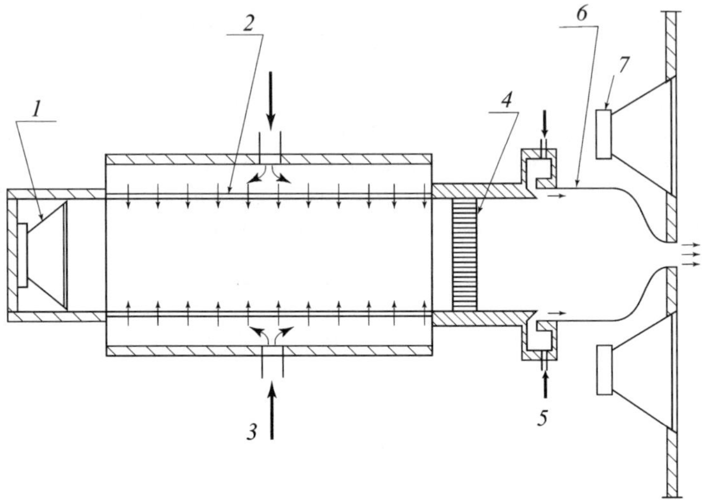

Fig. 2.56. The experimental apparatus for study of a turbulent jet excited by the axial and azimuthal modes. *1)* the radiator, *2)* the porous cylinder, *3)* the air, *4)* the honeycomb, *5)* the smoke, *6)* the nozzle 25:1, *7)* the radiators

The two-frequency acoustical excitation of jets at low frequencies, as well as the mixing intensification effect, is considered in [2.2, 2.38, 2.43]. We dwell on the results of two latter papers. The low-frequency acoustical excitation of a turbulent jet by the axial and azimuthal modes is studied in [2.38] where the azimuthal mode frequency is half the frequency of the axial mode. The experimental apparatus layout is shown in Fig. 2.56. The azimuthal disturbances are generated by four or two radiators with phase shift 90° and 180° correspondingly. The nozzle outlet section diameter is $d = 21.5$ mm, the issue speed is $u_0 = 7$ and 14 m/s with the corresponding Strouhal numbers Re = $u_0 d / \nu = 10^4$ and $2{\cdot}10^4$. The initial turbulence at the nozzle outlet section centre is $\varepsilon_0 = 1\%$. The Strouhal number $\mathrm{St}_s = 0.6$. The azimuthal disturbances are generated when switching on two opposite loudspeakers with the phase shift 180°.

Fig. 2.57 presents the photos of the jet flow for Re = $2{\cdot}10^4$, $\mathrm{St}_s = 0.6$, $\mathrm{St}_h = 0.3$, $u'_a / u_0 = u'_h / u_0 = 0.03\%$. Such combined excitation of the jet causes a considerable change of its opening angle. Still more dramatic expansion of the jet is observed at Re = 10^4, $\mathrm{St}_a = 0.55$, $u'_a / u_0 = 0.12\%$ (the zero mode) and $\mathrm{St}_a = 0.55$, $\mathrm{St}_h = 0.55/2$, $u'_a / u_0 = 0.12\%$ and $u'_h / u_0 = 0.04\%$ (for two-mode excitation) where the jet opening angle is mounted to 70°. The dependency of the jet opening

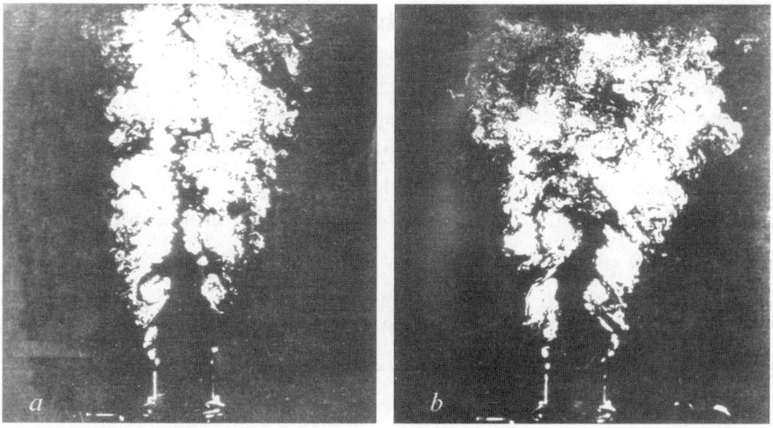

Fig. 2.57. The photos of the jet flow for single- and two-mode acoustical excitation (the laser method). (*a*) The longitudinal single-mode acoustical excitation Re = $2 \cdot 10^4$, $\mathrm{St}_a = 0.6$, $A_a = 0.03\%$. (*b*) The two-mode acoustical excitation Re = $2 \cdot 10^4$, $\mathrm{St}_a = 0.6$, $\mathrm{St}_h = 0.3$, $u'_a/u_0 = u'_h/u_0 = 0.03\%$.

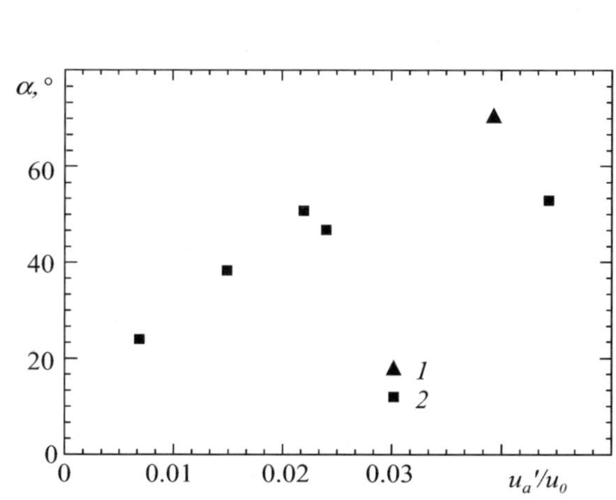

Fig. 2.58. The dependency of the jet opening angle, α, on the transversal excitation amplitude for two values of the axial excitation amplitude, A_a = 0.05 and 0.12%, Re = 10^4, $\mathrm{St}_a = 0.64$, $\mathrm{St}_h = 0.3$. *1)* $u'_a/u_0 = 0.12\%$, *2)* $u'_a/u_0 = 0.05\%$.

angle, α, on the azimuthal and axial excitation amplitudes is shown in Fig.2.58 for the above conditions. The angle α increases from 24° to 70°[1]!

Very interesting results for two-frequency low-frequency excitation of turbulent jets are obtained in [2.43]. In that paper two-frequency low-frequency excitation of a jet for Strouhal number $St_s = 0.4$ and 0.2, 0.6 and 0.3, 0.8 and 0.4 with $M_0 = 0.2$ and 0.45 for different amplitudes of the main frequency and its subharmonic $u'_{f/2}/u_e$ and the phase shift 0 – 360° changing with the step 45°. By way of example, Fig. 2.59 presents the changes in the axial velocity u/u_0 along the jet axis for an undisturbed jet, as well as for a jet under two-frequency excitation for Strouhal numbers 0.4 and 0.2, 0.6 and 0.3, 0.8 and 0.4 with the same amplitude $u'_f = u'_{f/2} = 0.03\,u_0$ and phase shift $\psi = 270°$. Similar changes along the flow in the momentum loss thickness, θ/d, are presented also. It follows that the two-frequency acoustical excitation is much more efficient than the single-frequency one. Another important conclusion of the cited paper is the fact that the subharmonic amplification does not depend on the initial phase shift for high levels of fundamental and subharmonic excitation.

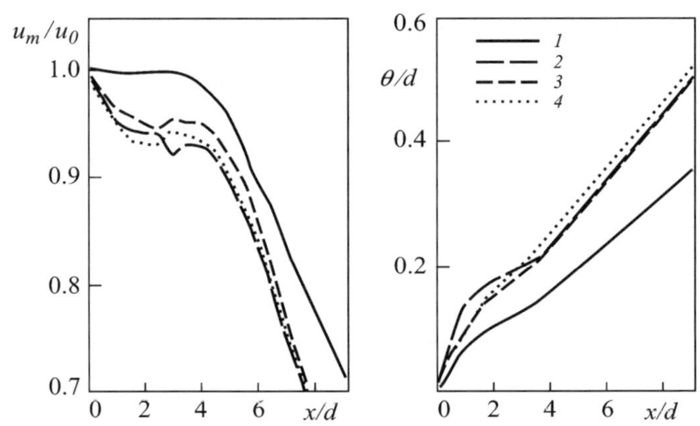

Fig. 2.59. The changes in the velocity along the jet axis and in the impulse loss thickness for a jet under two-frequency longitudinal acoustical excitation for $M_0 = 0.2$, $u'_f = u'_{f/2} = 0.03u_0$ and the phase shift $\psi = 270°$. *1)* the undisturbed jet, *2)* $St_s = 0.4$ and 0.2, *3)* $St_s = 0.6$ and 0.3, *4)* $St_s = 0.8$ and 0.4 [2.43].

One feature of the above papers should be noted. In the case of two-frequency acoustical excitation of a turbulent jet at the main frequency and its subharmonic, the considerable effect of the aerodynamic characteristic control could be achieved only at low frequencies. Here two-frequency acoustical excitation, as compared to

[1] Subsequent development of these studies is contained in the paper Cho SK, Yoo JY, Choi H (2000) Resonance in axisymmetric jet under controlled helical, fundamental, and axisymmetric subharmonic forcing. AIAA J 38, # 3: 434 – 441.

one-frequency excitation, causes considerable mixing intensification. The detectable attenuation of turbulent mixing in a jet under high-frequency two-frequency acoustical excitation, as compared to one-frequency excitation, was not revealed in the above papers. A number of experimental studies on such excitation were performed only at the jet segments $x = (0 - 0.8)d$ and $x = (0 - 1.5)d$.

Thus, the authors have performed a special study [2.70] to reveal possibilities of increase in efficiency of two-frequency transversal acoustical excitation of jets as compared to single-frequency excitation both at low frequencies (the mixing intensification) and at high frequencies (the mixing attenuation). Some results of that study are presented below for $\mathrm{Re} = 2.8 \cdot 10^4$, $d = 0.02$ m, $u_0 = 20$ m/s for low-frequency ($\mathrm{St}_s = 0.4$ and 0.8) and high-frequency ($\mathrm{St}_s = 3$ and 6, 3.9 and 7.8) excitation with the phase shift $\psi = 0 - 360°$. The sound pressure levels for low-frequency (L_1) and high-frequency (L_2) excitation were identical: $L_1 = L_2 = 125$ dB.

Fig. 2.60 presents oscillograms (*I*) and spectra (*II*) of the pressure pulsations measured without the jet near the nozzle outlet edge for $f_1 = 400$ Hz, $f_2 = 800$ Hz and $f_1 = 3000$ Hz, $f_2 = 6000$ Hz for the phase shift $\psi = 0$ and $\psi = 145°$. As is obvious from the presented spectra, such an excitation can be considered as the two-frequency one because the intensity of two main frequencies exceeds their harmonic intensities approximately by 25 dB.

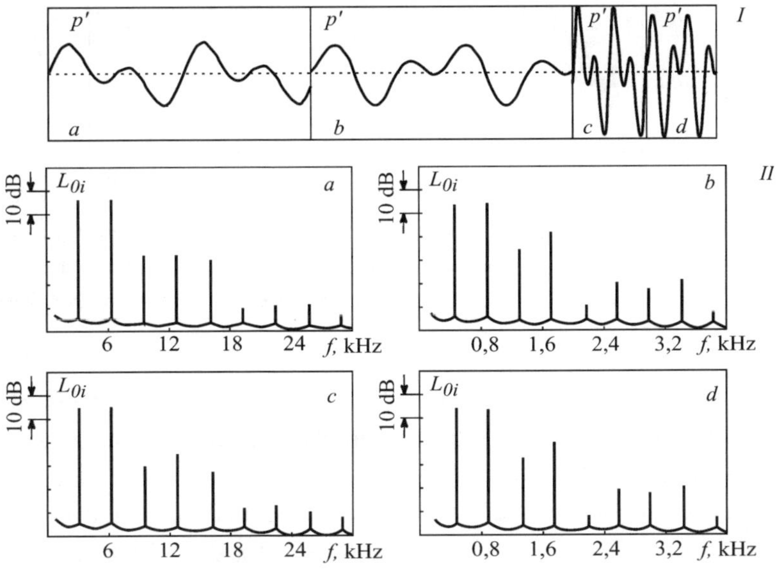

Fig. 2.60. The oscillograms (*I*) and spectra (*II*) of the pressure pulsations for the two-frequency excitation at the frequencies (*a, c*) $f_1 = 3000$ Hz and $f_2 = 6000$ Hz, $f = 400$ Hz and $f_2 = 800$ Hz with the phase shift (*a, b*) $\psi = 0$ and (*c,d*) $\psi = 145°$.

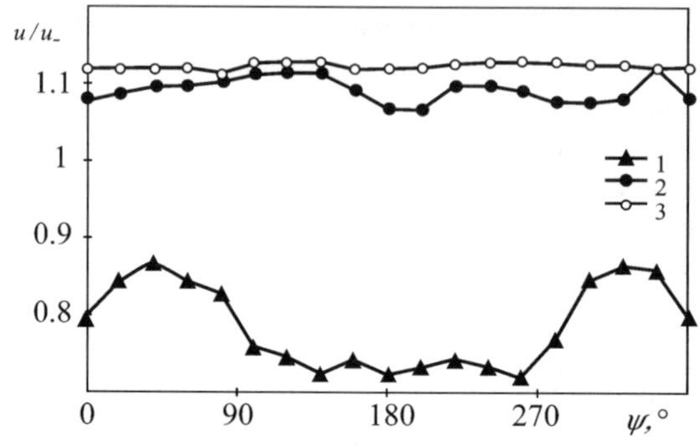

Fig. 2.61. The velocity ratio at the jet axis point $x/d = 8$ as a function of the phase for two-frequency acoustical excitation with L = 125 dB. *1)* St_s = 0.4 and 0.8, *2)* St_s = 3 and 6, *3)* St_s = 3.9 and 7.8.

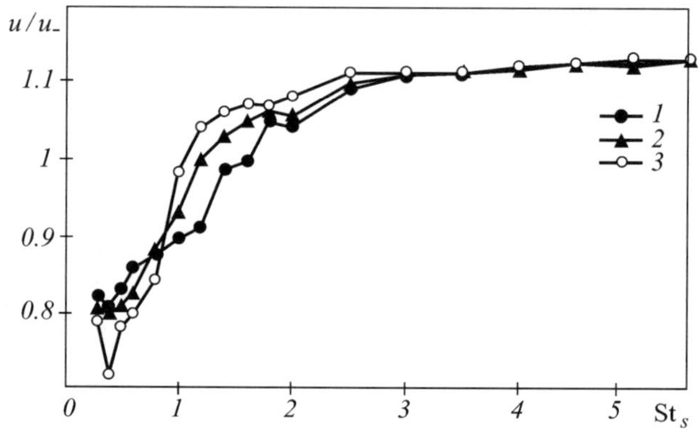

Fig. 2.62. The velocity ratio at the jet axis point $x/d = 8$ for two-frequency acoustical excitation as a function of the least value of two Strouhal numbers. *1)* the single-frequency excitation, *2)* two-frequency excitation for $\psi = 0$, *3)* two-frequency excitation for $\psi = 145°$.

The level of the phase shift effect for St_s = 0.4 and 0.8, St_s = 3 and 6, St_s = 3.9 and 7.8 can be judged from Fig. 2.61 where the ratio u/u_- at the jet axis point $x/d = 8$ is presented. It follows that the phase shift is an essential parameter only for low frequencies of excitation (St_s = 0.4 and 0.8). In the case of two-frequency high-frequency excitation (St_s = 3 and 6, St_s = 3.9 and 7.8), the change of the phase shift in the range $\psi = 0 - 360°$ does not have a pronounced excitation effect.

Fig. 2.62 presents the dependencies of u/u_- at the jet axis point $x/d = 8$ on the least value of the Strouhal numbers, St_s, in the case of two-frequency acoustical excitation for L = 125 dB and the phase shift ψ = 0 and 145°, as well as in the case of single-frequency excitation. It follows that the two-frequency acoustical excitation at low frequencies is more efficient than the single-frequency one. However, the two- and single-frequency excitations give practically identical results at high frequencies. Fig. 2.63 presents the change in the average velocity and longitudinal velocity pulsations along the jet axis without acoustical excitation, for single-frequency excitation, as well as for two-frequency excitation with phase shift $\psi = 0$ and 145° for the low-frequency ($St_s = 0.4$ and 0.8) and the high-frequency excitation ($St_s = 3.9$ and 7.8). The results clearly demonstrate the efficiency of the two-frequency excitation of a jet compared with the one-frequency excitation for low frequencies and, on the contrary, its inefficiency for high frequencies.

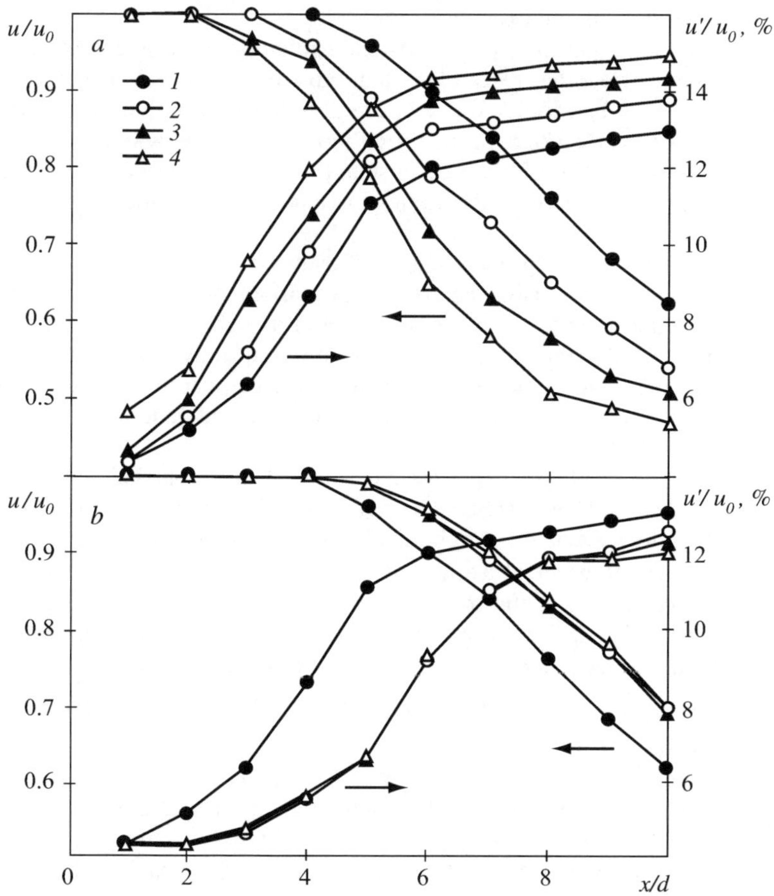

Fig. 2.63. The change in the average velocity and mean-square velocity pulsations along the jet axis for a jet *1*) without acoustical excitation, *2*) for the single-frequency excitation, for the two-frequency excitation with the phase shift *3*) $\psi = 0$ and *4*) $\psi = 145°$. (*a*) $St_s = 0.4$, (*b*) $St_s = 3.9$.

2.15. Multi-Frequency Acoustical Excitation of Turbulent Jets

As is well known, when generating sound of pure tones by radiators of different kinds, components with harmonic frequencies, in addition to the main tone component, appear due to nonlinearity of acoustic characteristics of the radiator (the loudspeaker). The relative level of harmonic components of higher orders increases, as a rule, with the excitation level.

In the authors' experiments on acoustical excitation of turbulent jets, as in similar experiments of another researchers that have used electrodynamic radiators of sound, the level of harmonics was considerably lower than that of the main component. Hence the effect of acoustical oscillations on the jet was almost unaffected by these harmonics. Therefore, we can state that the tone harmonic excitation was in the most part of experimental studies on acoustical excitation of turbulent jets and mixing layers.

As noted in Section 2.14, to raise the efficiency of control of turbulent jet aerodynamic characteristics using acoustical excitation, one may implement multi-frequency excitation at the main frequency and its subharmonics with the appropriate phase shift. Another approach is related to the study of the nonharmonic effects. The changes in the acoustical signal shape can cause changes in the front slope of the pressure wave acting on the nozzle edge and generating vortex disturbances (vortex rings).

Below are the results of experimental study of the acoustical signal effect on the jet aerodynamic characteristics. The object of investigation was the change in the average velocity and the longitudinal velocity pulsations at the jet axis point ($x/d=8$) under transversal acoustical irradiation of a jet for different frequencies, levels and shapes of the sound signal [2.69]. The experimental apparatus is described in [2.63]. Its main parameters are as follows: the nozzle diameter is d = 0.02 m, the issue speed is $u_0=10$ and 20 m/s, the corresponding Reynolds numbers are Re = $u_0 d/\nu = 1.4\cdot 10^4$ and $2.8\cdot 10^4$. The initial boundary layer is almost laminar.

The acoustical excitation is performed by means of electrodynamic radiator (loudspeaker). The loudspeaker coil is supplied by harmonic signals (necessary for generation of harmonic acoustical excitation) or by periodical rectangular electric impulses that are transformed into periodical acoustic signals of the appropriate shape. Each of the positive rectangular electric impulses being supplied to the electrodynamic radiator coil generates pressure impulses with different signs in the acoustical field. The changes in the signal shapes are achieved through the changes in parameters of the periodical rectangular electric impulses characterizing by the period, the corresponding level (voltage) and the ratio between the impulse duration, T_0, and the period, $k=T_0/T$ = 0.1, 0.3, 0.5, and 0.9 (Fig. 2.64).

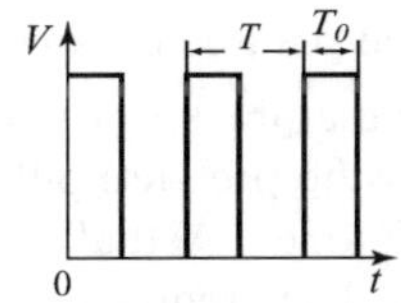

Fig. 2.64. The rectangular periodical electric impulses, the voltage, *v*, as a function of time, *t*; $k = T_0 / T$.

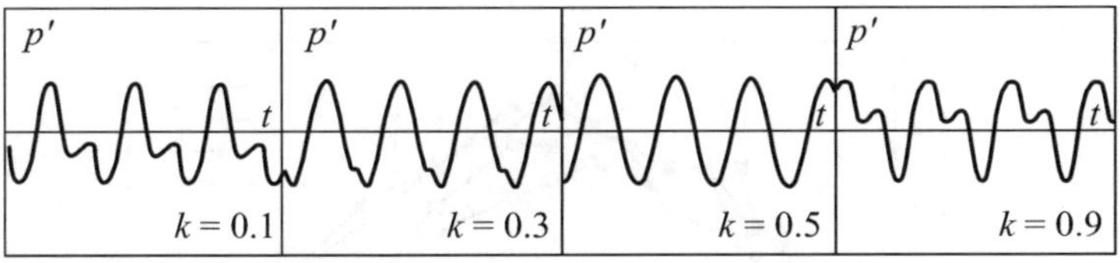

Fig. 2.65. The oscillograms of pressure pulsations near the nozzle edge for $k = 0.1 - 0.9$, $f = 500$ Hz and $L_0 = 120$ dB.

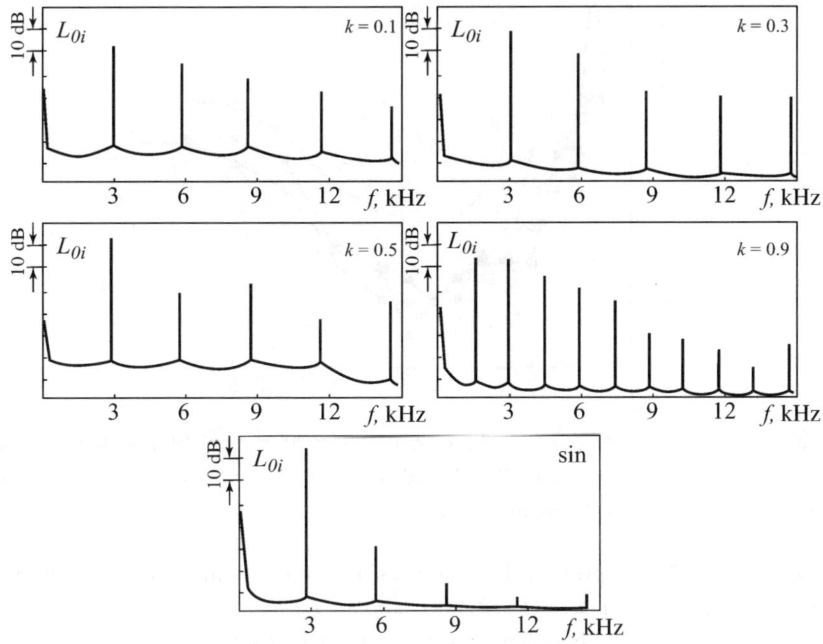

Fig. 2.66. The narrow band spectra of pressure pulsations for $k = 0.1$-0.9 and the reference frequency $f = 3000$ Hz .

Figs. 2.65 and 2.66 present for $f = 3000$ Hz and $L_0 = 120$ dB the oscillograms of pressure pulsations near the nozzle edge and the corresponding narrow band spectra for different values of the parameter $k = T_0 / T = 0.1 - 0.9$ and for a harmonic signal. It should be noted that the spectra have more relatively intense harmonics when the oscillograms of the pressure pulsations near the nozzle edge noticeably differ from the harmonic ones. Actually, in this case the multi-frequency acoustical excitation is realized. The frequency of rectangular electric impulses that were supplied to the loudspeaker was used as the reference frequency, *f*. As may be seen from Fig. 2.66 in the case of harmonic signals, the harmonic level is negligibly small (it is less than the main tone level by 40 – 50 dB).

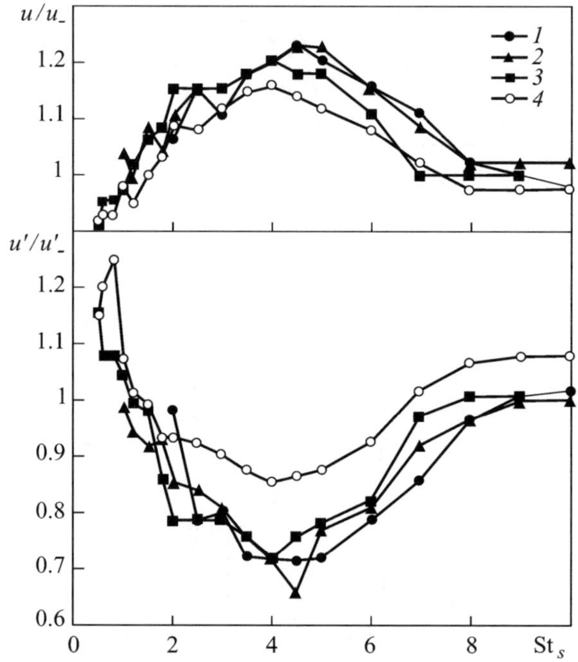

Fig. 2.67. The functions $u/u_{-} = F_1(k, \mathrm{St}_s)$ and $u'/u'_{-} = F_2(k, \mathrm{St}_s)$ at the jet axis point $x/d = 8$ for $u_0 = 10$ and 20 m/s and $L_0 = 130$ dB, $v'_s/u_0 = 1.5\%$. *1)* $k = 0.9$, *2)* $k = 0.5$, *3)* $k = 0.1$, *4)* harmonic signals.

Figs. 2.67 and 2.68 present the results of hot-wire anemometer measurements of the dependencies $u/u_{-} = F_1(k, \mathrm{St}_s)$ and $u'/u'_{-} = F_2(k, \mathrm{St}_s)$ at the jet axis point $x/d = 8$ for $u_0 = 10$ and 20 m/s (Re = $1.4 \cdot 10^4$ and $2.8 \cdot 10^4$). The corresponding results for harmonic excitation are presented there also. The most interesting results are obtained for high-frequency excitation ($\mathrm{St}_s = 3 - 8$). Here for $k = 0.1 - 0.9$ the values of decrease in u'/u'_{-} and increase in u/u_{-} are considerably greater than that for harmonic excitation.

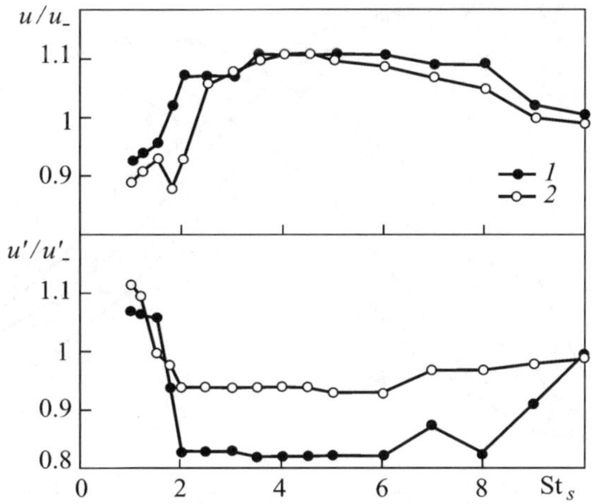

Fig. 2.68. The functions $u/u_- = \psi_1(\mathrm{St}_s)$ and $u'/u'_- = \psi_2(\mathrm{St}_s)$ for $k = 0.9$ (*1*) and for harmonic signals (*2*) at the jet axis point $x/d = 8$ for $u_0 = 20$ m/s and $L_0 = 130$ dB, $v'_s/u_0 = 0.75\%$.

It should be noted that in the case of equal values of sound pressure near the nozzle edge, L_0 = 130 dB, the efficiency of the jet high-frequency excitation by a nonharmonic signal decreases as the velocity increases from $u_0 = 10$ m/s to $u_0 = 20$ m/s. For $u_0 = 10$ m/s at the jet axis point $x/d = 8$ the minimum of u'/u'_- for $\mathrm{St}_s = 4$ equals to 0.9 and 0.7 for a harmonic and a nonharmonic signal correspondingly, whereas for $u_0 = 20$ m/s these values are 0.92 and 0.82. This fact could be explained by different values of v'_s/u_0 that are equal to 1.5% and 0.75% correspondingly. To gain the identical effect in the both cases when increasing twice the issue speeds (from 10 to 20 m/s), one should increase the sound pressure level by 6 dB, i.e. to set 136 dB instead of 130 dB.

Fig. 2.69 presents the dependencies of u/u_0 and u'/u_0 on the longitudinal coordinate for a turbulent jet ($u_0 = 10$ m/s) under harmonic and nonharmonic ($k = 0.9$) acoustical excitation ($\mathrm{St}_s = 6$, $v'_s/u_0 = 1.5\%$), as well as without any excitation. These functions illustrate efficiency of the nonharmonic and nonharmonic excitation aimed at the turbulence attenuation.

2.16. Acoustical Excitation of Noncircular Jets

Aerodynamic characteristics of turbulent acoustically excited submerged jets issuing from nozzles with noncircular cross-sections have been studied. The equivalent diameter, i.e. the diameter of the circular nozzle with the area that is equal to the area of the noncircular nozzle, is taken as the linear size.

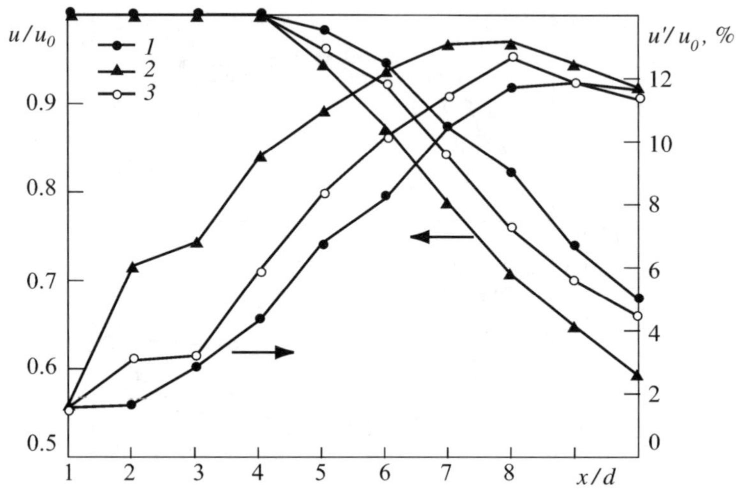

Fig. 2.69. The change of the averaged velocity, u/u_0, and the mean-square velocity pulsations, u'/u_0, along the jet axis for the transversal acoustical excitation $\mathrm{St}_s = 6$, $u_0 = 10$ m/s, $L_0 = 130$ dB, $v'_s/u_0 = 1.5\%$. *1)* $k = 0.9$, *2)* the unexcited jet, *3)* the harmonic excitation.

The most interesting results for low-frequency excitation of noncircular jets are obtained for the elliptic outlet cross-section of the nozzle, as applied to the both laminar and turbulent initial boundary layers [2.20, 2.22]. By way of example in the case of the nozzle elliptic cross-section with the axis ratio 2 : 1, Fig. 2.70 presents the longitudinal velocity pulsations along the jet axis, x/d_e, for the Strouhal numbers of acoustical excitation, $\mathrm{St}_s = 0 - 0.9$, and the excitation level $u'_s/u_0 = 2\%$ for the initial laminar (Fig. 2.66, *a*) and turbulent (Fig. 2.66, *b*) boundary layers. The Reynolds numbers are Re = $u_0 d_e / \nu = 10^5$ and $3.2{\cdot}10^4$. The cited papers demonstrate among other things the possibility for the control of the reorientation (along the flow) point position of the major and minor axes of the jet cross-section by means of acoustical excitation of elliptic jets. Fig. 2.71 shows the increase in the effective thickness, $\delta_{1/2}$, of an elliptic jet in the longitudinal direction without and with acoustical excitation (St = $fd_e/u_0 = 0.4$, $u'_s/u_0 = 2.5\%$) along the minor and major symmetry axes for the cases of the laminar (*a*) and turbulent (*b*) initial boundary layers. It follows that the acoustical excitation shifts substantially the cross-section where the reorientation of the major and minor axes begins.

Aerodynamic characteristics of jets issuing from nozzles of different cross-sections (circular, equilateral triangular, 6-lobed) under transversal acoustical excitation (low- and high-frequency) are studied in [2.72]. The boundary layer at the outlet section was laminar, the Reynolds number is Re = $u_0 d_e / \nu = 2.8{\cdot}10^4$, the sound pressure level near the nozzle edges is $L = 125$ dB.

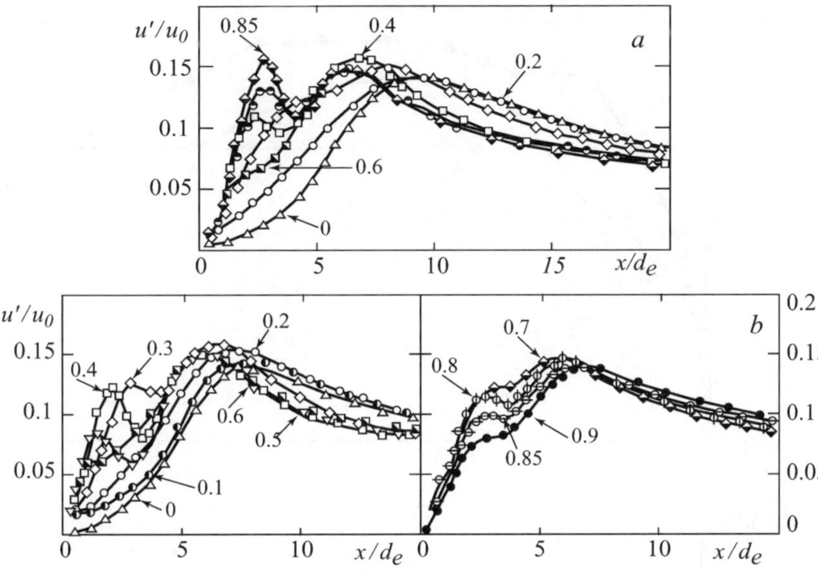

Fig. 2.70. The distribution of the longitudinal velocity pulsations along the elliptic jet axis for the axis ratio 2 : 1 at the nozzle cut-off and $u_s'/u_0 = 2\%$ for different values of the Strouhal number, $\mathrm{St}_s = fd_e/u_0$. The initial boundary layer is (*a*) laminar and (*b*) turbulent.

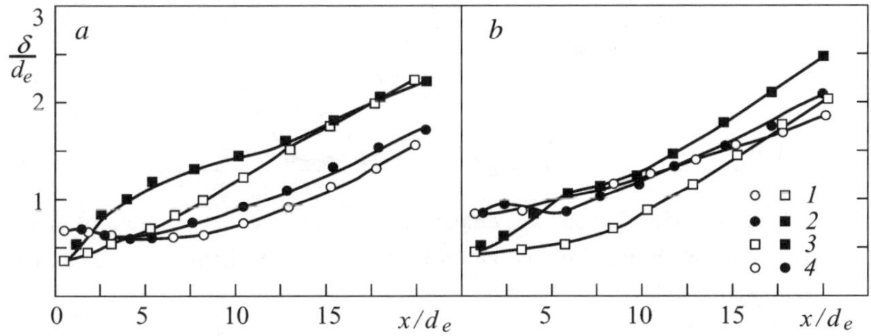

Fig. 2.71. The influence of acoustical excitation ($\mathrm{St}_s = 0.4$) on the reorientation along the flow of the minor and major axes of the jet cross-section for the (*a*) laminar and (*b*) turbulent initial boundary layers. The elongation of the nozzle elliptic outlet cross-section is 2 : 1. *1)* without acoustical excitation, *2)* with acoustical excitation, *3)* along the minor symmetry axis, *4)* along the major symmetry axis. The light marks correspond to the unexcited jet, the dark marks correspond to the excited jet.

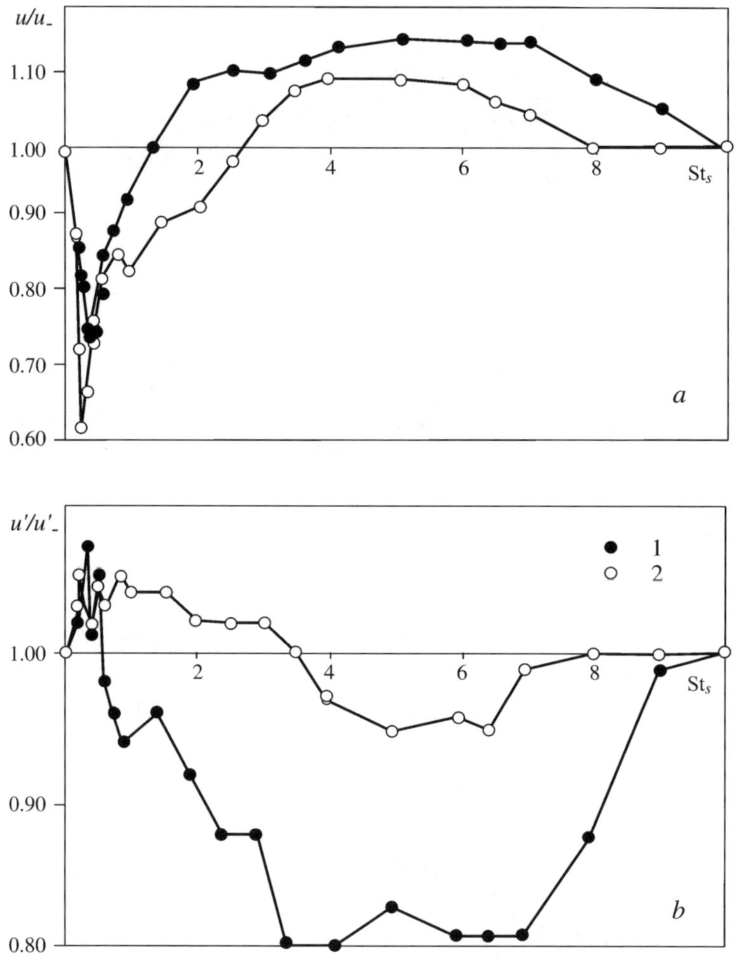

Fig. 2.72. The dependencies of the averaged velocity and velocity pulsations at the jet axis for $x/d_e = 8$ on the Strouhal number of acoustical excitation for (*1*) circular and (*2*) triangular nozzles.

Fig. 2.72 presents the functions $u/u_- = F_1(\mathrm{St}_s)$ and $u'/u'_- = F_2(\mathrm{St}_s)$ illustrating the influence of the acoustical excitation frequency for triangular and circular nozzles on the averaged velocity and longitudinal velocity pulsations at the jet axis section $x/d_e = 8$. Fig. 2.73 presents the changes in the averaged velocity and longitudinal velocity pulsations along the jet axis for $\mathrm{St}_s = 0.3$ and 5.0 for triangular and circular nozzles. These functions show that a jet issuing from a triangular nozzle is more sensitive to low-frequency excitation ($\mathrm{St}_s = 0.3$) and less sensitive to high-frequency excitation ($\mathrm{St}_s = 5.0$) as compared to a circular jet.

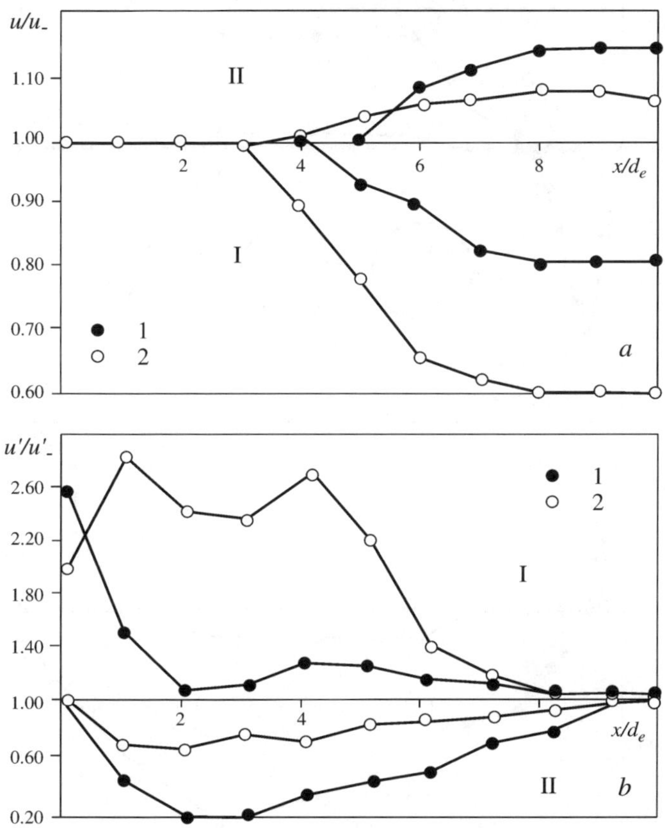

Fig. 2.73. The changes in the (*a*) averaged velocity and (*b*) velocity pulsations along the jet axis for $St_s = 0.3$ and 5.0 for (*1*) circular and (*2*) triangular nozzles.

It is interesting to compare the dependencies of the averaged velocity at the jet axis for $x/d_e = 8$ on the Strouhal number for jets issuing from a circular nozzle, a nozzle with 6 generators of longitudinal vortices, and a 6-lobed nozzle (Fig. 2.74). It follows, among other things, that a jet from the nozzles of two latter shapes is almost insensitive to both low- and high-frequency excitation. This conforms with [2.34] where acoustical characteristics of jets issuing from nozzles with different shapes are studied. This phenomenon could be explained by corruption of the flow azimuthal homogeneity in the initial region of the jet issuing from the nozzle of complex configuration and by the corresponding attenuation of the coherent structures.

The acoustical excitation of turbulent jets issuing from nozzles having complex cross-sections with sharp angles (equilateral and isosceles triangles, star-shaped nozzles) gives also a possibility to attenuate large-scale coherent structures and, hence, amplify the relative role of small-scale mixing [2.16, 2.17].

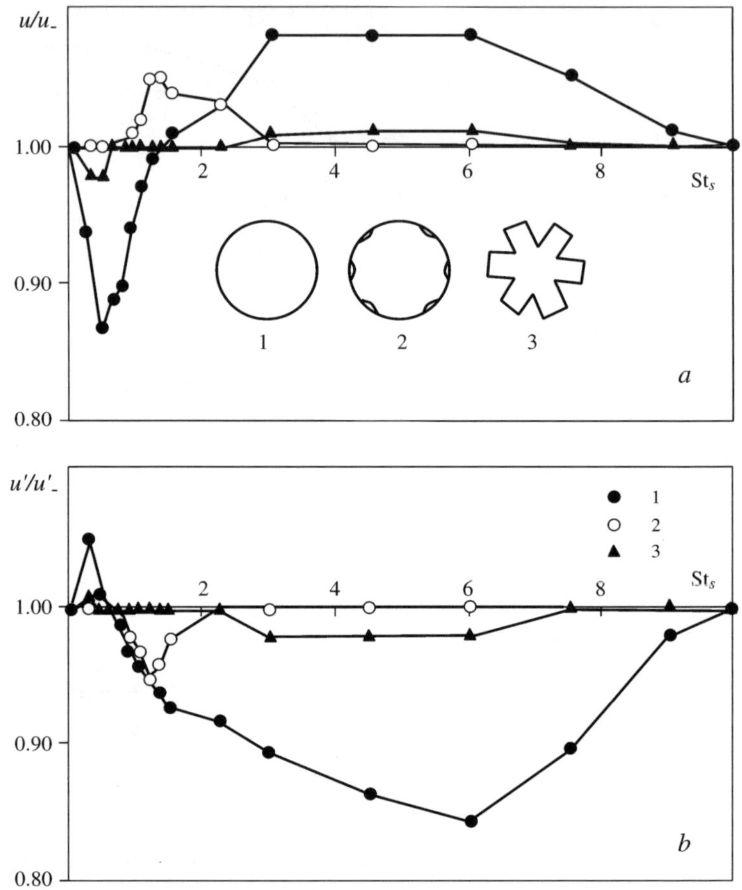

Fig. 2.74. The dependencies of the averaged velocity and longitudinal velocity pulsations at the jet axis for $x/d_e = 8$ on the Strouhal number of acoustical excitation for (*1*) a circular nozzle, (*2*) a nozzle with 6 generators of longitudinal vortices, and (*3*) a 6-lobed nozzle; Re = $4.2 \cdot 10^4$.

2.17. Effect of the Nozzle Edge Shape on Sensitivity of Jets to Acoustical Excitation

As noted above, the jet acoustical excitation is governed by transformation of the acoustical disturbances into the vortex ones at the nozzle edge. Thus, it is interesting to study the effect of the nozzle edge shape on the jet acoustical excitation, i.e. on the dependency of the jet characteristics on the acoustic field parameters. Such study is described in [2.29] where the change in the nozzle edge configuration caused the change in the jet acoustical characteristics.

The experiments are performed for the transversal acoustical irradiation of jets. The main parameters are as follows: the Reynolds number is $\mathrm{Re} = u_0 d / \nu = 5.8 \cdot 10^4$, the initial turbulence at the nozzle outlet section is $\varepsilon_0 = 0.5\%$, the issue speed is u_0 = 17 m/s, the sound pressure level is L = 120 dB, the boundary layer at the nozzle outlet section is laminar and, upon the turbulator mounting, turbulent. The experiments were performed at low and high frequencies corresponding to St_s = 0.29 and 2.6.

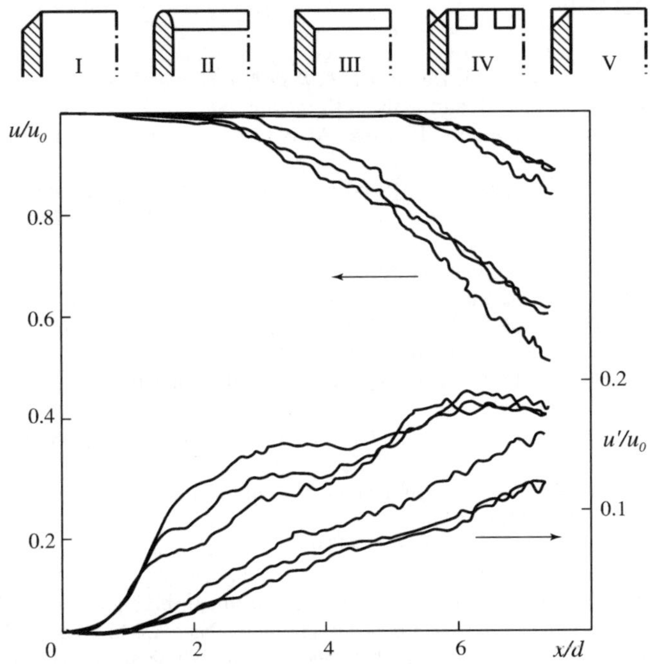

Fig. 2.75. The changes in the average velocity and intensity of the longitudinal velocity pulsations at the jet axis along the longitudinal coordinate for two values of the Strouhal number and for three types of the nozzle edges. St_s = 0.29: *1 – 3)* the nozzle edges I, V, IV; *4)* without outer acoustical excitation. St_s = 2.6: *4, 5)* the nozzle edges IV and I.

Fig. 2.75 presents the results for three kinds of nozzle outlet edges (from seven kinds under investigation) in the form of functions $u / u_0 = F_1(x/d)$ and $u' / u_0 = F_2(x/d)$ for St_s = 0.29 and 2.6. It follows that the change in the nozzle edge shape is convincingly reflected on the form of the functions for both low and high frequencies. In these cases both effects (the mixing intensification and attenuation) are observed for the initial laminar and turbulent boundary layers.

Fig. 2.72, *a* shows 5 types of the studied edges (*I* – *V*). Case *IV* corresponds to a saw-tooth edge formed by splitting the nozzle edge into 36 parts and their subsequent sharpening at an angle of 45° alternately on the outside and on the inside.

In full accordance with Section 2.3, the boundary layer turbulization for the fixed sound pressure level and high frequencies of the acoustical action causes the decrease or even annihilation of the mixing attenuation effect. The initial boundary layer turbulization for low frequencies of the acoustical excitation scarcely affect the mixing intensification in a jet.

REFERENCES

2.1. Ahuja KK, Lepicovsky J, Burrin RH (1981) Acoustic and turbulence measurements of a tone excited jet with and without flight simulation. AIAA Pap 2007

2.2. Arbey H, Ffowcs Williams JE (1984) Active cancellation of pure tones in an excited jet. J Fluid Mech 149: 445 – 454

2.3. Beloglazov BP, Ginevsky AS (1985) Influence of initial turbulence and initial turbulence scale on characteristics of accompanying jets (in Russian). Industrial Aerodynamics 1 (33): 195 – 212

2.4. Belotserkovsky SM, Ginevsky AS (1995) Simulation of turbulent jets and wakes using the method of discrete vortices (in Russian). Fizmatlit, Moscow

2.5. Borisov YuYa, Gynkina NM (1975) Excitation of high-speed jets by acoustical oscillations. Soviet Physics Acoustic 21, # 3: 230 – 233

2.6. Crighton DG (1981) Acoustic as a branch of fluid mechanics. J Fluid Mech 106: 261 – 298

2.7. Crow SC, Champagne FH (1971) Orderly structure in jet turbulence. J Fluid Mech 3: 547 – 591

2.8. Furletov VI (1969) Effect of Sound Vibrations on a Turbulent Gas Jet. Fluid Dynamics. 4, #5: 111 – 114

2.9. Gertsenshtein SYa, Sukhorukov AN (1998) Hard excitation of three-dimensional finite-amplitude modes in streamflows. Doklady Physics 43, # 4: 245 – 247

2.10. Ginevsky AS (1992) Methods of control of turbulent mixing in jets (in Russian). In: Scientific fundamentals of turbulence phenomena. Nauka, Moscow, pp 212 – 228

2.11. Ginevsky AS (1994) Acoustic, vibrational, and aerodynamical methods of turbulent mixing control. In: Active control of vibration. Symposium, London, pp 203 – 207

2.12. Ginevsky AS (1994) Acoustic Methods of Turbulent Mixing Control in Jet Flows. In: The Third Int. Congress on Air- and Structure-Borne Sound and Vibration. Montreal, pp 1191 – 1198

2.13. Ginevsky AS, Pochkina KA (1967) The influence of initial turbulence on characteristics of an axisymmetric immersed jet. J Eng Phys and Thermophys 12, # 1: 15 – 19

2.14. Ginevsky AS, Vlasov YeV, Kolesnikov AV (1978) Aeroacoustical interactions (in Russian). Mashinostroyenie, Moscow

2.15. Gurevich MI (1979) Theory of ideal fluid jets (in Russian). Fizmatlit, Moscow

2.16. Gutmark E, Schadow KC, Parr DM, Harris CK, Wilson KJ (1985) The mean and turbulent structure of noncircular jets. AIAA Pap 543

2.17. Gutmark E, Schadow KC, Wilson KJ, Parr DM (1986) Small-scale mixing enhancement in acoustically excited jets. AIAA Pap 1885

2.18. Ho CM, Huerre P (1984) Perturbed free shear layers. Ann Rev Fluid Mech 16: 365 – 424

2.19. Hussain AKMF (1983) Coherent structures – reality and myth. Phys Fluids 26, # 10: 2816 – 2859
2.20. Hussain HS, Hussain AKMF (1985) Excited elliptic jets. AIAA Pap 544
2.21. Hussain HS, Hussain F (1988) Subharmonic resonance in a shear layer. In: Adv. Turbulence 2. Proc. 2nd Eur Turbulence Conf, Berlin, pp 96 – 101
2.22. Hussain F, Hussain HS (1989) Elliptic jets. Part 1. Characteristics of unexcited and excited jets. J Fluid Mech 208: 257 – 320
2.23. Isataev SI, Tarasov SB (1971) On the Effect of an Acoustic Field, Directed Along a Jet Axis, on the Jet. Fluid Dynamics 6, #2: 332 – 334
2.24. Isataev SI, Tarasov SB (1972) On velocity pulsation characteristic frequencies and spectra in initial sections of axisymmetric jets (in Russian). In: Applied and theoretical physics 4, Alma-Ata, pp 247 – 252
2.25. Isataev SI, Tarasov SB (1975) On velocity pulsation characteristic scales in the transition area of free axisymmetric jets (in Russian). In: Applied and theoretical physics 7, Alma-Ata, pp 140 – 146
2.26. Kibens V (1980) Discrete noise spectrum generated by an acoustically jet. AIAA J 18, □ 4: 434 – 441
2.27. Kudryashov AV, Mansfel'd AD, Rabinovich MI, Sushchik MM (1984) Modulation mechanism of suppression of turbulence in shear flows. Soviet Physics Doklady 29, # 7: 530 – 532
2.28. Laufer J, Zhang X (1983) Unsteady aspects of a low Mach number jet. Phys Fluids 7: 1740 – 1750
2.29. Lebedev LL, Pavelev AA (2002) Effect of the nozzle edge shape on the acoustic sensitivity of a jet. Fluid Dynamics 37, #1: 21 – 26
2.30. Lepicovsky J, Ahuja KK, Salikuddin M (1986) An experimental study of tone-excited heated jets. J Propulsion 2: 149 – 154
2.31. Lepicovsky J, Brown WH (1989) Effects of nozzle excite boundary-layer conditions on excitability on heated free jets. AIAA J 6: 712 – 718
2.32. Long TA, Petersen RA (1992) Controlled interaction in forced axisymmetric jets. Part 1. The distortion of the mean flow. J Fluid Mech 235: 37 – 55
2.33. Michalke A (1971) Instabilität eines kompressiben runden Freisttrahls unter Berücksichtigung des Strahlgrenzschichtdicke. Z Flugwiss 19, # 8-9: 319 – 328
2.34. Moore CJ, Brierley DH (1979) Shear Layer Instability Noise Produced by Various Jet Nozzle Configurations. In: Mech. Sound Gener. Flows. Joint Symp. Berlin, pp 48 – 54
2.35. Nallasamy M, Hussain AKMF (1989) Effects of excitation on turbulence levels in a shear layer. Trans ASME J Fluids Eng 111: 102 – 104
2.36. Ng TT, Bradley TA (1987) Effect of multi-frequency forcing on the near field development of a jet. AIAA Pap 54
2.37. Ogava N, Maki H, Kuroda K (1993) Studies on a tone-excited jet. Trans Jap Soc Mech Eng 566: 41 – 47
2.38. Parekh DE, Reynolds WC, Mungal MG (1987) Bifurcation of a round air jet by dual-mode acoustic excitation. AIAA Pap 164
2.39. Paschereit CO, Wygnanski I, Fielder HE (1995) Experimental invextigation of subsonic resonance in an axisymmetric jet. J Fluid Mech 283: 365 – 407
2.40. Pavel'ev AA, Tsyganok VI (1982) Influence of Acoustic and the Flow Regime in the Boundary Layer on Nozzle Walls on the Mixing Layer of an Immersed Jet. Fluid Dynamics. 17, #6: 844 – 849
2.41. Petersen RA, Kaplan RE, Laufer J (1974) Ordered structures and jet noise. NASA Contractor Rep 134733

2.42. Pimshtein VG (1999) Aeroacoustical interactions. Structure and noise of turbulent jets (in Russian). Album. TsAGI
2.43. Raman G, Rice EJ (1991) Axisymmetric jet forced by fundamental and subharmonic tones. AIAA J 7: 1114 – 1122
2.44. Raman G, Rice EJ, Mankbadi RR (1988) Saturation and a limit of jet mixing enhancement by single frequency plane wave excitation: experiment and theory. AIAA/ASME/SIAM/APS, 1st Nat Fluid Dyn Congr, Cincinnati, Ohio. Part 2, pp 1000 - 1007
2.45. Raman G, Zaman KBMQ, Rice EJ (1989) Initial turbulence effect on jet evolution with and without tonal excitation. Phys Fluids A 7: 1240 – 1248
2.46. Rockwell DO (1972) External excitation of planar jets. Trans ASME J Appl Mech 4: 883 – 890
2.47. Schmidt C (1978) Aerodynamic characterization of excited jets. J Sound and Vibr 1: 148 – 152
2.48. Schadow KC, Wilson KJ, Parr DM, Bicker CJ, Gutmark E (1985) Reduction of flow coherence in forced, subsonic jets. AIAA Pap 1109
2.49. Van Dyke M (1982) An Album of Fluid Motion. The Parabolic Press, Stanford, California
2.50. Vlasov YeV, Ginevsky AS (1967) Acoustical modification of the aerodynamic characteristics of a turbulent jet. Fluid Dynamics 2, #4: 93 – 96
2.51. Vlasov YeV, Ginevsky AS (1973) Generation and suppression of turbulence in an axisymmetric turbulent jet in the presence of an acoustic influence. Fluid Dynamics 8, #6.
2.52. Vlasov YeV, Ginevsky AS (1973) Vibration action on aerodynamic characteristics of a turbulent jet (in Russian). Industrial Aerodynamics 33: 145 – 149
2.53. Vlasov YeV, Ginevsky AS (1975) Action of sound oscillations on characteristics of a turbulent submerged jet (in Russian). Proc. TsAGI 1655: 23 – 32
2.54. Vlasov YeV, Ginevsky AS (1986) Coherent structures in turbulent jets and wakes (in Russian). In: Advances in Science and Technology. Series "Mechanics of Fluids and Gases" 20, VINITI, Moscow, pp 3 – 84
2.55. Vlasov YeV, Ginevsky AS (1993) Acoustical excitation of a turbulent jet at the main frequency and its subharmonics (in Russian). In: Problems of mechanics and some current aspects of science. Nauka, Moscow, pp 99 – 107
2.56. Vlasov YeV, Ginevsky AS, Karavosov RK (1979) Coherent structures in turbulent jets with coflowing stream (in Russian). In: Turbulent jet flows. Part 1. Tallinn, pp 20 – 27
2.57. Vlasov YeV, Ginevsky AS, Karavosov RK (1982) Study of influence of nozzle vibrations on jet aerodynamic characteristics (in Russian). In: Jet flows of fluid and gas. Part 2. Novopolotsk, pp 9 – 16
2.58. Vlasov YeV, Ginevsky AS, Karavosov RK (1983) Aero-acoustical characteristics of acoustically excited jets. In: Acoustics of turbulent flows (in Russian). Nauka, Moscow, pp 14 – 21
2.59. Vlasov YeV, Ginevsky AS, Karavosov RK (1986) Influence of the mode composition of acoustical disturbances on jet aerodynamic characteristics. Soviet Physics Acoustic 32, # 4: 329 – 330
2.60. Vlasov YeV, Ginevsky AS, Karavosov RK (1987) Study of aerodynamic and acoustical characteristics of acoustically excited jets (in Russian). In: Modern problems of aeromechanics. Mashinostroyenie, Moscow, pp 154 – 168
2.61. Vlasov YeV, Ginevsky AS, Pimstein VG (1989) Multifrequency acoustic excitation of coherent structures of subsonic turbulent jets. Proc 5th EPS Liquid State Conf, Moscow, pp 133 – 136

2.62. Vlasov YeV, Ginevsky AS, Pimstein VG (1992) Control of coherent structures and aero-acoustic characteristics of subsonic and supersonic turbulent jets. DGLR/AIAA 14th Aeroacoustics Conf, pp 672 – 678
2.63. Vlasov YeV, Ginevsky AS, Makarenko TM (1997) Influence of transversal acoustical irradiation on mixing intensification and section deformation of circular jets (in Russian). Sci. Trans. TsAGI, XVIII 2: 82 – 86
2.64. Vlasov YeV, Ginevsky AS, Makarenko TM (1998) Influence of flow mode in the initial boundary layer on the effect of turbulence attenuation under acoustical irradiation of a jet (in Russian). J Eng Phys and Thermophys 71, # 1: 81 – 85
2.65. Vlasov YeV, Ginevsky AS, Karavosov RK, Makarenko TM (1999) Acoustical excitation of turbulent jets issuing from orifices with sharp edges (in Russian). In: Modern problems of aero-hydromechanics. Ve 2. Nauka, Moscow, pp 156 – 161
2.66. Vlasov YeV, Ginevsky AS, Karavosov RK, Makarenko TM (1999) Turbulence Suppression in Subsonic Jets by High-Frequency Acoustic Excitation. Fluid Dynamics 34, #1: 23 – 28
2.67. Vlasov YeV, Ginevsky AS, Karavosov RK, Makarenko TM (1999) Experimental investigation of turbulent jet expiring from the diaphragm. In: Experimental and computational aerothermodynamics of internal flows. Proc Conf, V II, Dresden, pp 256 – 260
2.68. Vlasov YeV, Ginevsky AS, Karavosov RK, Makarenko TM (2001) On change of the sign of high-frequency acoustical action on a turbulent jet with excitation increase (in Russian). J Eng Phys and Thermophys 1: 8 – 9
2.69. Vlasov YeV, Ginevsky AS, Karavosov RK, Makarenko TM (2001) Action of a nonharmonic acoustical signal on a turbulent jet (in Russian). J Eng Phys and Thermophys 5: 33 – 35
2.70. Vlasov YeV, Ginevsky AS, Karavosov RK, Makarenko TM (2001) Peculiarities of high-frequency acoustical excitation of turbulent jets under change of acoustical field intensity (in Russian). Proc. TsAGI 2647: 27 – 32
2.71. Vlasov YeV, Ginevsky AS, Karavosov RK, Makarenko TM (2002) Study of two-frequency excitation of turbulent jets by fundamental and subharmonic tones. Fluid Dynamics 37, # 3: 505 – 507
2.72. Vlasov YeV, Ginevsky AS, Karavosov RK, Makarenko TM (2002) Acoustical excitation of noncircular turbulent jets (in Russian). J Eng Phys and Thermophys 76, # 1:
2.73. Zaman KBMQ, Hussain AKMF (1981) Turbulence suppression in free shear flows by controlled excitation. J Fluid Mech 103: 133 – 159
2.74. Zaman KBMQ, Rice EJ (1992) On the mechanism of turbulence suppression in free shear flows under acoustic excitation. AIAA Paper 65

CHAPTER 3 CONTROL OF ACOUSTIC CHARACTERISTICS OF SUBSONIC TURBULENT JETS

3.1. Acoustic Characteristics of Near and Far Fields of Turbulent Jets under Acoustical Excitation

The changes in averaged and pulsation aerodynamic characteristics of jets under acoustical excitation should be accompanied by the corresponding changes in the jet natural acoustical characteristics determined by the flow aerodynamic parameters (cf. Chap. 1). A study of this phenomenon is not only of scientific but also of practical interest as it offers possibilities of the purposeful control of jet acoustical characteristics. Consider the effect of a harmonic acoustical signal on the field of pressure pulsations in a jet itself and in its near and far acoustical fields.

3.1.1. The Near Acoustical Field and Pressure Pulsations in a Jet. Detailed measurements of pressure pulsations in an acoustically excited jet were performed by Chan [3.4]. He studied the spatial evolution on the pressure pulsations in a circular jet (u_0 = 65 m/s, Re = $2.6 \cdot 10^5$) under longitudinal low-frequency excitation. The initial boundary layer at the nozzle cut-off was turbulent, the sound pressure level at the nozzle outlet section near its edge is L_0 = 125 dB, the Strouhal number, St_s, was varied within 0.26 – 0.85.

Fig. 3.1 shows the changes of the pressure pulsations in a jet along the flow at the jet axis (r = 0) and in the mixing layer (along the line □ = 1°). Here the values of $\mathrm{St}_x = (x/d)\mathrm{St}_s$ are plotted along the abscissa axis and the values of $20\lg(p'/p'_0)$ along the ordinate axis. The mean-square values of the pressure pulsations p'_0 correspond to L_0 = 125 dB. The analysis of these functions demonstrates that the pressure disturbances p' are propagated from the shear layer to the jet axis because at the axis they begin to grow not immediately upon their issue from the nozzle but at some distance downstream from it. The maximum of the disturbances is observed at some distance from the nozzle and then the disturbances are damped out. Fig. 3.2 presents the maximum values of the pressure pulsations at the jet axis, in the mixing layer and at the jet outer border as functions of the Strouhal number, St_s. The functions show that low-frequency longitudinal

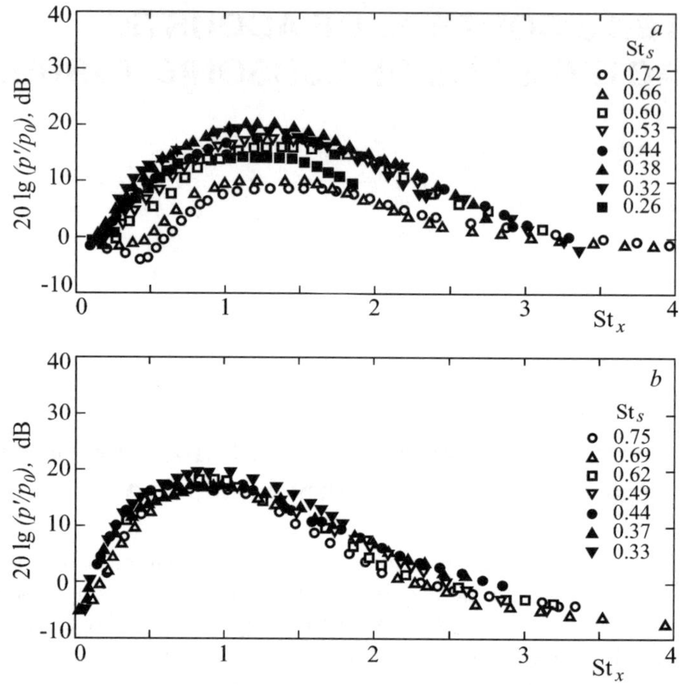

Fig. 3.1. The pressure pulsations at the jet axis and in the mixing layer for low-frequency longitudinal acoustical excitation: (*a*) at the jet axis ($r = 0$), (*b*) in the mixing layer ($\varphi = 1°$).

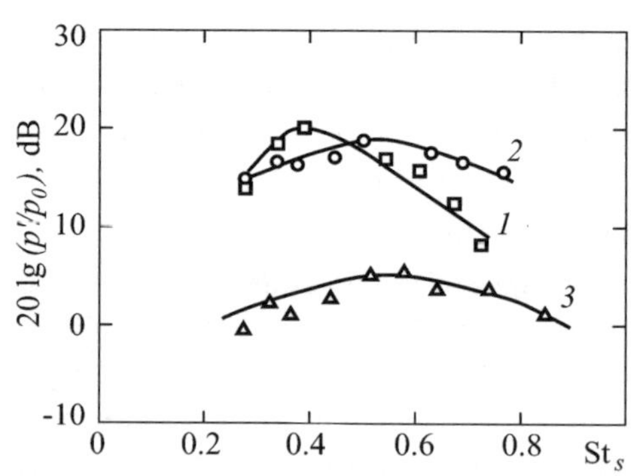

Fig. 3.2. The maximum values of the pressure pulsations (*1*) at the jet axis, (*2*) in the mixing layer and (*3*) at the jet outer border as functions of the Strouhal number, St_s .

acoustical irradiation of a jet causes changes in the pressure pulsation field inside the jet.

The near sound field characteristics for a jet under acoustical excitation are changed similarly. It follows from the pressure pulsation spectra measured in one-third octave bands [3.21, 3.23] in the near acoustical field of a jet under low- and high-frequency acoustical excitation and presented in Fig. 3.3, *a*, *b* that in the first case the excitation causes the increase of the broad-band noise (Fig. 3.3, *a*) and in the second case its decrease (Fig. 3.3, *b*). The same results (Fig. 3.3, *c*) were obtained also for measurements of the jet noise spectra in narrow bands ($\Delta f = 32$ Hz).

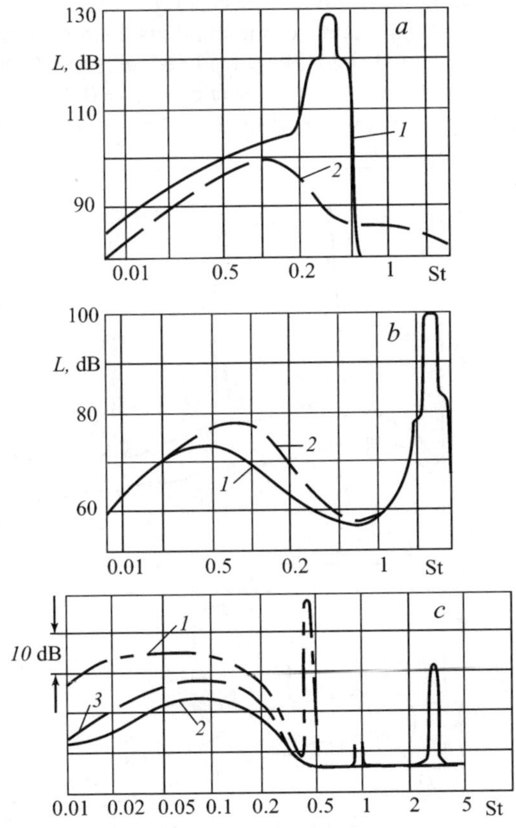

Fig. 3.3. The pressure pulsation spectra in the near acoustical field of a jet under low- and high-frequency acoustical excitation. (*a*) one-third octave spectra; Re = $2.1 \cdot 10^5$; M = 0.29. *1)* acoustical excitation with $St_s = 0.36$, *2)* the unexcited jet. (*b*) one-third octave spectra; Re = $1.75 \cdot 10^4$; M = 0.073. *1)* acoustical excitation with $St_s = 2.8$, *2)* the unexcited jet. (*c*) narrow-band spectra ($\Delta f = 32$ Hz). *1)* low-frequency acoustical excitation with $St_s = 0.46$, *2)* high-frequency acoustical excitation with $St_s = 3.5$, *3)* the unexcited jet.

3.1.2. The Far Acoustical Field of a Jet. Consider the acoustical excitement effect on the far sound field of turbulent jets. In 1975 Bechert and Pfizenmaier discovered amplification of the jet broad-band noise in the far field for low-frequency longitudinal acoustical excitation ($M_0 = 0.6$, $St_s = 0.48$, $\langle u_s'^2 \rangle^{1/2} / u_0 = 0.35\%$) [3.2]. The amplification could be as much as 6 – 7 dB. The same authors obtained in 1977 a similar result [3.3] for jet excitation by spiral waves ($n = 2$). Fig. 3.4 presents the corresponding narrow-band spectra for (*a*) plane and (*b*) spiral waves. In the first case the jet overall noise contains for the most part discrete components (the noise at the excitation frequency and at its harmonics) that far exceed the amplified broad-band noise of the jet. In the second case the noise at the excitation frequency and at its harmonics stand out less perceptibly.

Moore [3.18] obtained important experimental results for the jet acoustical excitation effect on the jet far sound field. He considered subsonic turbulent jets with turbulent boundary layer at the nozzle outlet section ($M_0 = 0.2 – 1.0$, $Re = 3 \cdot 10^5$) under longitudinal excitation by plane waves at the both low- and high- frequencies.

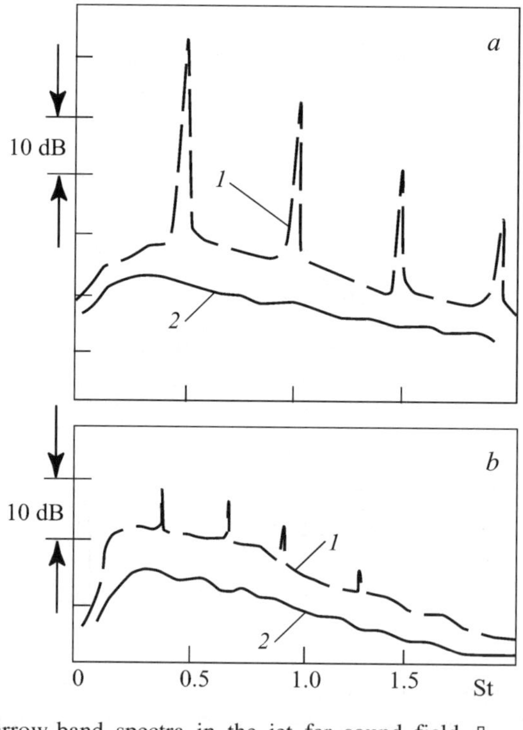

Fig. 3.4. Narrow-band spectra in the jet far sound field, θ = 45°, M_0 = 0.6. (*a*) excitation by longitudinal acoustical waves, $St_s = 0.48$; (*b*) excitation by spiral waves, $St_s = 0.30$. *1)* a jet under acoustical excitation, *2)* an unexcited jet.

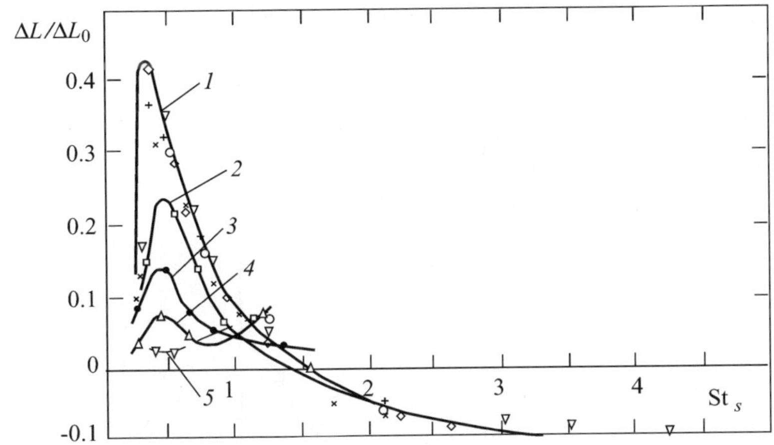

Fig. 3.5. The relative changes in the broadband noise level for the far sound field of a jet under irradiation. *1)* $M_0 = 0.2 - 0.6$, *2)* $M_0 = 0.7$, *3)* $M_0 = 0.8$, *4)* $M_0 = 0.9$, *5)* $M_0 = 1.0$.

The maximum increase in the broadband noise was gained at $St_s = 0.35$. However, some decrease in the broadband noise for $St_s > 1.5$ and $M_0 = 0.2 - 0.7$ was discovered also. The mean-square values of pressure pulsations in the sound wave at the nozzle cut-off did not exceed 0.08% of the dynamic pressure. Moore pointed out, referring to [3.20], the common character of the mechanisms for the broad-band noise reduction in the jet far field and the turbulent mixing attenuation in a jet under high-frequency acoustical excitation.

Moore visualized the flow for a jet under low-frequency acoustical excitation ($St_s < 1$) and demonstrated that the considerable reconstruction of coherent structures occurs in the jet initial region at the $St_s = 0.35$. Fig. 3.5 presents the relative changes in the wide-band noise level as functions of the Strouhal number, St_s, for different values of the Mach number, M_0 [3.18]. Here the relative increments of the jet broadband noise (□*L*, dB) due to the acoustical action with respect to the noise pure tone (□L_0, dB) acting on the jet are plotted along the ordinate axis. According to these functions the increment $\Delta L / \Delta L_0$ decreases with increasing the Mach number, for low-frequency acoustical excitation M_0 and becomes negligible for $M_0 \to 1$. The quantity $\Delta L / \Delta L_0 \to 0.1$ for high-frequency excitation ($St_s = 2.0 - 4.5$). It was also demonstrated [3.18] that the well-known "law of the eighth power" (Fig. 3.6) that is typical for turbulent jets free of acoustical excitation (cf. Chap. 1). The noise of an unexcited jet is proportional approximately to the sixth power of the issue speed.

The increase in the jet broadband noise is realized when the exciting acoustical signal levels exceed a certain threshold. The increase in the inherent broadband noise of a jet under excitement by a pure tone signal is more than in the case of excitement by a signal with a sufficiently wide frequency band.

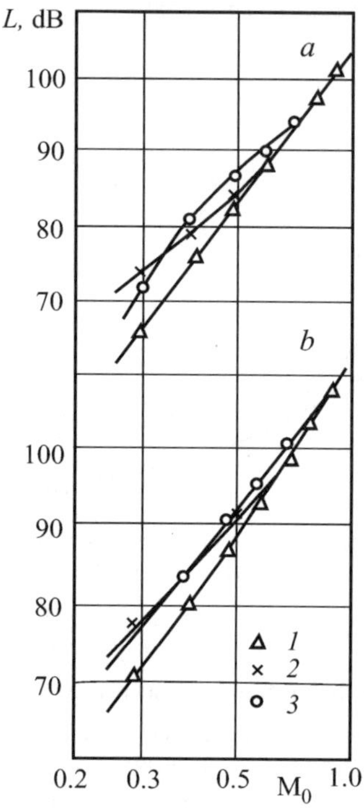

Fig. 3.6. The violation of the "eighth power law" for a jet under acoustical irradiation; (*a*) $\varphi = 90°$, (*b*) $\varphi = 30°$. *1)* an irradiation-free jet, *2)* a jet under irradiation by the pure tone sound at frequency $f = 1281$ Hz, *3)* the same at frequency $f = 2050$ Hz .

Similar results were obtained in [3.11, 3.14] for a low-speed subsonic jet under high-frequency acoustical excitation. The decrease in the broadband noise was recorded at 4 – 8 dB for $St_s = 3.77$ (Fig. 3.7, *a*) [3.14] and at 3 dB for $St_s = 7.74$ (Fig. 3.7, *b*) [3.11].

Hussain and Hasan [3.11] studied the influence of acoustical excitation on low-speed turbulent jets for low-and high-frequency excitation ($M_0 < 0.2$, $Re \le 10^5$). The boundary layer at the nozzle outlet section was laminar. The results are presented in Fig. 3.8 for $M_0 = 0.15$ in the form of dependencies of ΔL of St_θ . The increase in the jet broadband noise of $\approx$ 3 dB was observed in the range $St_\theta = 0.005 - 0.008$ and the decrease of $\approx$ 1.6 dB for $St_\theta = 0.01 - 0.017$. Here St_θ is the dimensionless frequency of the acoustical signal acting on the jet and calculated from the momentum loss thickness of the boundary layer at the nozzle cut-off (the Strouhal number).

Fig. 3.9 presents the results of measurement for noise spectra in the far field of turbulent jets under low- and high-frequency excitation by axisymmetric ($n = 0$)

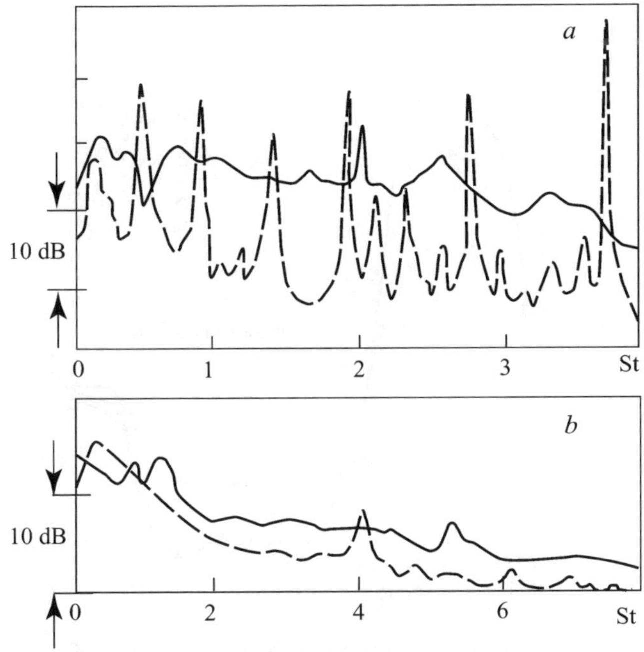

Fig. 3.7. Noise narrow-band spectra in the jet far sound field. (*a*) $\varphi = 31°$, $M_0 = 0.09$, $St_s = 3.7$, (*b*) $\varphi = 90°$, $M_0 = 0.15$, $St_s = 7.74$.

and spiral ($n = 2$) acoustical waves [3.23]. The broadband noise increase for $St_s < 1$ and its decrease for $St_s > 1$ are recorded here also.

The effect of the sound signal level (for low-frequency excitation of a subsonic turbulent jet) on the increase of the wide-band noise in the far acoustical field ($\varphi = 30°$, $M_0 = 0$, $St_s = 0.5$, $L_0 = 130 - 141$ dB) is illustrated by one-third octave spectra [3.1] presented in Fig. 3.10.

It was demonstrated in Chap. 2 that the turbulence attenuation effect is achieved for a distinct range of the sound pressure levels of high-frequency acoustical signals. When exceeding some limiting level, the action sign changes and the turbulence generation occurs instead of the turbulence attenuation. The jet broadband sound appears to increase also.

To study this case, a high-frequency gas-jet radiator with the sound source in the form of a slot jet issuing under supercritical pressure differential and leaking on a resonator [3.22] is used. The radiator axis was perpendicular to the axis of a jet under study, i.e. the jet was under transversal irradiation. The sound level was $L = 140$ dB for the frequency $f_s = 16000$Hz corresponding to the Strouhal number $St_s \approx 7.3$ for the jet issue speed $u_0 = 250$ m/s and the nozzle diameter $d = 0.115$ m. As follows from the jet noise one-third octave spectra measured for $\varphi = 30°$,

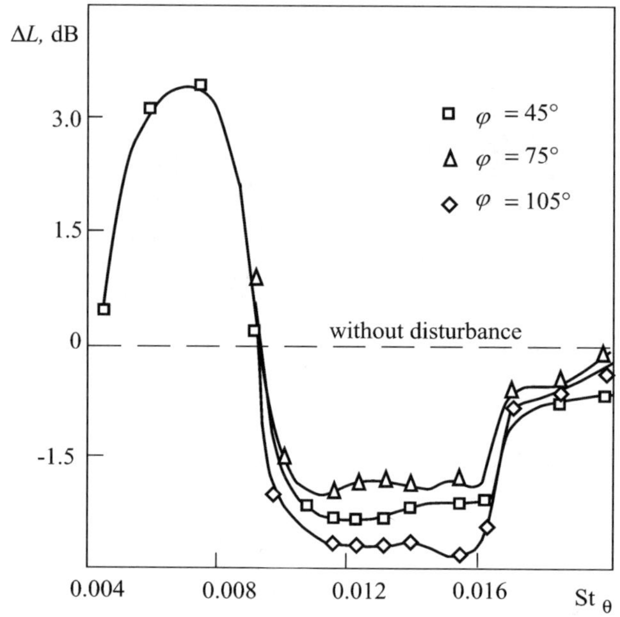

Fig. 3.8. The dimensionless noise spectra in the far sound field of a jet under low- and high-frequency excitation.

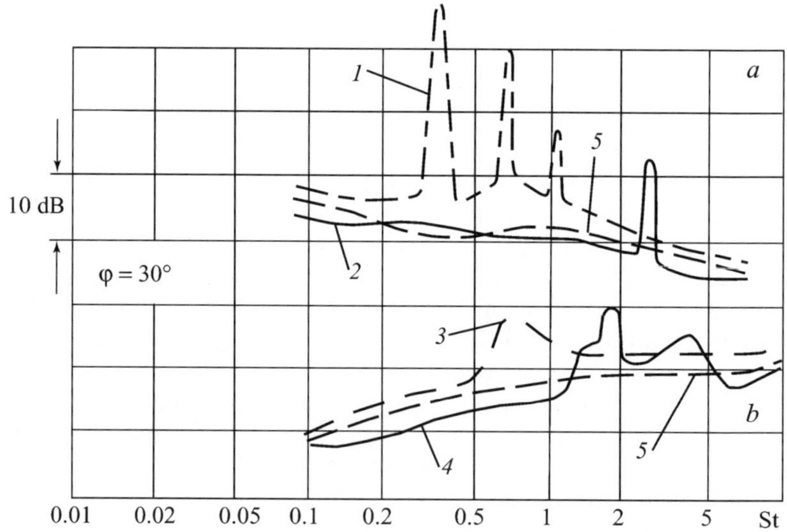

Fig. 3.9. The narrow-band noise spectra in the far sound field of a jet, $\varphi = 30°$. (*a*) excitation by longitudinal plane waves, (*b*) excitation by spiral acoustical waves. *1)* $St_s = 0.6$, *2)* $St_s = 3.5$, *3)* $St_s = 0.32$, *4)* $St_s = 1.65$, *5)* an excitation-free jet.

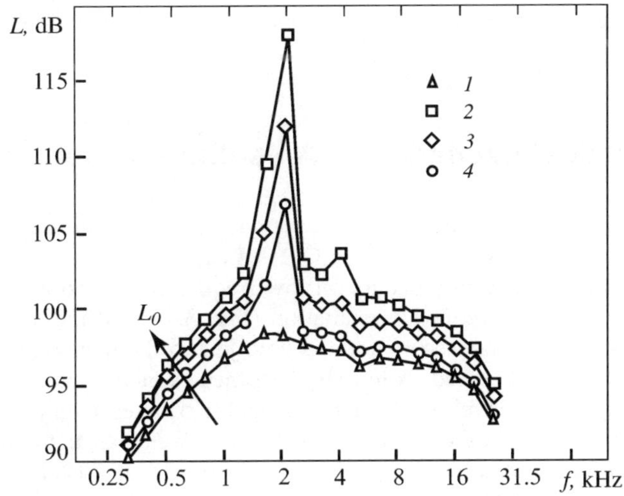

Fig. 3.10. The pressure pulsation spectra in the jet far field. *1)* an excitation-free jet, *2)* $L_0 = 141$ dB, *3)* $L_0 = 136$ dB, *4)* $L_0 = 130$ dB.

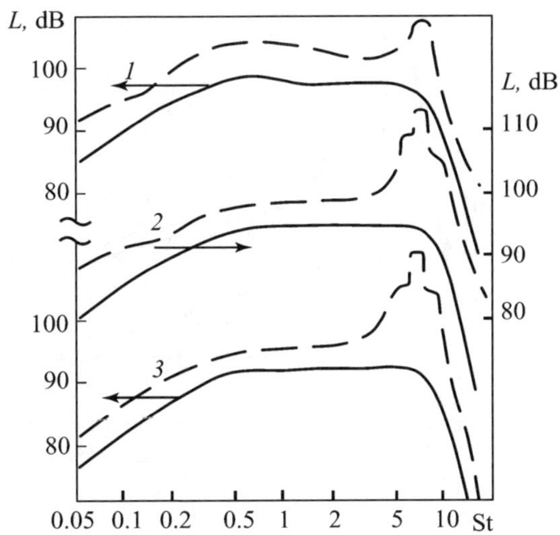

Fig. 3.11. The jet noise one-third octave spectra in the far sound field. *1)* $\varphi = 30°$, *2)* $\varphi = 60°$, *3)* $\varphi = 90°$. Solid lines correspond to an irradiation free jet and dashed lines to irradiation with $St_s = 7.3$.

60° and 90° and presented in Fig. 3.11, the high-frequency irradiation has caused the increase in the jet broadband noise in the far sound field and not its decrease.

Therefore, the changes in the broadband noise of turbulent jets under acoustical excitation were determined trustworthily. The subsequent studies along this line were devoted to refinements of the obtained rules as well as to their extension to the case of anisothermic and annular jets and jets propagating in coflowing streams.

3.2. Acoustical Excitation of Anisothermic Jets

The effect of acoustical excitement on hot jets with temperatures up to 900 K and Mach numbers M_0 = 0.47 is studied [3.12]. As in the case of cold jets, the jet broadband noise amplification occurs for low-frequency excitation (Fig. 3.12) and its attenuation for high-frequency excitation (Fig. 3.13). At the same time some distinctions between responses of hot and cold jets to the exciting low-frequency acoustical signal are observed. A hot jet "becomes excited" initially easier than a cold one, i.e. for lower levels of the acting signal. However, later on the hot jet responds to the increase in the acting sound level slower than the cold one. These conclusions are illustrated by data presented in Fig. 3.14. It was also observed that the increase in the broadband noise of a hot jet under low-frequency excitation occurs to a greater extent in the forward area of the far sound field (at large angles, φ, to the axis) and is most pronounced at frequencies near that of the exciting signal (cf. Fig. 3.12).

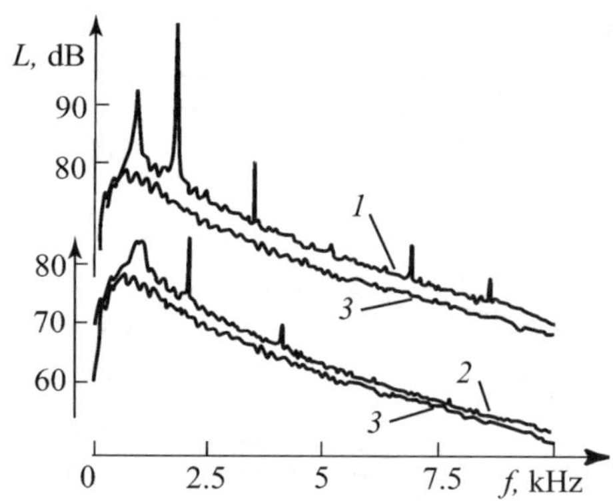

Fig. 3.12. The narrow-band noise spectra in the far sound field of a jet under low-frequency excitation, φ = 45°, M_0 = 0.47, T_{00} = 900 K. *1)* St_s = 0.5, L_0 = 126 dB; *2)* St_s = 0.6, L_0 = 112 dB, *3)* an excitation-free jet.

The authors of the paper [3.7] point out that the broadband noise for a hot high-speed jet (M_0 = 0.85, T_{00} = 600 K) under low-frequency irradiation increases generally in the same manner as for a cold jet (Fig. 3.15). It is pertinent to note that the author of the paper [3.12], on the basis of his own experiments and his analysis of the data from [3.18], concluded that the attenuation of the broadband noise for a hot jet under high-frequency excitation is more evident than that for a cold jet. The same fact was noted also by Crighton [3.5].

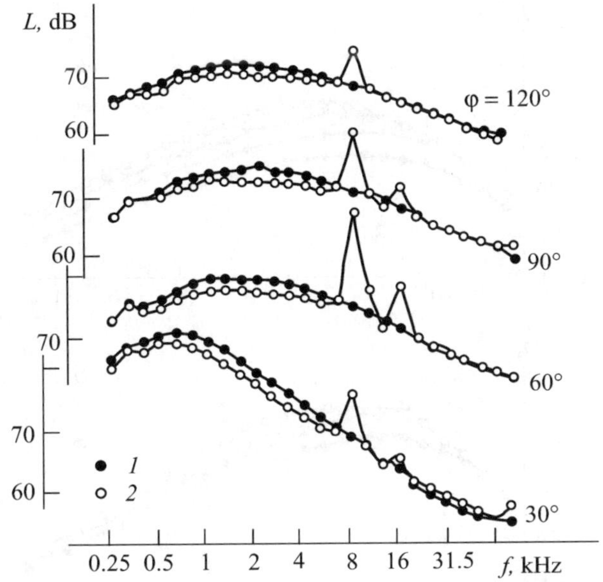

Fig. 3.13. The one-third octave spectra of the pressure pulsations in the far sound field of a jet under high-frequency excitation, $M_0 = 0.47$, $T_{00} = 600$ K. *1)* an excitation-free jet; *2)* an excited jet, $\mathrm{St}_s = 2.9$.

3.3. Acoustical Excitation of Jets in Coflowing Streams and of Coaxial Jets

It is worthy noting that effects of the broadband noise increase are characteristic not only for submerged jets but also for jets in coflowing streams, as well as for coaxial jets. For example, the paper [3.17] describes the jet broadband noise increase for acoustical excitement of both central jet and secondary annular flow. It turns out that the mixing layer of the outer jet is much more sensitive to acoustical excitement as compared to the mixing layer of the central jet. The review [3.5] contains the experimental results obtained by 'Rolls-Royce'. The experiments show that the noise broadband amplification of a jet under low-frequency excitation decreases as the accompanying flow speed increases. The jet excitation is not practically observed for $u_\infty / u_0 \approx 0.5$. High-speed filming shows that the vortex pairing does not occur in the main jet. These facts correlate completely with the results in [3.3, 3.4].

Paper [3.1] contains data on the influence of low-speed coflowing streams on the jet acoustical excitation ($\mathrm{St}_s = 0.5$, $L_0 = 134$ dB, $M_0 = 0.785$, $\varphi = 90°$, $u_\infty = 45$ m/s). It follows from Fig. 3.16 that in that case the coflowing stream does not

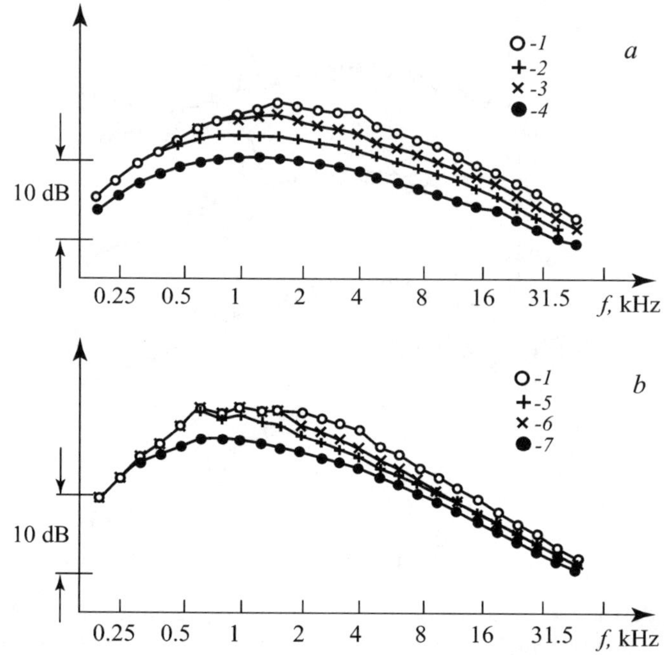

Fig. 3.14. The sound power spectra of a jet under low-frequency excitation, $St_s = 0.6$, $M_0 = 0.47$. (*a*) $T_{00} = 900$ K; *1)* $L_0 = 112.8$ dB, *2)* $L_0 = 106.6$ dB, *3)* $L_0 = 110.0$ dB, *4)* an excitation-free jet. (*b*) $T_{00} = 600$ K; *4)* – $L_0 = 119.5$ dB, *5)* $L_0 = 104.6$ dB, *6)* $L_0 = 110.7$ dB, *7)* an excitation-free jet.

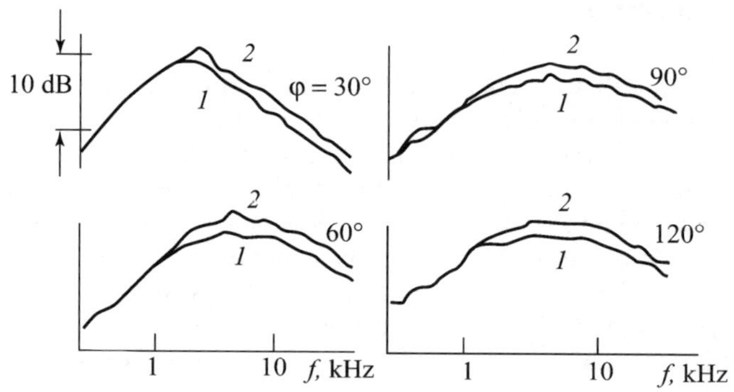

Fig. 3.15. The one-third octave sound spectra in the far sound field of a jet, $M_0 = 0.85$, $T_{00} = 600$ K. *1)* an excitation-free jet; *2)* an excited jet, $f = 5$ kHz.

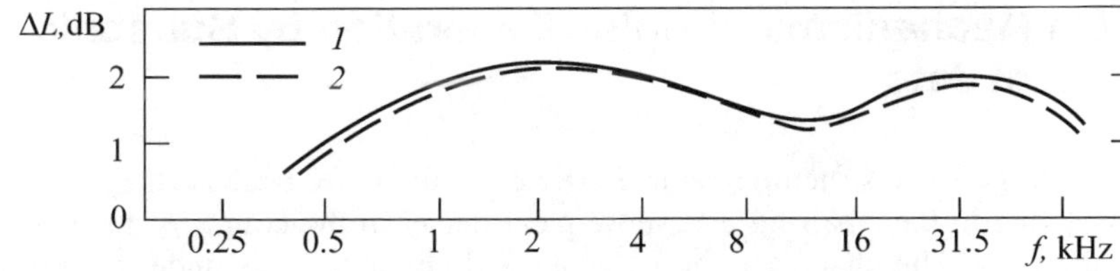

Fig. 3.16. The one-third octave sound spectra in the far sound field of a jet in the accompanying flow. *1)* $u_\infty = 0$, *2)* $u_\infty = 45$ m/s.

practically influence the main jet excitation. Here □*L* is the amplification factor (in dB) for the jet wide-band noise under acoustical excitation.

The influence of the coflowing stream generated by a high-speed cold annular jet ($M_{02} = 0.41$, $T_2 = 330$ K) on the acoustical excitation of the inner hot jet ($M_{01} = 0.47$, $T_1 = 600$ K) for the ratio between the areas of the issuing flows $F_2 / F_1 = 3.5$ is studied in [3.12]. The availability of the outer jet causes almost complete annihilation of the inner jet excitation effect due to the jet irradiation by the sound signal at frequency corresponding to the Strouhal number $\mathrm{St}_s \approx 0.5$ (Fig. 3.17).

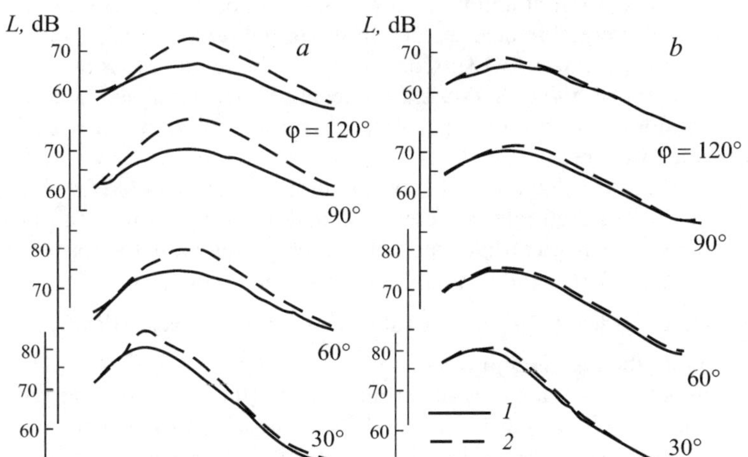

Fig. 3.17. The one-third octave sound spectra in the far sound field of coaxial jets (*a*) a single inner jet, $M_{01} = 0.47$, $T_1 = 600$ K, (*b*) coaxial jets, $M_{01} = 0.47$, $T_1 = 600$ K; $M_{01} = 0.41$, $T_1 = 330$ K. *1)* an excitation-free jet; *2)* an excited jet, $\mathrm{St}_s \approx 0.5$.

3.4. On Mechanisms of Noise Generation by Subsonic Turbulent Jets

Coherent large-scale structures radiate efficiently the noise because they vary considerably due to their pairing and subsequent decay in the course of their motion along the flow. The change in the jet acoustical characteristics under tone excitation is determined, primarily, by susceptibility of the coherent structures to periodical excitation. The localization of the pairing and decay points for the coherent structures occurs in the jet initial region and correspondingly the noise sources change their positions. There is no general agreement as to the noise source positions in the jet initial region.

According to one point of view [3.16, 3.18], the main noise sources are located at the points of annular vortex pairing (cf. Chap. 1). Another point of view [3.10] implies that precise decay of the azimuthal homogeneity of toroidal vortices, generation of azimuthally distributed substructures, and interaction between these structures are responsible for generation of most of the jet noise. The effects of changes in the jet broadband noise under acoustical irradiation could be explained in the framework of this hypothesis. Since low-frequency excitation amplifies the zero mode and blocks the development of higher azimuthal modes and, consequently, the generation of azimuthal lobes, more intensive generation of the structures and their interaction near the end of the initial region favour amplification of the jet broad-band noise [3.13]. Conversely, high-frequency excitation amplifies the development of higher azimuthal modes in the jet initial segment. As a result, the azimuthal lobe generation intensity and, consequently, the jet wide-band noise are attenuated in general.

The methods for calculation of subsonic turbulent jet noise so far elaborated are based on the use of the Lighthill acoustical analogy. According to this analogy, the general nonhomogeneous wave equation can be presented in the form of the sound propagation equation in the rest medium under the outer stress field, T_{ij}. Lighthill proposed to consider T_{ij} as the equivalent distribution of acoustical sources radiating sound into the rest medium.

Experimental characteristics of turbulent flow (the averaged and pulsation velocity profiles, the Reynolds normal and shear stresses, spatial-time characteristics of the velocity pulsation field) or relations of the semi-empirical turbulence theory (algebraic and differential models of turbulence [3.19]) are used to simulate the Lighthill tensor for unexcited jets. The coherent structures are not taken explicitly into calculations, though the empirical formula (Chap. 1) is used for the reference frequency of velocity pulsations in the mixing layer equivalent to the assumption that the Strouhal number $\mathrm{St} \approx 0.2 - 0.5$. Several attempts to predict the noise of turbulent jets based on the study of the vorticity field in a jet using the method of discrete vortices are known also [3.6, 3.9].

It is believed that some of these methods based on the Reynolds equations that are closed by means of the turbulence differential theories or on the method of discrete vortices could be extended to the case of periodically excited jets. To this aim one has to study their unsteady characteristics under periodical excitement with separation of the coherent structures.

Ffowcs Williams and Kempton [3.8] attempted simulating the Lighthill tensor so that the effect of increase/decrease of the jet broadband noise under low- or high-frequency acoustical excitation could be explained. They considered two schemes of noise generation in the jet initial region based on the flow wave model or on the vortex-pairing model, incorporating a random quality into each of the models.

It was assumed that the noise is generated for the first model in the breakdown of the instability wave and for the second model in the course of the vortex pairing. Both models describe well the increase of the jet wide-band noise under low-frequency excitation and the first one also the reverse effect, i.e. the decrease of the jet broadband noise under high-frequency excitation. In the latter case the high-frequency waves decay nearer to the nozzle. When decaying, they retard the growth of the subsequent waves that are the potential noise sources.

Paper [3.15] studies the noise emitted by separate segments of an excited jet and the contribution of separate modes into the overall noise level using nontraditional approach to measurements. The microphone lattice moved along the jet axis enclosing a cylindrical surface that extended in the jet far field up to 50 calibers from the nozzle cut-off. The proposed approach admitted the isolation of three quadruple components in the isotropic middle sound field of an excited turbulent jet and localization of the sources in the initial segment. The analysis of separate azimuthal components in narrow frequency bands demonstrated that the noise from separate vortex clusters localized at different distances from the nozzle cut-off, depending on the frequency, could have a similar structure. This conclusion favours the view of the important role of large-scale structures in the overall noise of an excited jet.

REFERENCES

3.1. Ahuja KK, Lepicovsky J, Burrin RH (1981) Acoustic and turbulence measurements of a tone excited jet with and without flight simulation. AIAA Pap 2007
3.2. Bechert D, Pfizenmaier E (1976) On the amplification of broad-band jet noise by a pure tone excitation. AIAA Pap 489
3.3. Bechert D, Pfizenmaier E (1977) Amplification of jet noise by a higher-mode acoustical excitation. AIAA J 15, □ 9: 1268 – 1271
3.4. Chan YY (1974) Spatial waves in turbulent jets. Phys Fluids 17, □ 1: 46 – 53
3.5. Crighton DG (1981) Acoustic as a branch of fluid mechanics. J Fluid Mech 106: 261 – 298
3.6. Davies POAL, Hardin JC, Edwards AVJ (1975) A potential flow model for calculation of jet noise. AIAA Pap 441
3.7. Deneuville P, Jacques J (1977) Jet noise amplification: a practically important problem. AIAA Pap 368
3.8. Ffowcs Williams JE, Kempton AJ (1978) The noise from the large-scale structure of a jet. J Fluid Mech 84, № 4: 673 – 694

3.9. Gaidaenko VI, Pogrebnaya TV (2000) Use of the method of discrete vortices for calculation of acoustical characteristics for a circular subsonic jet (in Russian). In: Methods of Discrete Singularities in Problems of Mathematical Physics. IX Int Symp Proc, Oriol, pp 123 – 126
3.10. Hussain AKMF (1983) Coherent structures – reality and myth. Phys Fluids 26, # 10: 2816 – 2850
3.11. Hussain AKMF, Hasan MAZ (1985) Turbulence suppression in free turbulent shear flows under controlled excitation. P.2. Jet-Noise reduction. J Fluid Mech 150: 159 – 168
3.12. Jubelin B (1980) New experimental studies on jet noise amplification. AIAA Pap 961
3.13. Juve D, Sunyach M (1981) Near and far field azimuthal correlation for excited jets. AIAA Pap 2011
3.14. Kibens V (1980) Discrete noise spectrum generated by an acoustically excited jet. AIAA J 18, ▯ 4: 434 – 441
3.15. Kopiev VF, Zaitsev MYu, Chernyshev SA, Kotova AN (1999) The role of large-scale vortex in a turbulent jet noise. AIAA Pap 1839
3.16. Laufer J (1974) On the mechanism of noise generated by turbulence. In: Omaggio a Carlo Ferrari, Torino, pp 451 – 464
3.17. Lu HY (1981) Effect of excitation on coaxial jet noise as observed by an elliptic mirror. AIAA Pap 2044
3.18. Moore CJ (1977) The role of shear-layer instability waves in jet exhaust noise. J Fluid Mech 80 Pt. 2: 321 – 367
3.19. Munin AG, Kuznetsov VM, Leontyev EA (1986) Aerodynamic sources of noise (in Russian). Mashinostroyenie, Moscow
3.20. Vlasov YeV, Ginevsky AS (1973) Generation and suppression of turbulence in an axisymmetric turbulent jet in the presence of an acoustic influence. Fluid Dynamics 8, ▯ 6:
3.21. Vlasov YeV, Ginevsky AS (1980) A problem of aeroacoustical interactions. Soviet Physics Acoustic 26, # 1: 1 – 7
3.22. Vlasov YeV, Ginevsky AS, Karavosov RK (1983) Aero-acoustical characteristics of acoustically excited jets. In: Acoustics of turbulent flows (in Russian). Nauka, Moscow, pp 14 – 21
3.23. Vlasov YeV, Ginevsky AS, Karavosov RK (1987) Study of aerodynamic and acoustical characteristics of acoustically excited jets (in Russian). In: Modern problems of aeromechanics. Mashinostroyenie, Moscow, pp 154 – 168
3.24. Zaman KBMQ (1985) Effect of initial condition on subsonic jet noise. AIAA J 23, # 9, 1370 – 1373

CHAPTER 4 EFFECT OF INTENSIVE ACOUSTIC DISTURBANCES ON SUBSONIC JETS

The approach to the control of aerodynamic and acoustical characteristics of subsonic turbulent jets by means of weak acoustical disturbances was demonstrated in Chap. 2 and 3. In the present chapter we consider some results of experimental study for effects of intensive periodical and, particularly, acoustical disturbances on aerodynamic characteristics of turbulent jets. We do not concern energy benefits of such in the approach to control of turbulent jets. It should only be noted that a number of researchers have performed experimental studies of turbulent jet characteristics under periodical excitation of high intensity. However, the comparison of their results is rather difficult because the periodical in time law of jet expenditure modulation was determined by constructional features of devices (flow interrupters) generating pulsations in jets. This situation hampers generalization of results in the published papers or their correlation, as the flow structure in an excited jet appears to be dependent on the spectral composition of the periodical velocity pulsations and the turbulence scale at the nozzle outlet section. This fact is confirmed by essential distinctions in the propagation laws for highly excited turbulent jets deduced in different authors' papers [4.2, 4.3, 4.5, 4.6, 4.8].

4.1. High-Amplitude Low-Frequency Periodical Excitation of a Circular Jet and Plane Mixing Layer

In [4.10] the sinusoidal law of air expenditure modulation is implemented in the wide range of the velocity pulsation intensities at outlet of a round nozzle ($\varepsilon_0 = 10 - 45\%$). This is supported by the fact that the amplitudes of the periodic pulsation harmonics and subharmonics were less than the amplitude of the pulsation main tone approximately by 20 dB. The mean outflow velocity was sustained at a constant value (u_0 = 10 m/s, Re = $3.5{\cdot}10^4$) and the frequency was f = 30 – 100 Hz corresponding to $\mathrm{St}_d = 0.15 - 0.5$.

The experimental set-up is presented in Fig. 4.1. The geometrical parameters of the damping chamber (*1*) were chosen so that the whole inner hollow of the set-up could consider as a simple acoustical resonator (like the Helmholtz resonator) in the operating frequency range. All the studies were performed at the air expenditure pulsation frequencies near the resonator natural frequency. This provided, first, the suppression of the main tone higher harmonics and, second, the larger

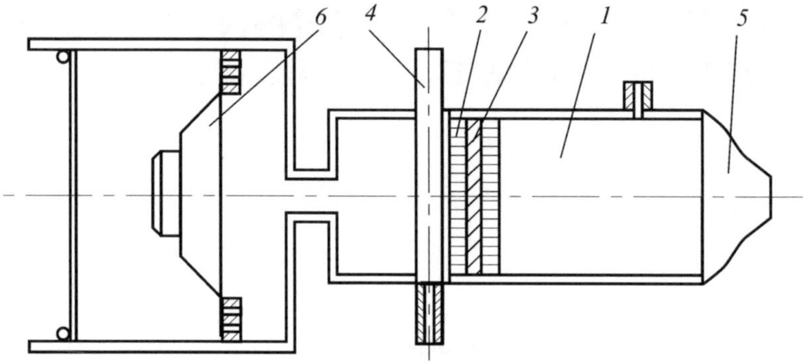

Fig. 4.1. The experimental apparatus. *1)* damping chamber, *2)* honey comb, *3)* deturbulizing lattice, *4)* air supply branch pipe, *5)* nozzle, *6)* electrodynamic speaker.

amplitudes of the sinusoidal velocity pulsations at the nozzle outlet (*5*) when using the electrodynamic loudspeaker (*6*) of small power.

The mean velocities along the jet axis and at the jet sections, the longitudinal velocity pulsations, their spectra and phase velocities of hydrodynamic wave propagation along the jet axis were measured by the hot-wire anemometer with the constant resistance. The hot-wire anemometer output linearization was not performed and all flow parameters were determined from the calibration characteristics of the hot-wire anemometer sensor. The hot-wire anemometer measurements for large amplitudes have considerable errors increasing with the distance from the jet axis.

Figs. 4.2 and 4.3 present the mean velocity decay and pulsations along the jet axis for different levels (ε_0) and frequencies (St_d) of periodical excitation. One can see that when ε_0 increases to 20 – 47%, a strongly excited jet decays initially (to $x/d = 4$) faster and then (for $x/d > 4$) slower than for an unexcited jet. Thus, for $x/d > 4$ and $\mathrm{St}_d = 0.15 - 0.25$ the mean velocity at the axis of an excited jet exceeds the velocity of an unexcited jet. A similar effect was discovered earlier in [4.2].

The analysis of the velocity profiles shows that a highly excited jet expands downstream more intensively than for an unexcited one. The vortex structure convection downstream velocity changes nonmonotonely and can exceed the flow mean velocity in some regions. Therefore, large vortices overtaking the flow must slow down and transmit their energy to the flow, i.e. the mechanism of energy transmission from the pulsation motion to the averaged one is realized in jets pulsating with high amplitudes. Here the pairings of annular vortices characteristic for unexcited and weakly excited jets are lacking. This is evident from power measurement of first harmonics and main tone subharmonics in the spectra of regular oscillations in a jet. For example, the subharmonic power for strong periodical excitation ($\varepsilon_0 > 20\%$) is very small and practically does not increase downstream. This fact validates the absence of the pairings.

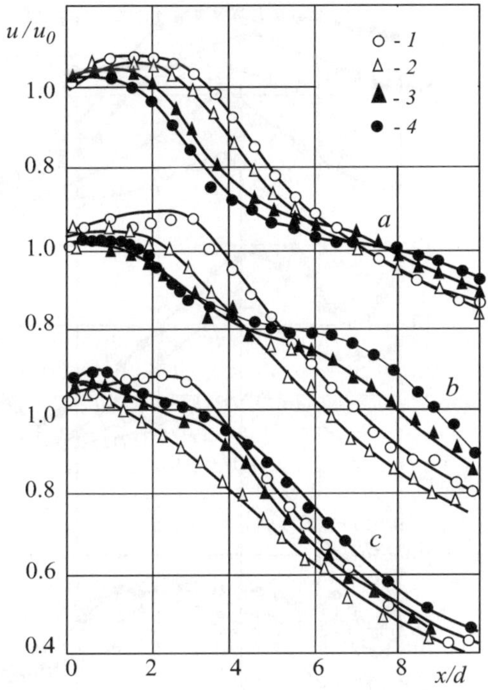

Fig. 4.2. The velocity decay along the jet axis for different frequencies and levels of periodical excitation. (*a*) $St_s = 0.15$; *1)* $\varepsilon_s = 0$, *2)* $\varepsilon_s = 18\%$, *3)* $\varepsilon_s = 30\%$, *4)* $\varepsilon_s = 37\%$. (*b*) $St_s = 0.25$; *1)* $\varepsilon_s = 0$, *2)* $\varepsilon_s = 14\%$, *3)* $\varepsilon_s = 36\%$, *4)* $\varepsilon_s = 47\%$. (*c*) $St_s = 0.50$; *1)* $\varepsilon_s = 0$, *2)* $\varepsilon_s = 11\%$, *3)* $\varepsilon_s = 21\%$, *4)* $\varepsilon_s = 28\%$.

It is interesting to note that the effect of nonmonotone change of velocity along the jet axis under low-frequency high-amplitude harmonic excitation is supported by calculations of plane pulsating turbulent jets on the basis of the method of discrete vortices [4.1, 4.10]. When analyzing the Reynolds shear stress profiles, it was demonstrated that in sections situated in a neighborhood of the abnormal segment of the velocity change near the jet axis, the Reynolds shear stresses change their sign, i.e. the pulsation motion energy is transmitted to the mean motion. The latter determines ultimately the nonmonotone change of the velocity along the jet axis (Figs. 4.4, 4.5). Here the velocity at the nozzle outlet section of width *h* changed according to the law $u = u_0(1 + A\sin 2\pi ft)$ where *A* and *f* are the pulsation amplitude and frequency correspondingly.

The paper [4.9] contains the study results for coherent structures in the highly excited turbulent mixing layer of two plane flows with velocities u_1 and u_2 satisfying the relation $(u_2 - u_1)/(u_2 + u_1) = 0.25$. The flow periodical excitation was performed by a flap mounted behind the plate dividing both flows. The flap

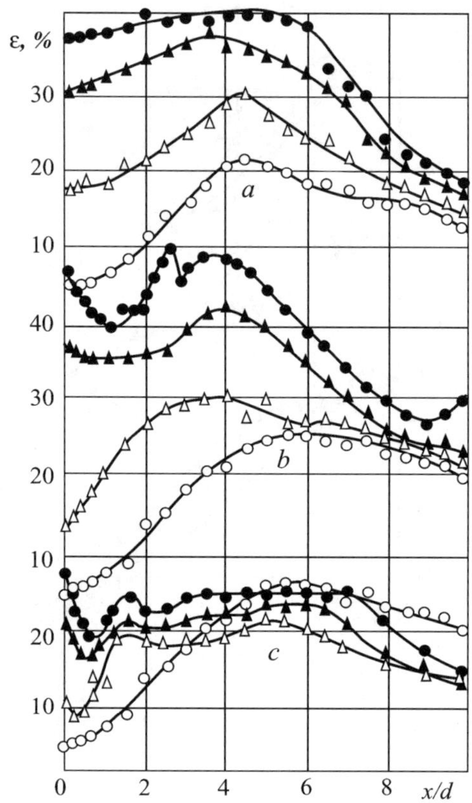

Fig. 4.3. The velocity pulsation intensity along the jet axis. The notation is the same as in Fig. 4.2.

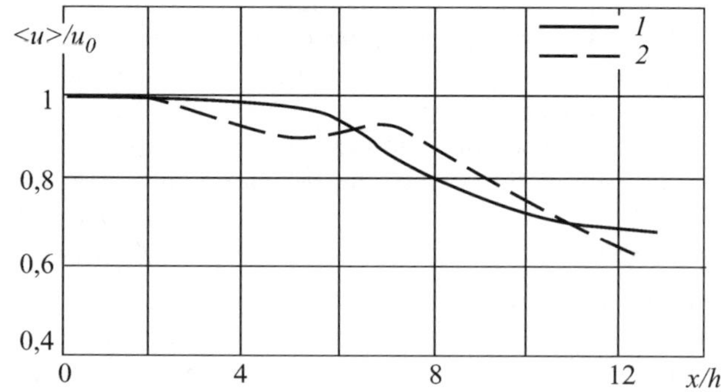

Fig. 4.4. The velocity decay along the axis of a plane turbulent pulsating jet (calculation). *1)* St = 0, *2)* St = 0.35, $A = 0.15$.

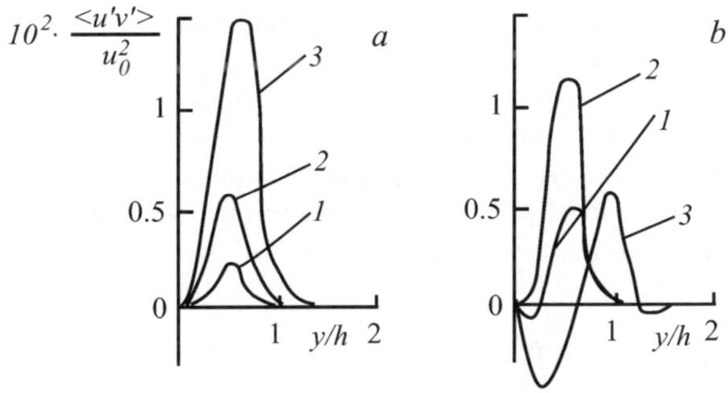

Fig. 4.5. The Reynolds shear stress profiles in a plane turbulent jet. (*a*) St = 0, (*b*) St = 0.35, *A* = 0.15. *1)* $x/h = 1$, *2)* $x/h = 2$, *3)* $x/h = 4$.

executed angular oscillations according to the sinusoidal law. Nonmonotone downstream change in the mixing layer thickness was discovered. Initially the layer is thickened, then it is thinned and, finally, it is slightly thickened. It was shown that the turbulence generation becomes negative at the segment of downstream reduction of the layer thickness (or of the momentum loss thickness, θ). This conclusion is correlated with the above-mentioned abnormal features of highly excited turbulent jets. It is significant that there are no coherent structure pairings as evidenced by the unimportance of the first subharmonic of the superimposed frequency. This conclusion is supported by the results of visual studies.

Described in [4.5] are experimental studies of a turbulent jet (the Reynolds number Re = 5700, the boundary layer at the nozzle cut-off is laminar) without periodical excitation and with axisymmetric periodical excitation of amplitude up to $u'/u_0 = 50\%$, and they have demonstrated that the annular vortex pairing process is realized ambiguously for high-amplitude low-frequency excitation. The pairing process is dampened for St = 0.34 and $\varepsilon_0 = 32\%$, whereas only one pairing is realized for St = 0.67 and $\varepsilon_0 = 17\%$; for $\varepsilon_0 = 0.5\%$ without excitation the first pairing occurs at the section $x/d = 1$, the second one occurs at the section $x/d = 1.50 - 2.56$, and for $x/d > 4$ the vortex rings collapse as a result of their azimuthal instability.

4.2. Flow Visualization in a Subsonic Circular Jet under Longitudinal and Transversal High-Amplitude Acoustical Excitation

To study the disturbance generation mechanism in a jet under the action of sound waves, gas-jet radiators of high intensity (L = 170 dB) were used. This has allowed to observation of (for shadow shootings) not only vortices generated by sound but also sound waves generating them [4.4, 4.7]. The Reynolds number determined by the nozzle diameter and the flow speed was Re = 10^6. The use of gas-jet radiators of high intensity has resulted in the change from the harmonic periodical excitation to the saw-tooth one (Fig. 4.6).

When analyzing the results of the study on effects of jet acoustical excitation where the Hartman whistle was used as the sound source, one should note the following: the signal generated by the Hartman whistle is nonharmonic; the spectrum contains, besides the main tone, the second and on occasion the third harmonics, so the jet is actually under multi-frequency excitation.

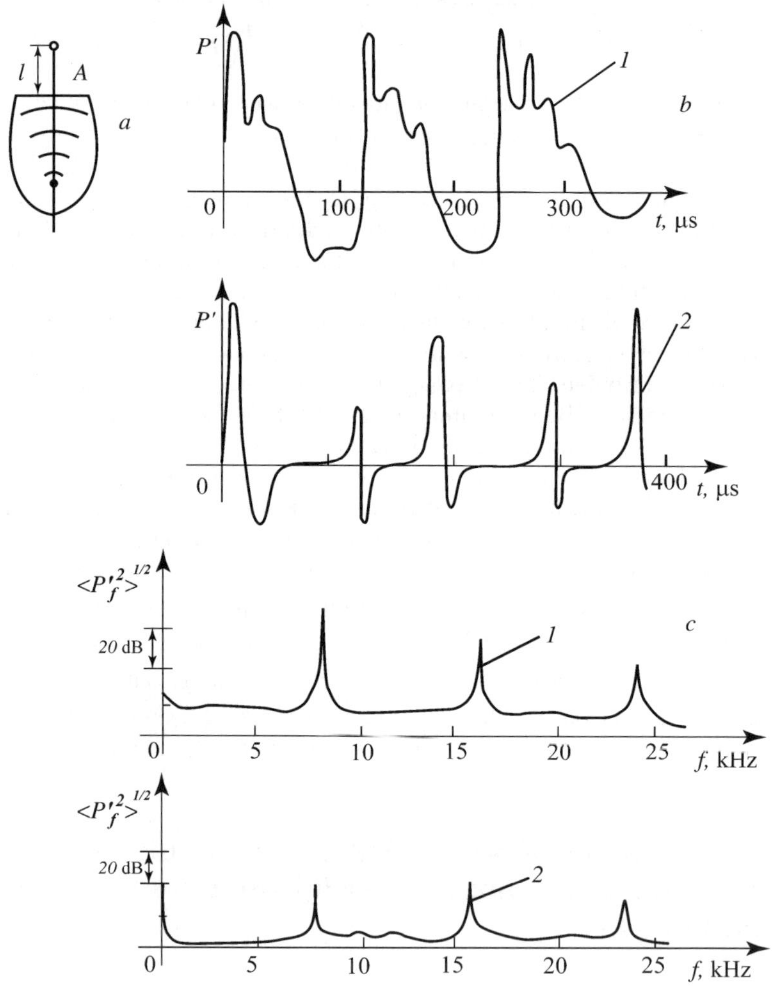

Fig. 4.6. The (*b*) oscillograms and (*c*) spectra of a gas-jet radiator with (*a*) an elliptic concentrator for different distances, *l*. *1)* $l = 15$ mm, $L = 169$ dB, *2)* $l = 300$ mm, $L = 150$ dB.

The irradiation directivity characteristics corresponding to different harmonics could vary. Therefore, the relationship among the noise intensities at different harmonics under changes of the relative position of the jet and the whistle may vary (sometimes the noise level at the second harmonic frequency was higher than that at the main tone frequency).

The acoustical wave generated by the Hartman whistle is a wave of finite amplitude due to the high noise level and the wave in the course of movement could change its shape (from a wave with the steep front near the radiator to an almost sinusoidal wave far away from the radiator) with the corresponding change in the relative level of the harmonic components. Hence in the adjustment of the irradiation level by way of changes in the distance between the jet and the radiator, the harmonic composition of the signal acting on the jet changes also even for the constant angle between the axes of the jet and the radiator.

The analysis of the interaction between high-intensity sound and a subsonic jet under *longitudinal* acoustical excitation demonstrates (Fig. 4.7) that the direct action of a longitudinal sound wave propagating form the damping chamber in a direction of the flow issue occurs in the mixing layer in the immediate vicinity of the nozzle outlet section. It is here that the toroidal vortex seen in the shadow photos is generated in the passage of the sound wave compression phase. This vortex is brought down by the flow, increases its size, and decays for $x/d = 3 - 4$. High-intensity sound waves outside of the mixing layer near the nozzle effects almost not at all on vortices that are brought downstream from the origin place.

The sound action on a jet under *transversal* irradiation, as in the case of longitudinal irradiation, is localized in the mixing layer in the immediate vicinity of the nozzle edge and causes generation of a vortex with azimuthal instability – its intensity is maximum at the jet border nearest to the radiator. This vortex is skew (Fig. 4.8).

When increasing the sound frequency, one can achieve generation of several vortices within the jet initial segment and study the sound effect on vortices at different stages of their development (Fig. 4.9). As was mentioned above, sound waves of intensity L = 170 dB do not influence on the generated vortices. At the same time, the sound wave front traveling through a vortex experiences considerable changes according to the velocity change at the vortex cross-section. Worthy of mention is an attempt of studying vortices under high-intensity sound from a radiator located at one of the focuses of an elliptic concentrator, which is used to localize the action within a small segment of the jet (near the second focus). The sound wave effect on the vortices is null here also.

The shadow photo analysis at high-amplitude acoustical irradiation makes it possible to conclude that the intensities of the generated vortices exceed considerably that of the annular vortices in the mixing layer of the radiation-free jet initial segment. The latter vortices could not be seen at the photos. It is worth noting that the pairing of the annular vortices generated by the sound is not observed.

The absence of the pairing of the annular vortices generated by the sound in the cases considered here probably could be explained by the high intensity of these vortices. As is well known, a similar effect is observed in the Karman vortex street for the separated flow over a cylinder where the pairings of the plane vortex cores are lacking also [4.6].

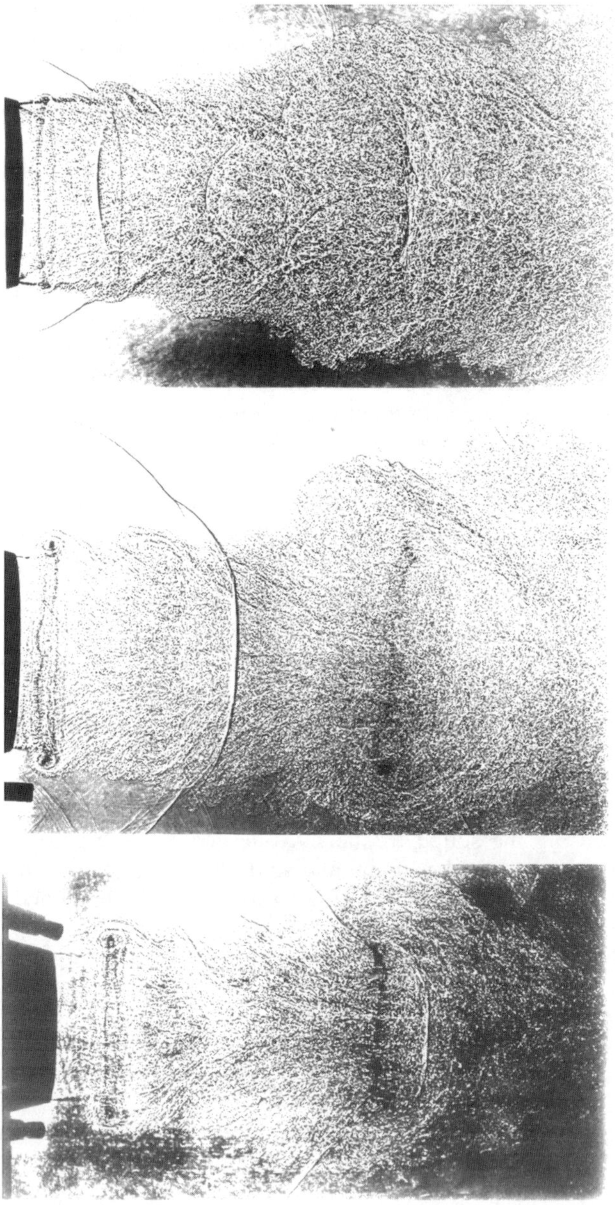

Fig. 4.7. Interaction of high-intensity sound and a subsonic jet (longitudinal irradiation, $M_0 = 0.75$, $St_s = 0.28$, $L = 170$ dB).

Fig. 4.8. A jet under transversal irradiation generating a skew vortex ($M_0 = 0.75$, $St_s = 0.23$, $L = 160$ dB).

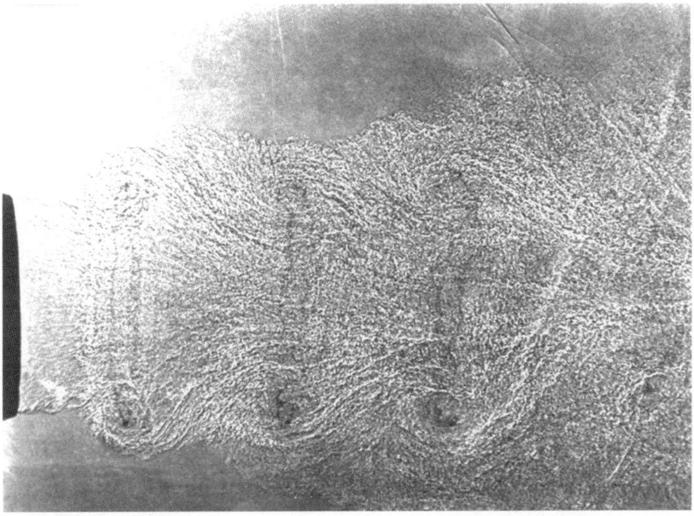

Fig. 4.9. Interaction of a sound wave and a chain of toroidal vortices under longitudinal irradiatio.

REFERENCES

4.1. Belotserkovsky SM, Ginevsky AS (1995) Simulation of turbulent jets and wakes using the method of discrete vortices (in Russian). Fizmatlit, Moscow

4.2. Krasheninnikov SYu, Liakhov VK, Migalin KV, Shaikin AP, Riaposov VB (1990) On the effect of the method of velocity pulsation shaping on the development of jet flows (in Russian). Izvestiya Vuzov, Aviation Technics, 3: 63 – 65

4.3. Migalin KV, Liakhov VK, Grabarnik SYa (1986) Some features of the pulsating jet development (in Russian). Izvestiya Vuzov, Aviation Technics, 3: 74 – 76

4.4. Pimshtein VG (1999) Album "Aeroacoustical interactions, structure and noise of turbulent jets" (in Russian). TsAGI

4.5. Reynolds WC, Bouchard EE (1981) The effect of forcing on the mixing-layer region of a round jet. In: Turbulent Shear Flows, Springer, pp 402 – 411

4.6. Vlasov YeV, Ginevsky AS (1986) Coherent structures in turbulent jets and wakes (in Russian). In: Advances in Science and Technology. Series "Mechanics of Fluids and Gases" 20, VINITI, Moscow, pp 3 – 84

4.7. Vlasov YeV, Ginevsky AS, Pimshtein VG (1991) On effect of the high intensity sound on a subsonic turbulent jet (in Russian). In: XI All-Union Acoustical Conference, Moscow. Ser G, pp 31 – 34

4.8. Vulis LA, Mikhasenko YuI, Khitrikov VA (1966) Effective control of propagation of a free turbulent jet (in Russian). Fluid Dynamics 1, # 6: 112 - 115

4.9. Weisbrot I, Wygnanski I (1988) On coherent structures in a highly excited mixing layer. J Fluid Mech 195: 137 - 159

4.10. Zosimov AV, Smirnykh YeA, Utenkova TA (1991) Experimental and theoretical study of turbulent pulsating jets under high levels of excitation (in Russian). Industrial Aerodynamics 4 (36): 105 – 123

CHAPTER 5 SELF-EXCITATION OF TURBULENT JET FLOWS

5.1. Self-excitation Schemes of Turbulent Jet Flows

The role of coherent structures in jets and mixing layers becomes deciding between aerodynamic and acoustical characteristics of jet flows. This is true for amplification of acoustical feedback in realization of different kinds of resonance. The simplest of such cases is the case of jet issue from a receiver (Fig. 5.1, *a*) that is a resonator with resonance frequencies belonging to the range of the jet sensitivity to periodical excitation. Here the jet is excited without any outer sound sources [5.1].

The best known effect of jet self-excitation is the so-called *wedge* tone realized when the plane jet or the mixing layer leaks on a wedge or a circular jet leaks on a coaxial ring (Fig. 5.1, *b*). Here the acoustical feedback is realized due to interaction of coherent structures and the wedge edge (the ring edge in the axisymmetric case). This feedback arises as follows. The pressure waves generated by collision of coherent structures and the wedge or ring edge propagate upstream and excite the mixing layer at the nozzle outlet section. The interaction of the shear layer and the wedge or the ring gives rise to intensive self-excited oscillations with frequencies determined by the flow speed, the mixing layer initial thickness, the wedge angle, and the distance x_0 from the nozzle edge or from the separating plate (in thc case of the mixing layer) downstream to the obstacle.

The self-excited oscillations give rise to discrete components in the spectrum of the velocity pulsations in the mixing layer and of the transversal force pulsations on the wedge. One can observe the jump changes in the frequency with accompanying hysteresis when the distance, x_0 gradually increases. The presence of the wedge with the shear layer inleakage causes the vortex structure changes that are similar to the changes in the case of outer periodical excitation of the shear layer at the main frequency of the large-scale vortex structure succession.

Similar resonance effects occur when a jet or, more precisely, a shear layer interacts with a cavity (Fig. 5.1, *c*). Here also the coherent structures, upon collision with the cavity trailing edge generate pressure waves propagating upstream with the sound speed and exciting the boundary layer near the cavity front space, resulting in the self-excited oscillations. This problem is discussed in Chapter 10. Another category of self-excited oscillations is realized when a jet flows in a tube

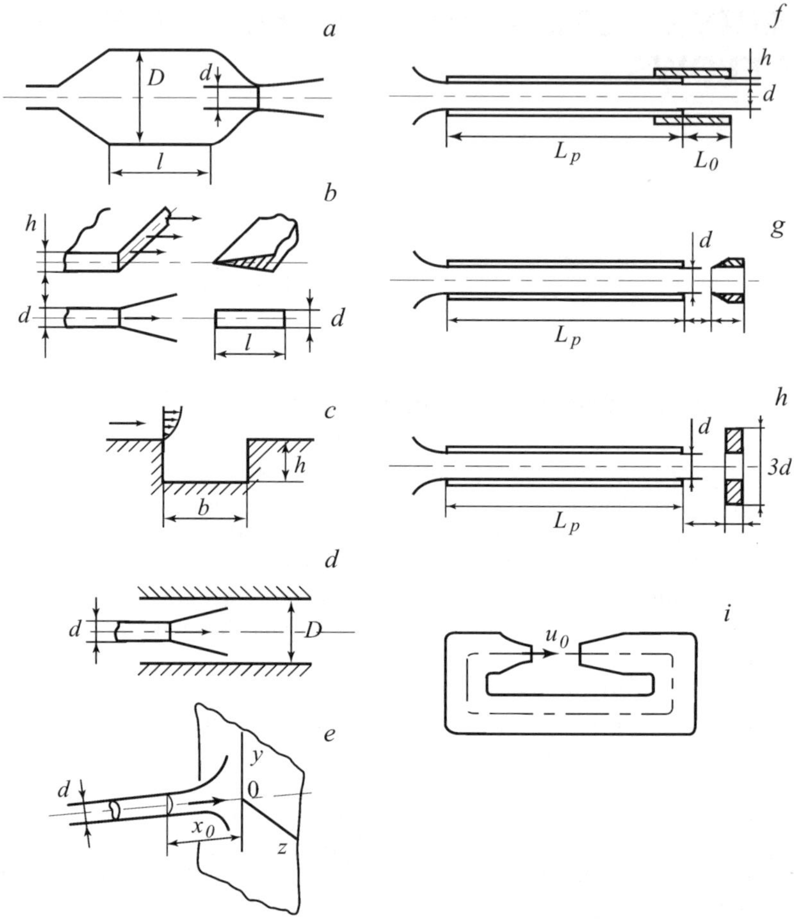

Fig. 5.1. The jet flow self-excitation schemes: (*a*) the jet issue from the receiver, (*b*) the plane jet leakage on the wedge or the axisymmetric jet on the coaxial tube, (*c*) interaction of the jet (the mixing layer) and the cavity, (*d*) the jet inflow in the tube, (*e*) the jet impingement on the baffle, (*f*) the 'whistle' nozzle, (*g*) the jet impingement on the ring, (*h*) the jet impingement on the washer with a hole, (*i*) the jet in the wind tunnel with the open test section.

(Fig. 5.1, *d*) open at the both sides. Here the interaction of the jet periodical coherent structures and the tube resonance characteristics occurs.

Strong self-excited oscillations are generated also when a turbulent jet with sufficiently high speed impinges on a baffle located not very far from the nozzle edge (Fig. 5.1, *e*). This problem is considered in Section 5.2.

Let us dwell on another prominent example of jet self-excitation. This is the so-called 'whistle nozzle' wherein the jet self-excitation is realized with controlled amplitude and frequency (Fig. 5.1, *f*). It consists of a tube with constant cross-section and successive coupling sliding over the tube. The length, L_0, of the tube broad end changes with the coupling longitudinal shift. The coupling provides the jump expansion of the flow from the tube of length L_p. This unit serves to excite the velocity harmonic oscillations of intensity $\varepsilon_u = \left(\left\langle u'^2 \right\rangle\right)^{1/2} / u_0$ at the tube outlet section without any energy input from the outside and, hence, to intensify considerably the mixing in the jet [5.5, 5.9].

The self-excited oscillations at the 'whistle nozzle' are generated under interaction of two independent resonance mechanisms: the characteristic tone of the shear layer due to the unsteady separated flow past the ledge (on the coupling) facing downstream and the resonance of the delivering ('organ') tube. The jump change in the self-excited oscillation frequency occurs for the smooth change in the 'whistle nozzle' geometric parameters. Here the adjacent steps are divided by 'dead' zones where the simultaneous realization of the two described resonance mechanisms is impossible. The frequency and the amplitude of the pure tone generated in the nozzle depends on the tube length, L_p, the ring length, L_0, the height, h, the issue speed, u_0, and the tube diameter, d. The shear layer tone frequency is a multiple of the frequency St_d in the frequency range $\mathrm{St}_d = 0.3 - 0.6$.

It should be stressed that the jet excitation besides the 'whistle nozzle' is possible with the ring (Fig. 5.1, *g*) or the washer (Fig. 5.1, *h*) in place of the coupling (Fig. 5.1, *f*). However, in that case the audible tone is less intensive than that for the 'whistle nozzle'. In addition, the mixing considerable intensification in the jet for these units is achieved only for the laminar mode of the flow at the tube outlet [5.9]. In summary, note another important case of the self-excited oscillations in wind tunnels with the open test section (Fig. 5.1, *i*) that is considered below in Sect. 5.3 and in Chap. 9.

Fig. 5.2 presents the change along the x axis in the mean velocity and in the longitudinal velocity pulsations in a turbulent jet issuing from a 'whistle nozzle'. The achieved mixing intensification occurs for Strouhal number $\mathrm{St}_d = 0.3 - 0.6$ and is observed for the both laminar and turbulent flow modes at the tube outlet. The excitation at the initial turbulent flow mode (at the tube end) turns out to be less intensive than at the initial laminar mode. As one could see in Fig. 5.2, here the complete analogy to the case of low-frequency acoustical excitation (cf. Sect. 3.3) takes place; the wide-band spectrum of the velocity pulsations in the whole frequency range increases for the same values of St_s in both cases.

The following empirical formula is derived for calculation of the pure tone frequencies, f, whereby a 'whistle nozzle' may generate the controlled excitation of a circular jet [5.5]:

$$\frac{f}{a_0 n}\left(L_p + 1.65 \frac{L_c}{j} + 0.75d \right) = 1 \,,$$

where a_0 is the sound speed and $n = ½, 1, 3/2$; $j = 1, 2, 3, \ldots$

The case $\mathrm{St}_d = 0$ in Fig. 5.2 corresponds to an unexcited jet when $L_0 = 0$.

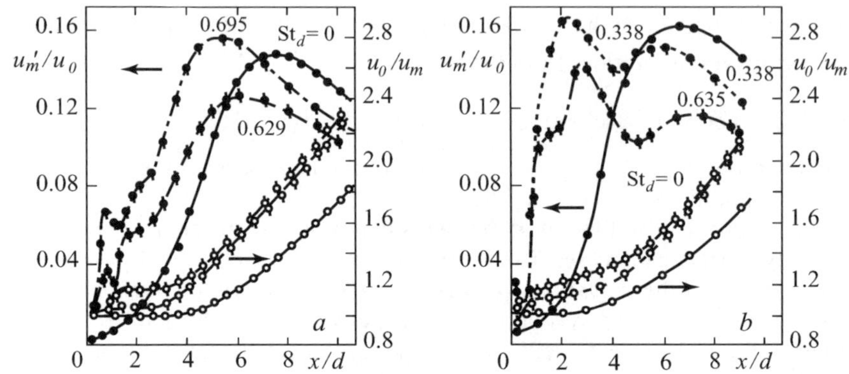

Fig. 5.2. The changes in the mean velocity and the longitudinal velocity pulsations along the axis of a jet issuing from the 'whistle nozzle': d = 2.54 cm, h = 0.3175 cm, u_0 = 36 m/s, $\varepsilon_m = u'_m / u_0 = 3\%$; (*a*) L_p = 15.24 cm, *1)* St_d = 0, *2)* St_d = 0.695, *3)* St_d = 0.629; (*b*) L_p = 30.40 cm, *1)* St_d = 0, *2)* St_d = 0.338, *3)* St_d = 0.335, light circles - u_0 / u_m, dark circles - u'_m / u_0.

The study of acoustical characteristics for a jet issuing from a "whistle nozzle" [5.6] demonstrates that there are intensive discrete components corresponding to the "whistle nozzle" and their harmonics in the pressure pulsation spectra for the far acoustical field at any angles φ with respect to the excited jet axis. The broadband noise level increases at any frequencies almost without the spectrum shape distortion as related to the excitation-free jet noise spectrum. The excess for the Mach numbers M = 0.25 – 0.57 and φ = 90° could be as much as 10 dB (Fig. 5.3). The mechanisms of this phenomenon considerably differ from the corresponding mechanisms for jets under acoustical excitation by artificial means [5.6].

5.2. Normal and Oblique Impingement of a Transonic Jet on a Baffle

5.2.1. Coherent Structures in Impact Jets. The presence of coherent structures in jets is most pronounced at the leakage of a subsonic turbulent jet on a baffle. Here for sufficiently large subsonic issue speeds and not very large distances of the nozzle from the baffle ($x_0 / d < 7.5$), intense self-excited oscillations occur at frequencies close to the frequency of the preferred mode in the free jet (St = 0.3 – 0.4). These oscillations are perceived in the jet far acoustical field as a sharp sound. Therefore, the coherent structures in this case could not only be seen (when visualizing the flow) but also heard. This phenomenon is studied in detail in [5.7, 5.8]. These papers contain a vast bibliography of the preceding studies.

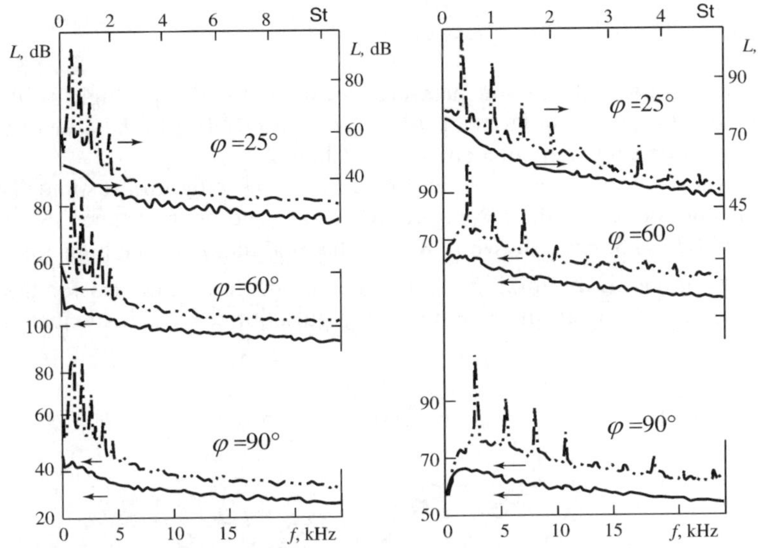

Fig. 5.3. The pressure pulsation spectra in the far acoustical field of a jet issuing from a "whistle nozzle" for Re = (2.1 – 5.0)·10^5 and the turbulent boundary layer at the nozzle outlet section. (*a*) $M_0 = 0.25, f = 874$ Hz; (*b*) $M_0 = 0.57, f = 2600$ Hz.

The acoustical feedback discovered in free jets (cf. Chap. 1) is considerably amplified here. Two feedback loops are established: the coherent structures carried downstream, and propagating upstream the pressure waves that are generated by collisions of the large-scale coherent structures and the baffle. These waves of frequency corresponding to the frequency of the large-scale coherent structures excite the mixing layer near the nozzle causing the abrupt amplification of the large-scale coherent structures. The characteristic period of the self-excited oscillations is determined by the convection velocity of the coherent structures, u_c, the velocity of the pressure waves propagating upstream (the sound speed), as well as by the distance between the nozzle and the baffle.

The necessary condition for the resonance realization is the fulfillment of the phase relation (1.2) where the distance between the nozzle edge and the baffle, x_0, should be substituted for x_i [5.1, 5.7, 5.8]:

$$\frac{x_0}{\lambda_1(f_r)} + \frac{x_0}{\lambda_2(f_r)} = N\,. \tag{5.1}$$

This relation characterizes the most fundamental property of flows having the feedback loop. When the phase relation (5.1) is satisfied, the resonance condition is a sufficiently high value of the Mach number M_0, since the pressure wave intensity in the mixing layer near the nozzle increases with this value. The corresponding condition for the pressure pulsation amplitude is as follows [5.15]:

$$p' = c\mathrm{M}_0 \,. \tag{5.2}$$

The availability of the sound pressure threshold for the jet high excitement gives an indication of a similarity of the considered problem to the case of the jet periodical excitement by an outer source (cf. Chap. 2).

The pressure pulsation measurements near the nozzle edge and at some distance from it outside the jet for the resonance ($\mathrm{M}_0 = 0.9$) and nonresonance ($\mathrm{M}_0 = 0.5$) modes at $x_0 / d = 4.5$ have shown [5.8] that the modulation of the high-frequency pressure pulsations in the mixing layer near the nozzle occurs also for the non-resonance mode, however, the low-frequency pulsation level is inadequate to excite the resonance.

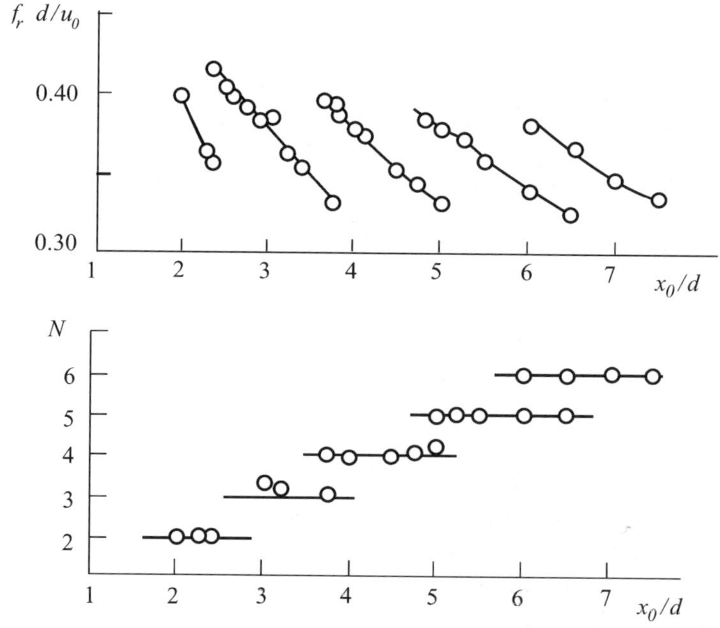

Fig. 5.4. The dimensionless resonance frequency and the parameter N as functions of x_0 / d for $\mathrm{M}_0 = 0.9$.

Fig. 5.4 presents the characteristic frequencies ($\mathrm{St}_r = f_r d / u_0$) of self-excited oscillations and of integer values of N for $\mathrm{M}_0 = 0.9$ and x_0 / d in the range from 2 to 7.5. The number of waves, N, is constant in each of the frequency step. When increasing the distance from the nozzle to the baffle, the resonance frequency decreases down to the minimum value of St_r; the subsequent increase of x_0 / d causes the jump increase of the frequency accompanying by the increase of the number of waves, N, in the feedback loop by 1. The lower bound $\mathrm{St}_r \approx 0.33$ corresponds to the most unstable mode of a free turbulent jet. In this case one can observe the hysteresis in experiments [5.8].

The downstream increase of the vortex scale in the mixing layer for relatively small distances between the nozzle and the screen and the corresponding decrease (approximately by one order) of the vortex frequency could not occur due to the vortex pairing as it takes place for the free jet. However, the so-called collective interaction of vortices is realized where the vortex multiple merging occurs instead of the vortex pairing [5.8]. The characteristic frequency of the large-scale structure succession in the radial near-wall jet upon the collective interaction is the same as it was before. This fact is supported in particular by measurements of the spectra of the surface friction pulsations on the baffle [5.16].

Usually, the realization of the resonance (self-oscillating) mode in the system '*jet – baffle*' is judged from the availability of the discrete component in the pressure pulsation spectrum for the near or far acoustical field or, alternatively, on the baffle surface. From the pressure pulsation spectrum for the far acoustical field (at the radius R/d = 37.5 making an angle $\varphi = 120°$ with the jet axis) presented in Fig. 5.5, *a* and *b* it is obvious that the discrete components appear or disappear as x_0/d increases from 1 to 9 for the fixed value of M_0 (*a*) and as M_0 increases from 0.63 to 0.95 (*b*) for the fixed distance $x_0/d = 4$. It follows that the discrete components in the pressure pulsation spectra for $x_0/d > 2$ and $M_0 > 0.77$ appear at the frequency St = 0.35. The self-oscillating modes are realized also for the hot jet impingement on a baffle, as illustrated by the pressure pulsation spectra in the far field (Fig. 5.5, *d*) for $x_0/d = 4$ and various T_{00}.

The possibility of realization of self-oscillating modes for oblique impingement of a jet on a baffle is of interest also. The corresponding pressure pulsation spectra in the far acoustical field of such a jet are presented in Fig. 5.5, *c* for $M_0 = 0.95$, $x_0/d = 4$ and different values of the baffle slope angle, ▯. The resonance loss for large values of the baffle slope angle ($\beta \geq 30°$) is explained by the perceptible decrease of the induced acoustical field intensity near the nozzle.

5.2.2. A Near-Wall Radial Jet. Coherent structures generated in the jet mixing layer and amplified duc to the acoustical feedback in a circular impact jet are conserved also in a near-wall radial jet spreading over the baffle. Presented in Fig. 5.6 are pulsation spectra of the near-wall pressure, p'_w, the surface friction, τ'_w, and the velocity pulsations, u', at the distance 0.25*d* from the baffle for $M_0 = 0.95$ and $x_0/d = 4$ which demonstrate the coherent structure degeneration with the distance from the baffle center [5.16].

The disappearance of the discrete component in the spectra of τ'_w and u' testifies also for the disappearance of these structures. The discrete components in the spectra of p'_w remain for all *r/d* because the microphones flush-mounted at the baffle measure the pressure pulsations in the jet far field. It is significant that the discrete outliers in the spectra of p'_w, τ'_w and u' correspond to approximately equal values of the Strouhal numbers St = 0.35. What this means is the coherent structures (generated upstream) moving along the radius do not interact with each other in the area of the near-wall radial jet, i.e. the coherent structures pairings accompanying by two-fold decrease in frequency are lacking.

Fig. 5.7 presents distributions along the radius for the mean pressure, $p^\circ = (p_w - p_a)/q_0$, and for the mean surface friction coefficient, $\tau^\circ = \tau_w / q_0$, as well as for the mean-square values of their pulsations, $\varepsilon_p = \left(\left\langle p_w'^2 \right\rangle\right)^{1/2} / q_0$ and $\varepsilon_\tau = \left(\left\langle \tau_w'^2 \right\rangle\right)^{1/2} / q_0$ for $x_0 / d = 4$ at the resonance ($M_0 = 0.95$) and nonresonance ($M_0 = 0.29$) modes. The exception is the mean pressure distribution that is practically the same for both modes.

5.2.3. Suppression of Self-Excited Oscillations. The generation of self-excited oscillations on the jet impingement on a baffle could be undesirable. In that case the spatial correlation coefficient of the pressure pulsations at the baffle centre and along the radius up to $r/d = 8$ retains large values ($R_{pp} > 0.5$) causing large pulsation loads on the baffle. In addition, the piercing sound in the surroundings is also undesirable.

The suppression of self-excited oscillations could be performed attenuating the acoustical feedback that is the main reason of the self-excited oscillation modes [5.16]. As mentioned above, deterioration of the mixing layer azimuthal homogeneity near the nozzle causes weak attenuation of the coherent structures. Hence the attenuation for a jet impinging on a baffle ($M_0 = 0.95$, $x_0/d = 4$) suffices to deteriorate the acoustical feedback and eliminate the resonance as demonstrated by the pressure pulsation spectra in the jet far field (Fig. 5.8, *a*). Here $\varphi = 120°$, the continuous line corresponds to the resonance mode and the dashed line to the self-excited oscillation suppression by insertion of three cones into the nozzle mixing layer (at a depth of 1 – 2 mm).

Another approach to attenuation of the acoustical feedback and to the self-excited oscillation suppression is the disruption of the jet coherent structures at the point of their collision with the baffle. This is achieved by the baffle surface finning, i.e. by installation of rather low ($\Delta h / d = 0.1$ - 0.2) partitions producing the square lattice with the step $\Delta z / d = 0.5$. Fig. 5.8, *b* presents the corresponding pressure pulsation spectra in the far field of a jet leaking on a smooth baffle and a finned surface for $M_0 = 0.95$ and $x_0 / d = 4$. Here $\varphi = 90°$, the continuous line corresponds to the resonance mode for the smooth baffle and the dashed line to the resonance elimination by the baffle surface finning. The elimination of the discrete components in the noise spectrum for this method of the self-excited oscillation suppression, as opposed to the previous one (cf. Fig. 5.8, *a*), accompanies by some amplification of the broadband noise.

The considerable attenuation of the self-excited oscillations in subsonic hot jets impinging on a baffle could be achieved also by replacing the solid baffle by a

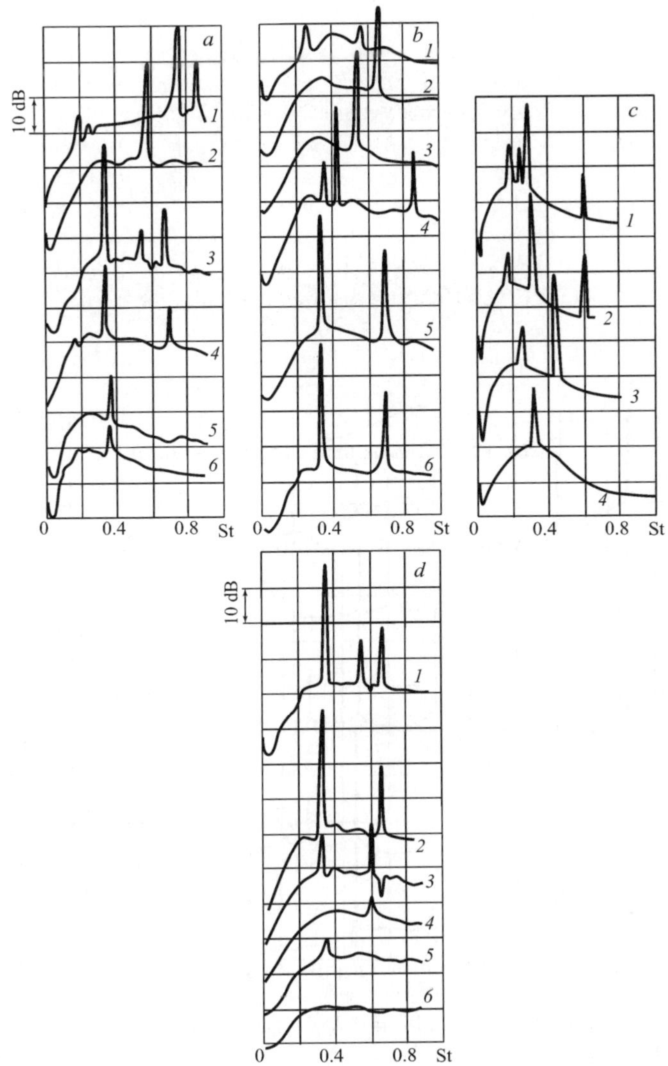

Fig. 5.5. The pressure pulsation spectra in the far field of an impact jet. (*a*) $M_0 = 0.95$; *1)* $x_0 / d = 1$, *2)* $x_0 / d = 2$, *3)* $x_0 / d = 4$, *4)* $x_0 / d = 6$, *5)* $x_0 / d = 8$, *6)* $x_0 / d = 8$; (*b*) $x_0 / d = 4$; *1)* $M_0 = 0.63$, *2)* $M_0 = 0.73$, *3)* $M_0 = 0.77$, *4)* $M_0 = 0.87$, *5)* $M_0 = 0.92$, *6)* $M_0 = 0.95$; (*c*) $M_0 = 0.87$, $x_0 / d = 4$; *1)* $T_{00} = 573$ K, *2)* $T_{00} = 523$ K, *3)* $T_{00} = 423$ K, *4)* $T_{00} = 293$ K; (*d*) $M_0 = 0.87$, $x_0 / d = 4$; *1)* $\beta = 0$, *2)* $\beta = 10°$, *3)* $\beta = 20°$, *4)* $\beta = 30°$, *5)* $\beta = 40°$, *6)* $\beta = 50°$.

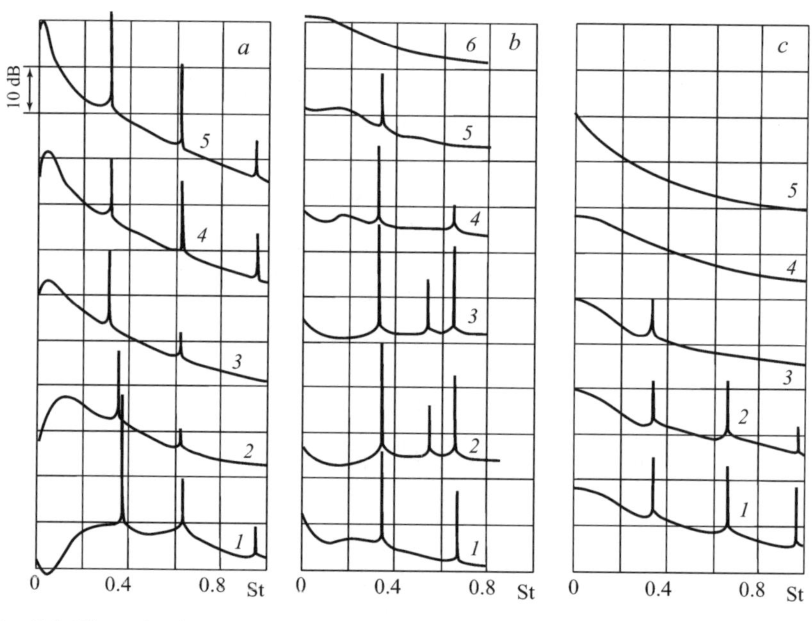

Fig. 5.6. The pulsation spectra at $M_0 = 0.95$, $x_0/d = 4$: (*a*) for the near-wall pressure on the baffle, $p'_w(r/d)$: *1)* $r/d = 2.5$, *2)* $r/d = 3$, *3)* $r/d = 4.5$, *4)* $r/d = 6$, *5)* $r/d = 7.5$; (*b*) for the surface friction, $\tau'_w(r/d)$: *1)* $r/d = 0.4$, *2)* $r/d = 0.75$, *3)* $r/d = = 1.5$, *4)* $r/d = 2.25$, *5)* $r/d = 3$, *6)* $r/d = 3.75$; (*c*) for the velocity, $u'(r/d)$ at $\Delta x/d = 0.25$: *1)* $r/d = 1.5$, *2)* $r/d = 2$, *3)* $r/d = 2.5$, *4)* $r/d = 3.25$, *5)* $r/d = 5$.

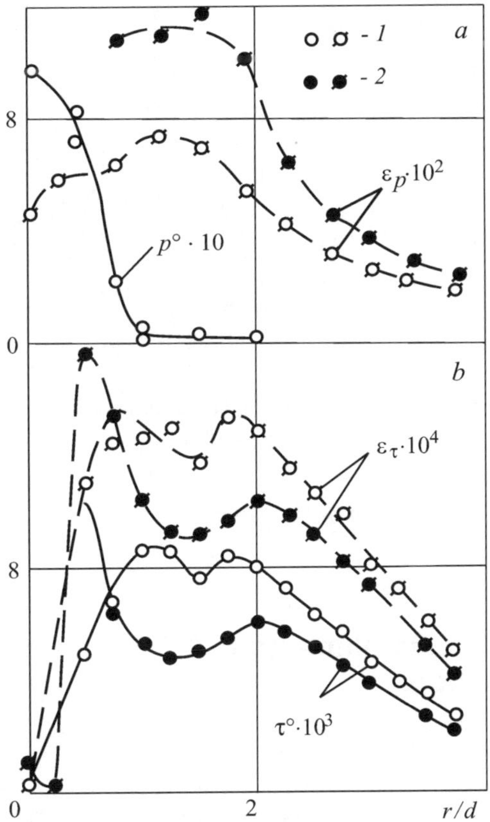

Fig. 5.7. The pressure and surface friction distributions along the radius of a near-wall jet and the corresponding mean-square distributions for $x_0 / d = 4$ at the resonance (*1*, $M_0 = 0.95$) and nonresonance (*2*, $M_0 = 0.29$) modes.

permeable, latticed or cellular one with organization of the transversal removal of the gas gone through the baffle when needed.

The approximate calculation of the normal impingement of a subsonic axisymmetric jet on a baffle is performed in [5.3]. The results are dependent upon the self-excited oscillation increment and frequency, as well as the integer parameter *N*, on the relative distance, x_0 / d, between the nozzle outlet section and the baffle for $M = 0.9$. The dependencies are similar to those obtained in experiments (cf. Fig. 5.4).

Of course, this simple theory does not pretend to give quantitative agreement with experiments. However, it describes all characteristic features of the phenomenon including hysteresis.

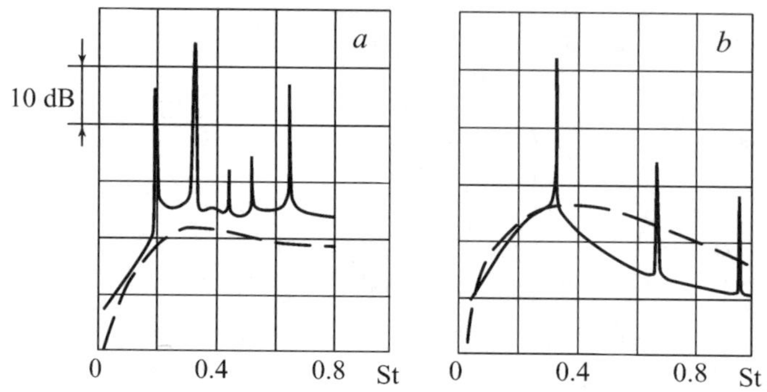

Fig. 5.8. The self-excited oscillation suppression in an impact jet (M_0 = 0.95, x_0 / d = 4) by deterioration of the mixing layer azimuthal homogeneity: (*a*) at the nozzle edge, (*b*) disruption of the axial symmetry of the circular vortices striking against the baffle by means of its surface finning.

5.3. Self-Excited Oscillations in Wind Tunnels with the Open Test Section

In closed-circuit wind tunnels with the open test section in a certain speed range self-excited oscillations occur causing considerable increase of the longitudinal low-frequency velocity pulsations in the test section and of the pressure pulsations in the overall air path of the wind tunnel [5.13, 5.14]. These oscillations are generated by the interaction of the flow hydrodynamic oscillations in the mixing layer of the free jet and the acoustical oscillations in the reverse flue (standing sound waves). The hydrodynamic oscillations in the jet mixing layer result in the excess of the velocity pulsation intensity in the flow core in the open test section over the corresponding pulsations in wind tunnels with the closed test section [5.11].

Another drawback of wind tunnels with the open test section is that, as the flow speed increases, if special care is not taken, the strong vibration of the wind tunnel elements bringing the threat to the completeness of the wind tunnel and building construction occurs, which threat increases for industrial wind tunnels of large diameter. However, even when special care is taken to reduce flow pulsation and wind tunnel shaking, the self-excited oscillations and the flow pulsation in the test section cannot be completely suppressed in the overall range of operating speeds by conventional means and the velocity pulsation intensity in the test section is too high exceeding 1.5 – 2.5% at some speeds. These speeds, as a rule, become unsuitable for experiments, resulting in constriction of the operating speed range.

The most popular approaches to the oscillation damping reduce to attenuation of regular vortex generation in the mixing layer of a free jet or attenuation of oscillation action in a free jet on oscillations in the return [5.14]. The first approach results if the boundary layer in the jet initial region is made azimuthally nonhomogeneous to attenuate or decay the annular vortices (coherent structures); the second is achieved by the use of holes in the diffuser walls near its inlet edge.

The nature of the self-excited oscillations in a wind tunnel is determined by two loops of the acoustical feedback: through the return downstream and through the test section upstream. The latter feedback corresponds to the case where a circular jet leaks on an axial ring or a tube, the so-called edge tone [5.2]. It follows that the self-excited oscillations are also possible in wind tunnels of unclosed type with the open test section surrounded by the Eiffel chamber [5.12].

The return is an acoustical resonator by virtue of which intense sound waves are excited in certain ranges of the flow speed. The submerged jet is realized in the open test section that, as noted above (cf. Sect. 1.1), is an amplifier of hydrodynamic pulsations. The association between the resonator and the amplifier is effected by way of mutual transformation acoustical and hydrodynamic waves that are most efficient near the resonator ends.

The vortices generated in the jet mixing layer while impinging on the diffuser edge generate the pressure impulses exciting sound waves that propagate in two directions: through the return downstream and through the test section upstream. When falling at the nozzle edge with the appropriate phase, the sound waves generate vortex disturbances playing the role of two feedback loops [5.13].

Fig. 5.9 presents the (*1*) pressure oscillation amplitude and (*2*) frequency in a free jet as functions of the flow speed illustrating generation of the self-excited oscillations in a wind tunnel with the nozzle diameter $d = 0.5$ m and the aerodynamic contour shown in Fig. 5.10.

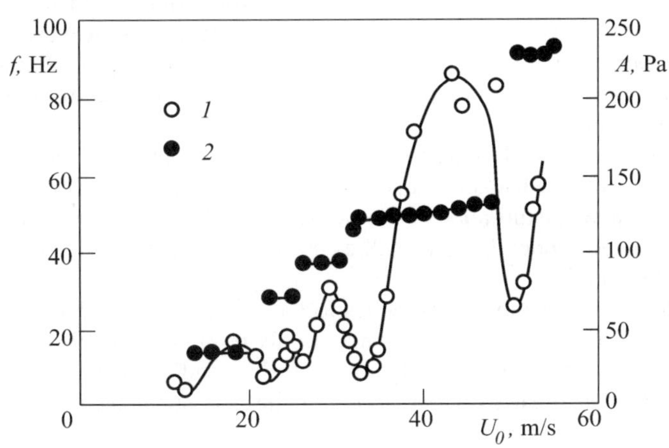

Fig. 5.9. The (*1*) pressure oscillation amplitude and (*2*) frequency in a free jet of a wind tunnel as functions of the flow speed illustrating generation of the self-excited oscillation modes.

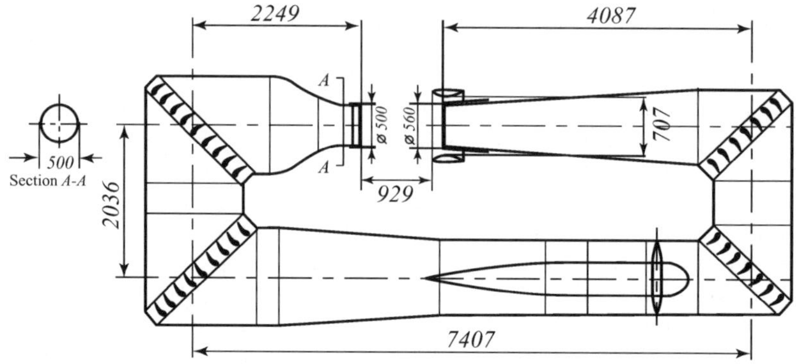

Fig. 5.10. The aerodynamic contour of the wind tunnel of Fig. 5.9.

We now consider the main features of the most commonly used units for damping self-excited oscillations in wind tunnels [5.14].

Firstly, organization of the flow entering into the diffuser. The flow entering into the diffuser is realized as a smoothly tapering funnel or, more often, a thin almost cylindrical part with the rounded edge. The cylindrical part is enclosed by an advanced annular wing. The elimination of the annular funnel results in considerable increase of the oscillation amplitudes (2 – 3 times), as well as in expansion of the periodical mode boundaries. As suggested in [5.14], the funnels are useful for the self-excited oscillation damping but their operation is quite insufficient.

Secondly, arrangement of holes at the diffuser entrance. It is usual to locate the holes (one or more rows) at a distance $(0.5 - 1.0)d$ from the diffuser front edge. The overall area of the holes is 30 – 35% of the diffuser entrance area. In spite of evident efficiency of this method of the oscillation damping, it should be noted that it does not provide the complete suppression of self-excited oscillations in the whole range of speeds.

Finally, generation of longitudinal vortices in the jet mixing layer. It is the reliable approach to the self-excited oscillation suppression. However, it is related to the boundary layer disturbance at the nozzle cut-off and accompanies by the increase in the wind tunnel drag, as well as by the undesirable distortion of the flow field in the test section.

We do not consider other approaches to the self-excited oscillation damping in wind tunnels with the open test section and only point to Chap. 9 where an acoustical approach to self-excited oscillation control is studied. The theory of such oscillations with due regard for two loops of the acoustical feedback is described in [5.10].

REFERENCES

5.1. Artamonov KI (1982) Thermo-hydro-acoustic instability (in Russian). Mashinostroyenie, Moscow

5.2. Curasava H, Yamanoue K, Obata T (1987) Self-excited oscillation in an axisymmetric jet with a coaxial circular pipe. Mem Nagano Techn Coll, 18: 21 – 30

5.3. Ginevsky AS, Landa PS, Zaikin AA (1994) Self-excitation of impinging jets with regard to acoustic feedback. In: Third Int Congress of Air- and Structure-borne Sound and Vibration, Montreal, pp 1207 – 1214

5.4. Gutmark E, Schadow KC, Parr DM, Harris CK, Wilson KJ (1985) The mean and turbulent structure of noncircular jets. AIAA Pap, № 543

5.5. Hasan MAZ, Hussain AKMF (1982) The self-excited axisymmetric jet. J Fluid Mech, 115: 59 – 89

5.6. Hasan MAZ, Islam O, Hussain AKMF (1984) Jet noise modification by the "whistler nozzle". AIAA J, 22, № 3: 340 – 347

5.7. Ho C-M, Nosseir NS (1980) Large coherent structures in an impinging jet. In: Turbulent Shear Flows 2, Berlin, pp 297 – 304

5.8. Ho C-M, Nosseir NS (1981) Dynamics of an impinging jet. Part 1. The feedback phenomenon. J Fluid Mech 105: 119 – 142

5.9. Hussain AKMF, Hasan MAZ (1982) The "whistler – nozzle" phenomenon. J Fluid Mech 134: 431 – 458

5.10. Landa PS (1997) Nonlinear oscillations and waves (in Russian). Fizmatlit, Moscow

5.11. Michel U (1989) Influence of shear layer induced fluctuations on the velocity unsteadiness in free-jet wind tunnels. In: Adv Turbulence 2. Proc 2nd Eur Turbulence Conf. Berlin, pp 480 – 483

5.12. Rebuffet P, Guedel A (1982) Etude sur maquette de differentes configurations de jet traversant une cavite (in French). Application a la soufferie CEPRA 19 du CEPr. La recherche Aerospatiale, 1: 11 – 22

5.13. Seiferth R (1946) Vorausberechnung und Beseitigung der Schwingungen von Freistrahl-Windkanalen (in German). In: Monographien űber Fortschritte der deutchen Luftfahrtforschung seit 1939. AVA Gőttingen, pp 11 – 22

5.14. Strelkov SP, Bendrikov GA, Smirnov NA (1946) Pulsations in wind tunnels and approaches to their damping (in Russian). Proc. TsAGI, № 593

5.15. Vlasov YeV, Ginevsky AS, Karavosov RK, Ukhanova LN (1986) Study of self-excited oscillation modes for a subsonic turbulent jet impinging on a baffle (in Russian). In: Industrial Aerodynamics, № 1 (33), pp 276 – 280

5.16. Vlasov YeV, Ginevsky AS, Karavosov RK, Ukhanova LN (1988) Resonant modes in impingement of a subsonic turbulent jet onto a baffle. Fluid Mechanics – Soviet Research 17, № 1: 14 – 24

CHAPTER 6 NUMERICAL SIMULATION OF PERIODICAL EXCITATION OF SUBSONIC TURBULENT JETS

The present chapter considers well-known approaches to mathematical simulation of subsonic turbulent jets under periodical harmonic excitation. Particular emphasis is placed on the description (in the framework of these approaches) of generation and interaction of large-scale coherent structures, as well as of their susceptibility to periodical excitation. The calculation results illustrating the turbulent mixing intensification (turbulence generation) in jets under low-frequency harmonic excitation and the mixing attenuation (turbulence suppression) under high-frequency excitation are presented. It is stressed that the mathematical simulation of turbulent jets under periodical excitation allows describing the laws of periodical (acoustical) excitation known from the experimental studies.

6.1. Direct Numerical Simulation of Turbulent Motion in the Initial Region of an Axisymmetric Jet under Low-Frequency Harmonic Excitation

The initial equation set comprises the nonstationary continuity, momentum and energy equations for subsonic flow of nonviscous gas (the Euler equations). The subgrid turbulence is disregarded. The numerical solution is obtained using the finite difference method and the corresponding approximation of the boundary conditions. The calculations are performed for the subsonic Mach numbers ($M_0 = 0.43$ and 0.57). The boundary layer effect at the nozzle cutoff is of course disregarded. However, the initial value of the momentum loss thickness at the first integration step across the mixing layer is given; two values are considered: $\theta_0 = d/70$ and $\theta_0 = d/140$. As a result of the accepted assumption about the flow axial symmetry, the reliable results are obtained in the jet segment of length not more than four calibers: $x = (0 - 4)d$.

To account for the jet periodical excitation, it is assumed that the outflow speed at the initial section changes in time according to the harmonic law

$$u = u_0[1 + a_1 \sin(2\pi f t)] \tag{6.1}$$

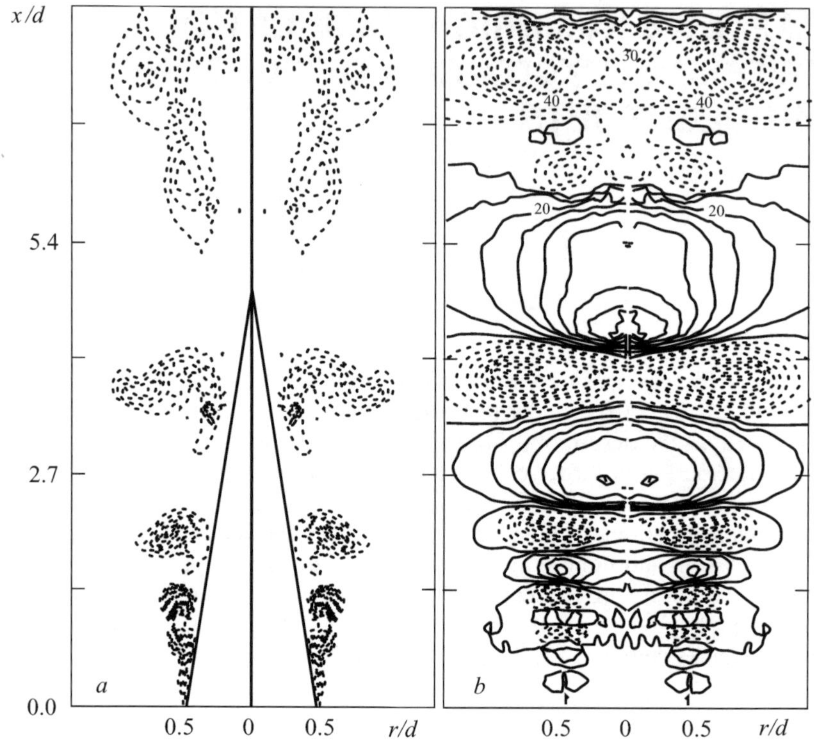

Fig. 6.1. The instantaneous fields of (*a*) vorticity and (*b*) pressure, $(p-p_\infty)/p_\infty$, at the meridian section of the circular excitation-free jet ($M_0 = 0.57$).

with the amplitude a_1 being in the range $a_1 = 0.01 - 0.10$. The calculations are performed for an excitation-free jet and for a jet under periodical excitation [6.10 – 6.12]. The results for an excitation-free jet are the fields of the instantaneous vorticity and pressure, as well as of the axial and radial velocities. In addition, there are calculated the fields of the averaged velocity, pressure and mean-square values of the longitudinal velocity pulsations, as well as the longitudinal velocity pulsation spectra in the mixing layer on the nozzle edge line. The data on forming and interaction (pairing) of coherent structures are presented. It is shown, in particular, that the vortex pairing may be accompanied by generation of local areas of the negative Reynolds stresses [6.12] corresponding to decrease in the turbulence kinetic energy (the turbulence producing is negative). This effect was observed earlier at experimental studies on acoustically excited circular jets.

By way of example, Fig. 6.1 presents the instantaneous pictures of the vorticity and pressure distributions at the meridian section of the jet over the segment $x = (0 - 8)d$ and Fig. 6.2 presents the longitudinal velocity pulsation spectrum on the nozzle edge line ($r/d = 0.5$) at the longitudinal coordinate x corresponding to the point behind the first pairing of the ring vortices as a function of the dimensionless frequency $\mathrm{St}_\theta = f\theta_0 / u_0$. The calculation data for the excitation-free jet agree satisfactorily with the experiment [6.10 – 6.12]. Consider now the numerical simulation results for the Strouhal number $\mathrm{St}_d = fd / u_0$ as a function of the longitudinal coordinate, x_i / d, exhibiting the changes along the jet axis in the intervals Δx_i that correspond to the first ($i = 1$), the second ($i = 2$), and the third ($i = 3$) pairings of the ring coherent structures [6.11]. Fig. 6.3 presents three functions: (*1*) the correlation measurements of Kibens [6.17], (*2*) the direct numerical simulation [6.11], and (*3*) the calculations of the function $\mathrm{St}_d(x_i / u_0)$ according to the relation (1.3) derived from the idea of the global mechanism of the feedback in the jet initial region. This relation is a quite satisfactory approximation to the both experimental and numerical data.

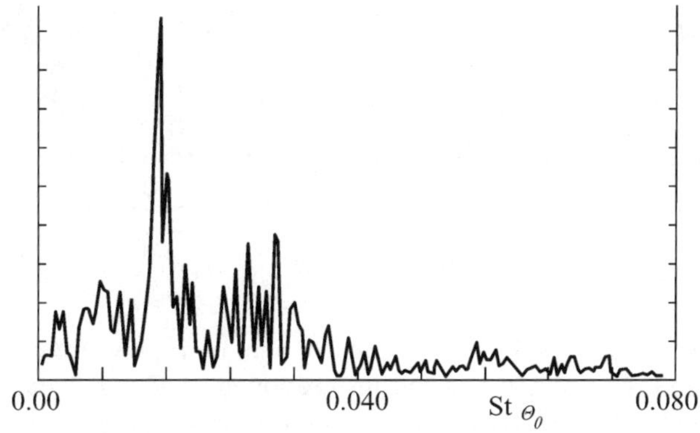

Fig. 6.2. The longitudinal velocity pulsation spectrum in the mixing layer of the turbulent jet.

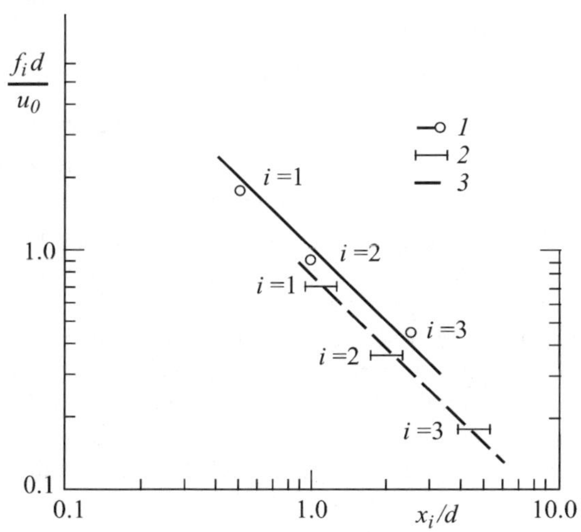

Fig. 6.3. The Strouhal number, St_d, along the jet axis: *1)* the experiments of Kibens [6.17], *2)* the direct numerical simulation [6.11], *3)* the calculations according to the relation (1.3).

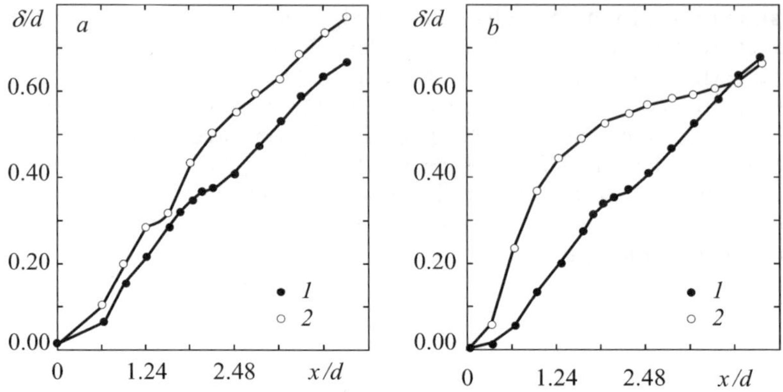

Fig. 6.4. The downstream thickening of the mixing layer in the initial region of (*1*) circular excitation-free and (2) excited turbulent jets. (*a*): *1)* $\langle u'^2 \rangle^{1/2} / u_0 = 0$, *2)* $\langle u'^2 \rangle^{1/2} / u_0 = 0.5\%$, $St_d = 0.26$; (*b*): *1)* $\langle u'^2 \rangle^{1/2} / u_0 = 0$, *2)* $\langle u'^2 \rangle^{1/2} / u_0 = 0.5\%$, $St_d = 0.35$.

Consider the numerical simulation data for a jet under low-frequency periodical excitation. It follows from the spectrum presented in Fig. 6.2 that the spectrum peak corresponds to $St_\theta = 0.015$. This conforms with the corresponding experimental results $St_\theta = 0.125 - 0.155$ [6.14] and slightly less than the theoretically predicted value $St_\theta = 0.017$, the difference being determined by the feedback effects [6.16]. Fig. 6.4 *a*, *b* presents the calculation results for the mixing layer thickness of excitation-free and excited jets at $St_d = 0.26$ and 0.35 and at the excitation levels $\langle u'^2 \rangle^{1/2} / u_0 = 0.5$ and 5%. Here the Strouhal number St_d corresponds to the second subharmonic, $St_{\theta 0}/4$, of the instability frequency, $\delta = r_{0.08} - r_{0.92}$ is the conventional mixing layer thickness determined from the difference of the ordinates where the dimensionless longitudinal velocity, u / u_0, equals to 0.08 and 0.92 respectively. The presented results illustrate the mixing layer thickening under low-frequency periodical excitation. Similar results are presented in Fig. 6.5 for distributions of the longitudinal velocity pulsations over the jet axis.

The presented results illustrate the possibilities of numerical simulation of the main effects observed at low-frequency harmonic excitation of a turbulent jet: the mixing layer thickening and the increase in the longitudinal velocity pulsations on the jet axis within the initial region depending on the excitation frequency and level. The important role of large-scale coherent structures in the mixing processes is taken properly into account.

Some scantiness of these results is determined by the assumption about the axial symmetry of the instantaneous stream.

Chap. 2 comprises the experimental results on dual-mode dual-frequency acoustical excitation of a turbulent jet at the main frequency and at its subharmonic [2.38]. The jet for the Reynolds numbers $Re = 10^4$ and $2 \cdot 10^4$ was excited by

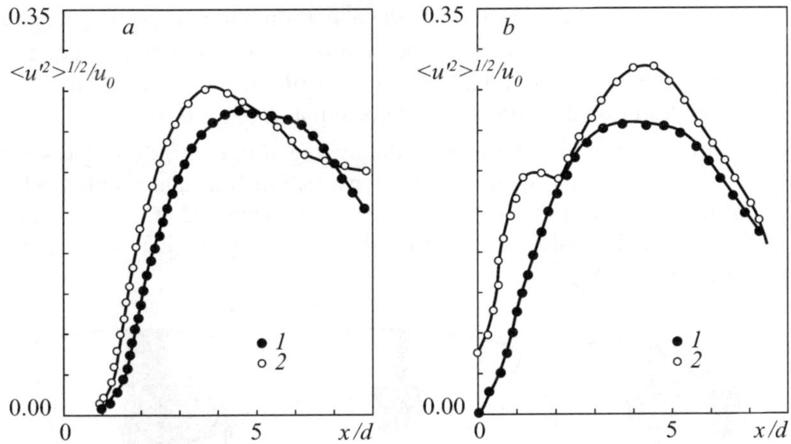

Fig. 6.5. The longitudinal velocity pulsations over the axis of circular (*1*) excitation-free and (*2*) excited turbulent jets. (*a*): *1)* $\langle u'^2\rangle^{1/2}/u_0 = 0$, *2)* $\langle u'^2\rangle^{1/2}/u_0 = 0.5\%$, $St_d = 0.26$; (*b*) *1)* $\langle u'^2\rangle^{1/2}/u_0 = 0$ (*1*), *2)* $\langle u'^2\rangle^{1/2}/u_0 = 0.5\%$, $St_d = 0.35$.

longitudinal acoustical waves with Strouhal number $St_a = 0.6$ and by azimuthal waves at the subharmonic frequency $St_h = 0.3$ with equal or different amplitudes. The azimuthal perturbations were generated by two loudspeakers placed on each side of the nozzle outlet section with the phase shift $\theta_* = \pi$. The maximum angle of the jet broadening ($\alpha = 70°$!) was achieved at the following parameters: $Re = 10^4$, $St_a = 2St_h = 0.55$ and the corresponding amplitudes $u'_a/u_0 = 0.12\%$ and $u'_h/u_0 = 0.04\%$.

To describe this phenomenon, dual-mode periodical excitation of a circular jet is studied [6.13] by direct numerical simulation on the basis of the nonstationary 3-D Navier – Stokes equations for the following parameters: $Re = 1500$, $St_a = 2St_h = 0.55$, $A_a = A_h = 0.15$. Here A_a and A_h are the amplitude values of the jet harmonic velocity pulsations under axial (the subscript *a*) and helical (the subscript *h*) excitation respectively. The calculations of the jet stream are performed using the method of determination of the scalar admixture transportation field. Figs. 6.6, *a*, *b* present the instantaneous pictures of the flow in two perpendicular planes $\psi = 0$ and $\psi = \pi/2$. Here $\psi = 0$ corresponds to the plane passing through the jet axis and the axis of two radiators placed on each side of the nozzle outlet section. Fig. 6.6, *c* presents the scalar admixture distribution over the plane $\psi = 0$ averaged in time.

To study dual-mode dual-frequency periodical excitement of a jet at higher values of the Strouhal number ($Re = 6\cdot10^3$), the Navier – Stokes equations together with the Large Eddy Simulation (LES) and the subgrid scale description using the Smagorinsky model are exploited in [6.13]. The calculations are performed for Strouhal numbers $St_a = 0.74$ and $St_h = 0.3$ and the amplitudes $A_a = 0.02$ and

$A_h = 0.075$. Here A_h is the amplitude of the actuation which is phase-locked in the plane $\psi = 0$. It corresponds to the superposition of two counter-rotating helical modes of the same frequency. The velocity profile at the nozzle cutoff is given and the momentum loss thickness is assumed equal to $\theta_0 = d/60$.

Fig. 6.7 presents the distributions of the averaged velocities and the scalar admixture concentrations along the axes of excitation-free and excited jets. It is shown that the maximum mixing intensification is achieved at the bifurcation regime. The presented results demonstrate the efficiency of dual-mode periodical excitation of a turbulent jet.

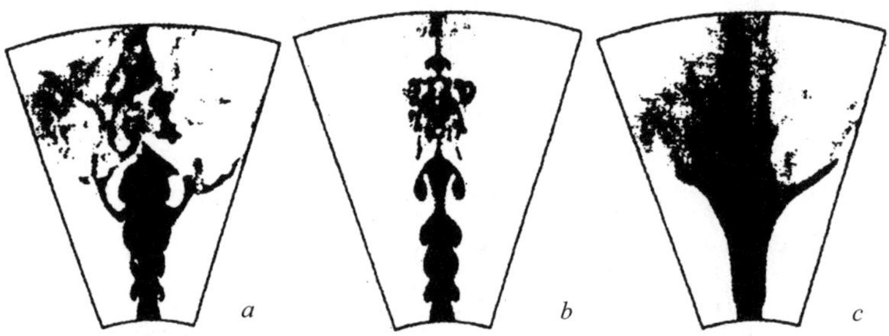

Fig. 6.6. The instantaneous pictures of the flow in the planes (*a*) $\psi = 0$ and (*b*) $\psi = \pi/2$ and (*c*) of the averaged flow in the plane $\psi = 0$.

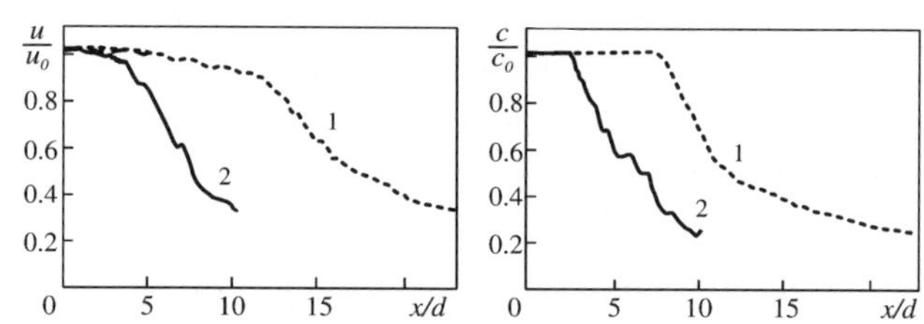

Fig. 6.7. The distributions of the averaged velocities and the scalar admixture concentrations in excitation-free and excited jets. *1)* an excitation-free jet, *2)* an excited jet.

6.2. Simulation of Plane and Circular Turbulent Jets under Low- and High-Frequency Harmonic Excitation Using the Method of Discrete Vortices

6.2.1. The Method of Discrete Vortices Simulates Plane and Circular Turbulent Jets for the Case of Ideal Incompressible Fluid. As applied to plane jets, two approaches may be used. As in the case of circular jets, the impermeability boundary conditions on the nozzle rigid surfaces are satisfied for the first approach only at a finite number of reference points. The impermeability boundary conditions for the second approach are satisfied precisely at all points of the rigid surfaces. The conformal image of the flow plane onto the upper half plane is used in the latter case. Both approaches for the first case with sufficiently compact arrangement of the reference points give practically identical results [6.5].

While modeling plane turbulent jets, the assumption about the availability of a symmetry plane for the instantaneous stream causes practically the total lack of mixing among vortex structures with opposite signs that are generated on both sides of the symmetry plane. By way of example, Fig. 6.8 presents vortex structures of a plane jet at a fixed time for symmetric and asymmetric cases. In strict conformity with the experiment, the second case demonstrates intensive mixing of vorticity blobs of opposite sign. The distributions of the averaged velocity over the jet axis calculated for symmetric and asymmetric arrangement of vortex structures presented in Fig. 6.9 show that the calculation results are consistent with the experiment in the both initial and main regions only in the second case [6.6].

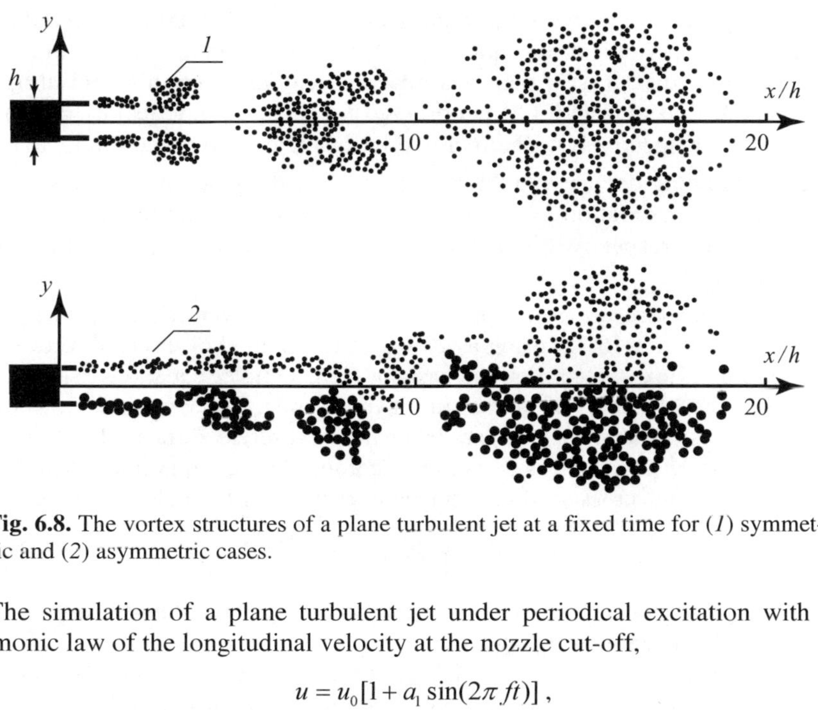

Fig. 6.8. The vortex structures of a plane turbulent jet at a fixed time for (*1*) symmetric and (*2*) asymmetric cases.

The simulation of a plane turbulent jet under periodical excitation with the harmonic law of the longitudinal velocity at the nozzle cut-off,

$$u = u_0[1 + a_1 \sin(2\pi ft)] ,$$

allows the study of the influence of the Strouhal number, $\mathrm{St} = fh/u_0$, and the amplitude, a_1, on the change in the averaged velocity, $\langle u \rangle / u_0$ and the longitudinal velocity pulsations, $\varepsilon_{u0} = \langle u'^2 \rangle^{1/2} / u_0$, along the jet axis at St = 0.35 and 2.0 (Fig. 6.10).

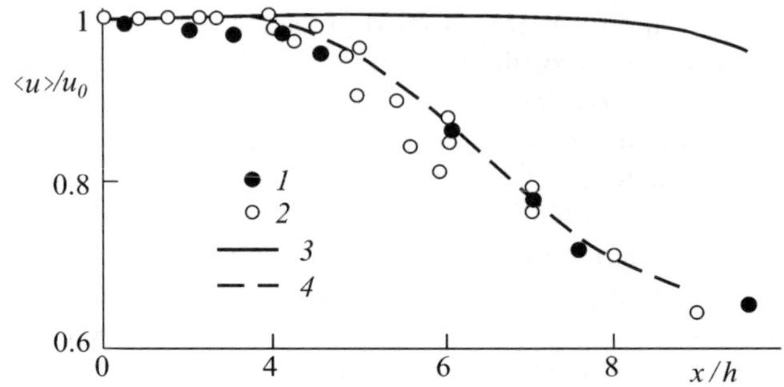

Fig. 6.9. The distributions of the averaged velocity over the axis of a plane turbulent jet; *1, 2)* experiments, *3)* the symmetric and *4)* the asymmetric approximations.

Here $\varepsilon_{u0} = a_1/\sqrt{2}$. The results are borrowed from [6.5, 6.33]. A certain tendency for a decrease in the initial region length (for $a_1 = 0.05$) is observed under low-frequency periodical excitation. The abnormal nonmonotone change in the velocity along the jet axis that was observed earlier in circular pulsating jets (cf. Chap. 4, Figs. 4.4 and 4.5) is observed for large amplitudes ($a_1 = 0.15$ and $\varepsilon_{u0} = 10.5\%$ respectively). The increase in the mean velocity at the jet axis and the corresponding decrease in the longitudinal velocity pulsations are stated for high-frequency excitation (St = 2) and for a very high excitation level ($a_1 = 0.3$ and $\varepsilon_{u0} = 21\%$ respectively). The high-frequency excitation effect is lower for smaller values of a_1.

Consider the simulation of a plane turbulent shear layer under high-frequency periodical excitation [6.31] using a version of the method of discrete vortices (the method of a vortex in a cell) on the basis of the 2-D Euler equations. The study is devoted to the evolution of the mixing layer in time. The finite thickness of the shear layer is simulated by four parallel chains of discrete vortices. The transverse distance between the vortices is chosen according to the requirement that the velocity profile in the cross section averaged over the longitudinal coordinate,

$$\langle u\rangle(y,t) = \int u(x,y,t)dt\,, \tag{6.2}$$

could be approximately expressed in the form of hyperbolic tangent that is inherent in plane turbulent mixing layers.

The objective of the study is the effect of the frequency (St_θ) and the amplitude (A) of periodical harmonic excitation on the ratio u'/u'_- where u' and u'_- are the mean-square velocity pulsations with and without periodical excitation respectively (Fig. 6.11). The ratio u'/u'_- is a function of the period, T, of harmonic perturbations.

The main conclusions of the study are as follows: 1) the turbulence suppression is lacking at small amplitudes of periodical excitation ($A = 0.003 - 0.0125$), 2) at

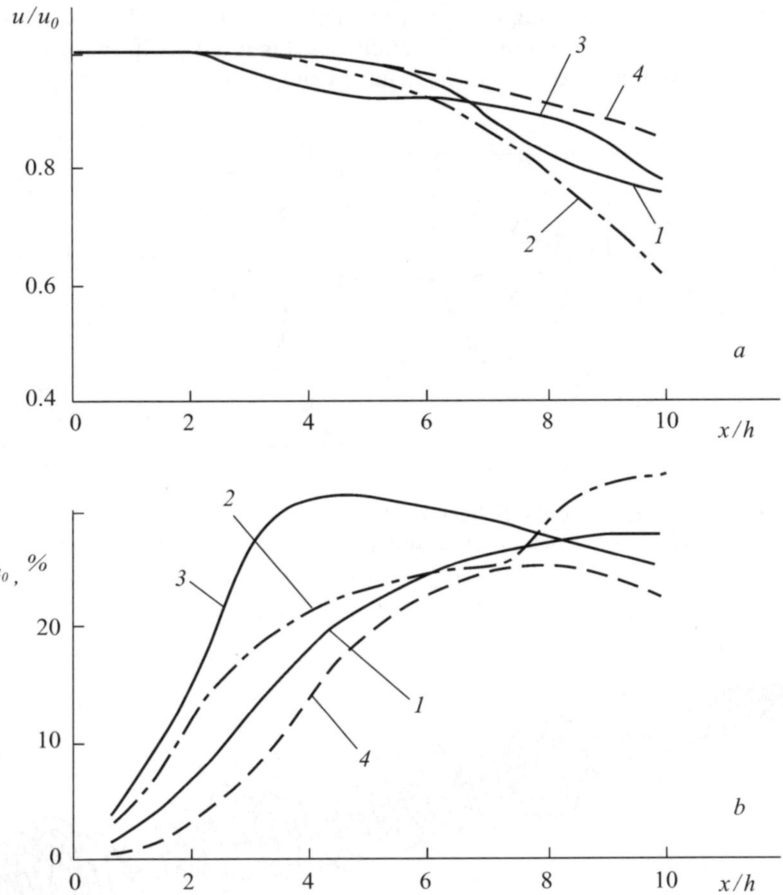

Fig. 6.10. (*a*) The mean velocity and (*b*) longitudinal velocity pulsations along the axis of a plane turbulent jet without and with periodical excitation. $St_h = 0.35$: *1)* $a = 0$, *2)* $a = 0.05$, *3)* $a = 0.15$; $St_h = 2.0$: *4)* $a = 0.3$.

the amplitude value 0.05, the maximum suppression occurs at the frequency corresponding to the maximum instability of the mixing layer ($St_\theta = 0.017$) (Fig. 6.11, *a*), and 3) at higher amplitudes ($A \geq 0.1$) the preference of this value ($St_\theta = 0.017$) is lost (Fig. 6.11, *b*).

6.2.2. The metod of Discrete Vortices Simulates Round Turbulent Jets. While modeling circular jets in the framework of the method of discrete vortices using a set of vortex rings, i.e. postulating the rigid condition of the axial symmetry, the calculation results come into conflict with the experimental results as the calculations do not realize the jet broadening and the velocity decrease downstream. While modeling a circular turbulent jet in the 3-D framework, vortex rings descending the nozzle edge are approximated by vortex polygons [6.5, 6.6] permitting description of deformation of these rings in their planes, as well as in the longitudinal direction along their motion. This is illustrated (Fig. 6.12) by comparison of the vortex structure of a circular submerged jet in two cases: where the solution is obtained in line with the assumption of the axial symmetry of

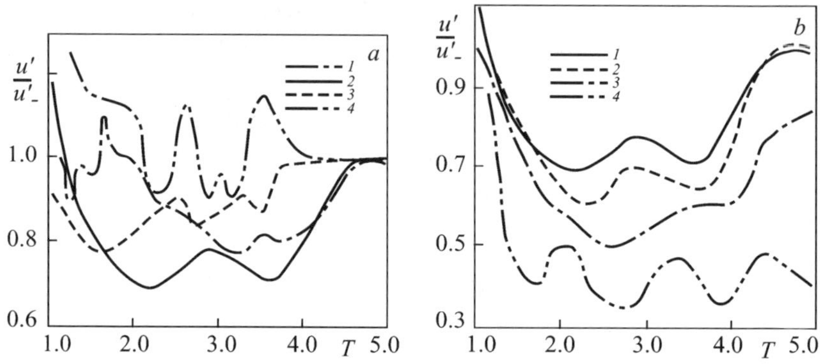

Fig. 6.11. The effect of the frequency (the Strouhal number, St_θ) and the amplitude (A) of periodical excitation of a turbulent mixing layer on the ratio u'/u'_-. (*a*) $A = 0.05$: *1)* $St_\theta = 0.006$, *2)* $St_\theta = 0.017$, *3)* $St_\theta = 0.034$, *4)* $St_\theta = 0.067$; (*b*) $St_\theta = 0.017$: *1)* $A = 0.05$, *2)* $A = 0.10$, *3)* $A = 0.20$, *4)* $A = 0.30$.

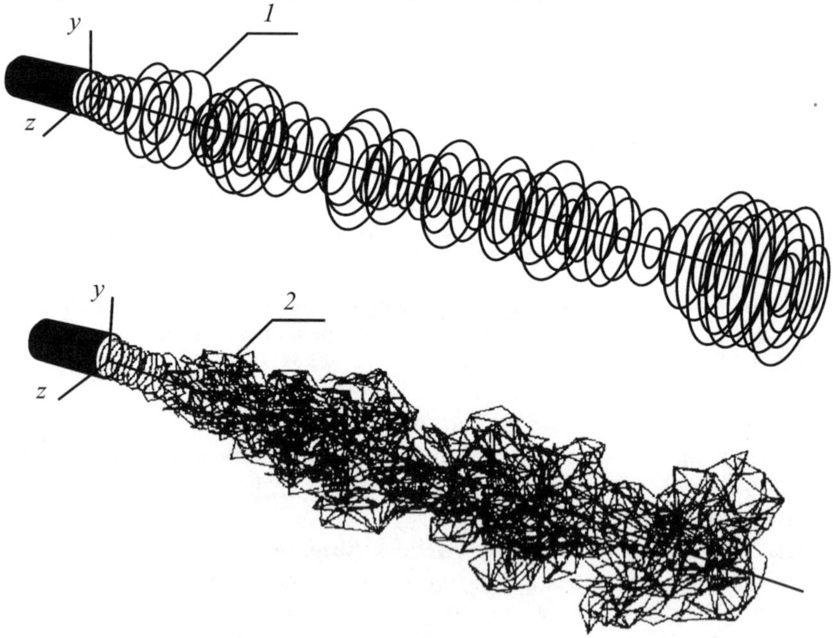

Fig. 6.12. The vortex structures of a circular turbulent jet at fixed time moments calculated in the axisymmetric (*1*) and 3-D (*2*) frameworks.

instantaneous stream (*1*) and where this assumption is invalid (*2*). In the second case the axial symmetry is retained only within 2.5 calibers ($x/d \le 2.5$). Further, we have the azimuthal and longitudinal deformation of the vortex polygons and the flow stochasticity for $x/d > 5$. The problem solution in the 3-D framework brings the calculation and experimental values of the velocity and other flow parameters (the profiles of the mean velocity, of three velocity pulsation components, and of the Reynolds shear stress) into coincidence for the both initial and main regions of the jet. Fig. 6.13 presents the changes in the axial velocity along the jet axis obtained in axisymmetric and 3-D calculations, as well as their comparison with the experimental results [6.7]. It is shown in the latter paper that the axisymmetric approximation yields too high values of the velocity pulsations even within the initial region ($x/d \le 4$).

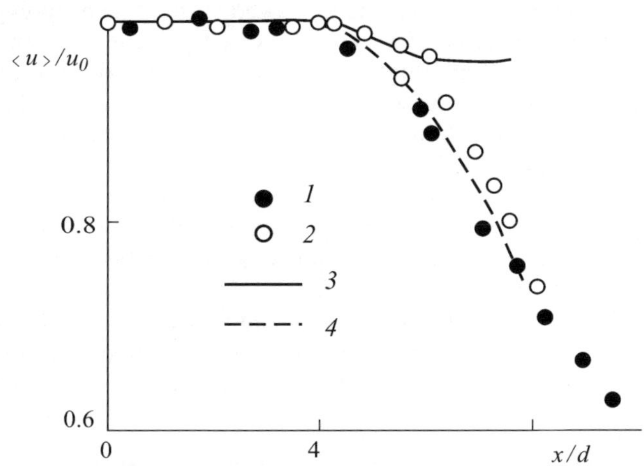

Fig. 6.13. The changes in the axial velocity along the axis of a circular jet with (*1*) and without (*2*) axial symmetry; (*3*, *4*) the experiments.

Unfortunately, the described approach is not realized yet for periodical excitation of circular turbulent jets. The attempt of the corresponding solution is undertaken in [6.1] under the assumption of axial symmetry of the instantaneous stream. There are a number of additional assumptions: the radii of the ring vortex cores carrying into the flow are given; two ring vortex chains slightly spaced along the radius are assigned instead of one ring vortex chain; the circulations of ring vortices descending from the circular cylindrical nozzle edge and their convection velocities are assigned in the form of harmonic functions of time:

$$\Gamma = \Gamma_0[1 + a_1 \sin(2\pi f t)], \quad u_c = u_{c0}[1 + a_1 \sin(2\pi f t)]. \tag{6.3}$$

As such, the approach to simulation for excitation-free jets ($a_1 = 0$) within the initial region ($x/d < 3-4$) qualitatively correctly describe the process of vorticity blob pairing, the mean velocity profiles, the ejection, and other statistical characteristics. The calculation of the periodically excited jet characteristics at the amplitude values $a_1 = 0.005 - 0.100$ and the Strouhal numbers $\mathrm{St} = fd/u_0 = 0.1 - 2.0$ has shown that the excitation in the range St = 0.3 – 2.0 of sufficient amplitude generates large-scale vortices in the jet. At St = 0.3 these vortices are very large and generated at $x/d = 2$; at St = 0.5 large vortices are generated already at $x/d = 1$ and conserved along the distance of several calibers downstream. At St > 1 vortices generated at the excitation frequency are small and quickly decay interacting with other vortices.

The calculation of ejection characteristics for a pulsating turbulent jet [6.1] does not reveal their distinct dependency on the Strouhal number.

The inclusion of deterioration of the instantaneous flow axial symmetry while modeling a circular pulsating submerged turbulent jet with the periodical throughput rate at the nozzle cut-off, $Q(t)/Q_0 = 1 + a\sin(2\pi ft)$, is attempted in [6.4]. The mixing layer is described by a finite set of closed vortex frames (polygons) just as in [6.5, 6.6] for the simulation of excitation-free jets. The jet ejection characteristics under low- (St = 0.4) and high-frequency (St = 1) periodical excitation are calculated using a version of the method of discrete vortices. The amplitude value is $a = 0.14$ and the following boundary conditions are satisfied: the impermeability on the nozzle walls; the equality of static pressures and normal velocities on both sides of the vortex sheet; the total decay of all perturbations at infinity; and the given throughput rate at the jet initial section. Moreover, the Chaplygin – Zhukovsky condition, i.e. the velocity finiteness, is satisfied at the points of the vortex sheet descend (on the nozzle edge).

It follows from Fig. 6.14 that the entrainment rate in the region of extension of four calibers ($x/d \le 4$) increases for the jet under low-frequency excitation and decreases under high-frequency excitation with respect to that for the excitation-free jet.

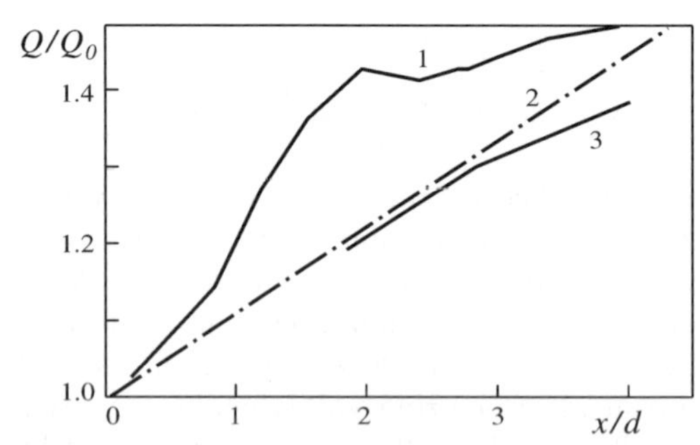

Fig. 6.14. The change in the entrainment rate along a circular jet under (*1*) low-frequency (St = 0.4) excitation, (*2*) without excitation, and (*3*) under high-frequency (St = 1) excitation.

6.3. Numerical Simulation of a Turbulent Mixing Layer on the Basis of the Nonstationary Reynolds Equations Closed by the Differential Model of Turbulence

The Reynolds equations together with the differential model of turbulence are used intensively for the calculation of hydrodynamic and thermal characteristics of various stationary and nonstationary turbulent flows. The description of a turbulent flow in a 2-D mixing layer in [6.9] uses the nonstationary Reynolds equations and the three-parameter model of turbulence [6.23]. The gas large-scale motions ($M^2 << 1$) are supposed to be two-dimensional and small-scale turbulent pulsations are three-dimensional; the phenomena of turbulence transit, generation, diffusion, and dissipation are incorporated. Consideration is given to subsonic flows of a perfect gas where the effects of molecular viscosity and heat conductivity are supposed to be inessential, i.e. $Re \to \infty$.

The simulation is performed for a mixing layer under periodical excitation at its initial region. The periodical harmonic excitation frequency corresponds to the most amplified perturbations according to the linear theory of hydrodynamic stability. In accordance to this requirement, the mean velocity profile is predetermined, as well as the periodical pulsations of the longitudinal and transversal velocity components. Moreover, stationary characteristics of small-scale pulsations, i.e. profiles of the turbulence energy, the Reynolds shear stress, and vorticity, are given at this section. The latter three functions correspond to some equilibrium state of turbulence.

A set of boundary conditions providing gating of large vortices with have the minimum effect of generating sound perturbations is given at the output section. The common conditions of vanishing derivatives with respect to the transversal coordinate of the velocity, pressure, turbulence energy, Reynolds shear stress, and vorticity are satisfied at the upper and lower borders of the rectangular area.

The initial value of the Mach number at the input section in the middle of the mixing layer is $M_0 = 0.2$. The main calculation series is performed for four values of the initial turbulence: $\varepsilon_0 = 0$, 3.26, 6.5, and 14.8%. All values yield the periodical regime of the solution for time $\tau \geq 80$.

The first version ($\varepsilon_0 = 0$) does not use any turbulence model, i.e. the starting position is the Euler equation set for perfect gas. All versions at the periodical regime establishment demonstrate the nonlinear process of the mixing layer rolling-up into large vortex structures carrying thereafter away from the calculated area through output sections.

Fig. 6.15 presents the lines of equal vorticity and pressure in the mixing layer at the fixed time for $\varepsilon_0 = 0$ and 14.8%, as well as the transversal velocity in the middle of the mixing layer as a function of the longitudinal coordinate for various ε_0. These functions clearly demonstrate attenuation of large-scale vortex structures in the mixing layer under the influence of small-scale turbulence.

We do not know any papers on similar simulations applied to plane and circular turbulent jets under periodical excitation.

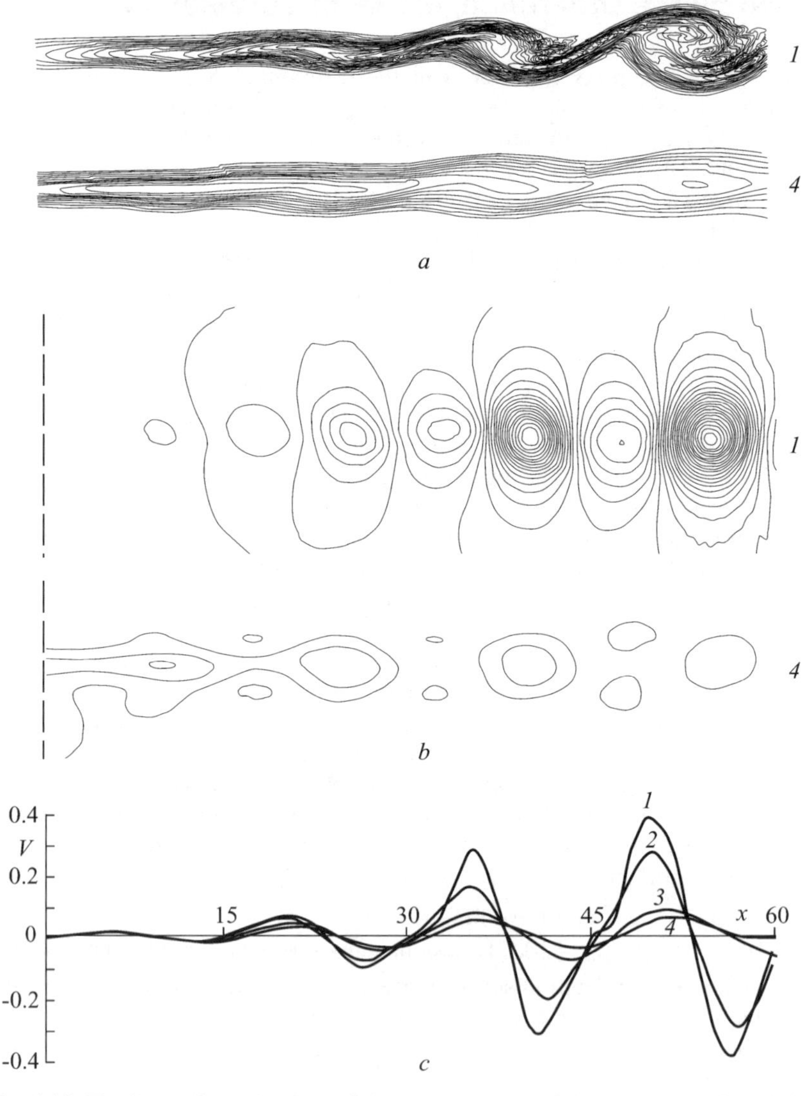

Fig. 6.15. The lines of equal values of the (*a*) vorticity and (*b*) pressure in the mixing layer for (*1*) $\varepsilon_0 = 0$ and (*4*) 14.8%. The change in the transversal velocity along the stream in the middle of the mixing layer (*c*): *1)* $\varepsilon_0 = 0$, *2)* $\varepsilon_0 = 3.26\%$, *3)* $\varepsilon_0 = 6.5\%$, *4)* $\varepsilon_0 = 14.8\%$.

6.4. Numerical Simulation of Turbulent Jet Flows on the Basis of the Generalized Reynolds Equations (the Three-Term Extension). The Effects of Low- and High-Frequency Harmonic Excitation

Assume that the variables (the velocity and pressure components) contained in the Navier – Stokes equations may be presented in the form of the triple extension [6.15]:

$$g(x,t) = \langle g(x,t) \rangle + \tilde{g}_c(x,t) + g'_r(x,t) \tag{6.4}$$

where $\langle g \rangle$ is the component independent of time, $\tilde{g}_c$ is the large-scale coherent component, and g'_r is the random small-scale component. It is assumed that the coherent and incoherent pulsations are uncorrelated. The equation set describing behaviour of each of three flow components is derived from the Navier – Stokes equation and the continuity equation expanding the quantities containing in these equations in the form (6.4) and applying the corresponding averaging operators: averaging in time and in phase. The latter operator is used for separation of the periodical coherent component. The resulting set of differential equations for the averaged, periodical and random motions contains, besides the Reynolds shear stresses $\langle u'v' \rangle$ caused by small-scale turbulent motion, the additional stresses $\langle \tilde{u}\tilde{v} \rangle$ resulting from periodical coherent structures. These equations also contain terms resulting from interactions between the periodical and random motions, as well as between each of them and the averaged motion [6.17, 6.20 – 6.22, 6.25 – 6.30]; the molecular viscosity is assumed to be negligibly small with respect to the turbulent one.

The presented approach is used to calculate turbulent jets and wakes in incompressible fluid under single- and dual-frequency periodical excitation [6.20, 6.21, 6.24, 6.26, 6.27]. The calculation uses approximation of a boundary layer, the jet is assumed to be isobaric and the fluid is incompressible and inviscid. Several approaches to the solution of such problems leading to a set of integral relations are well known. The evaluation of parameters of the small-scale turbulence uses various semiempirical turbulence models (ranging from algebraic models with the constant eddy viscosity across the mixing layer to differential models) or empirical formulae.

The calculation of periodical coherent components generally uses the linear theory of stability in inviscid approximation under assumption on local parallelism of the flow. The periodical components of velocity and pressure are presented in the form

$$[\tilde{u},\tilde{v},\tilde{w},\tilde{p}]=\{\overline{u}(y),\overline{v}(y),\overline{w}(y),\overline{p}(y)\}\times \\ \times\exp\left[i\left(\int\alpha(x)dx-\beta t+n\Theta\right)\right] \tag{6.5}$$

where $\alpha(x)$ is the complex function of x, its real part corresponds to the wave number and its imaginary part corresponds to the perturbation amplification coefficient, β is the circular frequency, and n is the azimuthal wave number. It is assumed that the periodical perturbations in (6.4) may be presented in the form of (6.5). The problem solution should incorporate the presence of nonlinear wave processes. For this purpose the energy integral method is used, i.e. the integral relations derived from the initial differential equations by integration across the mixing layer are considered instead of the differential equations for transition of kinetic energy of the mean, periodical large-scale and random small-scale streams.

By way of example consider the following set of integral relations for a circular jet [6.7]:

$$\begin{aligned}
&\frac{d}{dx}\int_0^\infty\frac{1}{2}\langle u\rangle^2\langle u\rangle y dy=\int_0^\infty\left(-\langle u'v'\rangle\right)\frac{\partial\langle u\rangle}{\partial y}ydy-\int_0^\infty\left(-\langle\tilde{u}\tilde{v}\rangle\right)\frac{\partial\langle u\rangle}{\partial y}ydy,\\
&\frac{d}{dx}\int_0^\infty\langle k\rangle\langle u\rangle y dy=\int_0^\infty\left(-\langle u'v'\rangle\right)\frac{\partial\langle u\rangle}{\partial y}ydy-\int_0^\infty\varepsilon y dy+\int_0^\infty\varphi y dy,\\
&\frac{d}{dx}\int_0^\infty\langle q\rangle\langle u\rangle y dy=\int_0^\infty\left(-\langle\tilde{u}\tilde{v}\rangle\right)\frac{\partial\langle u\rangle}{\partial y}ydy-\int_0^\infty\varphi y dy.
\end{aligned} \tag{6.6}$$

Here k and q are the kinetic energy of turbulent and periodical motions respectively, $k=\frac{1}{2}\left(\left\langle u'^2+v'^2+w'^2\right\rangle\right)$, $q=\frac{1}{2}\left(\left\langle\tilde{u}^2+\tilde{v}^2+\tilde{w}^2\right\rangle\right)$, and ε and φ are the kinetic energy dissipation rate of turbulence and periodical motion respectively. The dissipative function determines the energy transport from periodical perturbations to small-scale turbulent perturbations.

The solution of the set (6.6) requires the mean velocity profiles at different sections of the mixing layer and relations for the Reynolds stresses, $\langle u'v'\rangle$ and $\langle\tilde{u}\tilde{v}\rangle$. One may use the turbulence model to determine $\langle u'v'\rangle$ and the relation (6.5) that may be presented in the form $[\tilde{u},\tilde{v},\tilde{w},\tilde{p}]=q(x)\{\}\exp[()]$ (where $q(x)$ is the amplitude value of the corresponding periodical perturbation) to determine $\langle\tilde{u}\tilde{v}\rangle$. Hence, only relative profiles for the components of velocity and pressure are borrowed from the linear theory of stability and the set (6.6) describes the nonlinear process of interaction between periodical and random components.

The set (6.6) makes possible the evaluation of three parameters: the effective thickness of the mixing layer, the amplitude or average value of the turbulence kinetic energy and the amplitude value of periodical pulsations. It is necessary to set the initial values of these parameters and the characteristic frequency of periodical pulsations or the Strouhal number St. This offers possibilities of taking account of such parameters as the initial turbulence and the initial level of periodical excitation, as well as the excitation frequency and the azimuthal wave number, $n = 0, 1, 2, \ldots$

By way of example, Fig. 6.16 compares the calculated and experimental values of the pressure pulsations along the axis of the mixing layer and the transversal profiles of the pressure pulsations for a turbulent jet under acoustical excitation by plane ($n = 0$) and higher azimuthal ($n = 1$ and 2) modes [6.28]. Paper [6.26] contains the calculation of the momentum loss thickness broadening along the mixing layer of a circular jet under high-frequency excitation ($St_\theta = 0.017$). The calculation results (Fig. 6.17) adequately describe the corresponding experimental function obtained earlier for a jet under acoustical excitation (cf. Fig. 2.36).

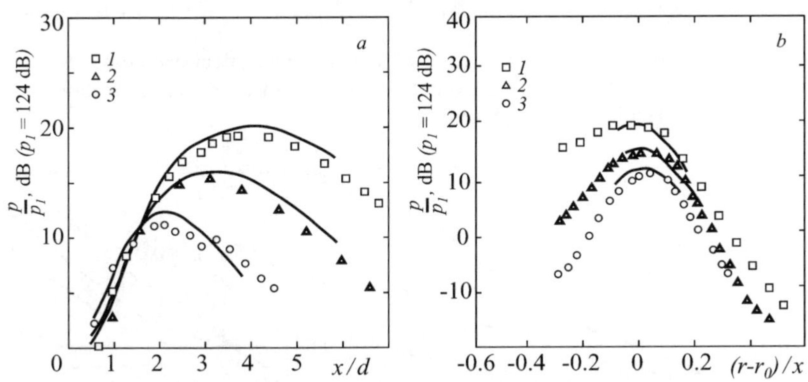

Fig. 6.16. The comparison of the calculated ($St_s = 0.48$) and experimental values of the pressure pulsations along the stream in the middle of the mixing layer (*a*) and the transversal profiles of the pressure pulsations (*b*) for a circular turbulent jet under low-frequency acoustical excitation ($St_s = 0.5$) at the zero ($n = 0$), first ($n = 1$) and second ($n = 2$) azimuthal modes. (*a*): *1)* $n = 0$, $St_s = 0.491$, *2)* $n = 1$, $St_s = 0.505$, *3)* $n = 2$, $St_s = 0.503$; (*b*): *1)* $n = 0$, $St_s = 0.513$, *2)* $n = 1$, $St_s = 0.501$, *3)* $n = 2$, $St_s = 0.502$.

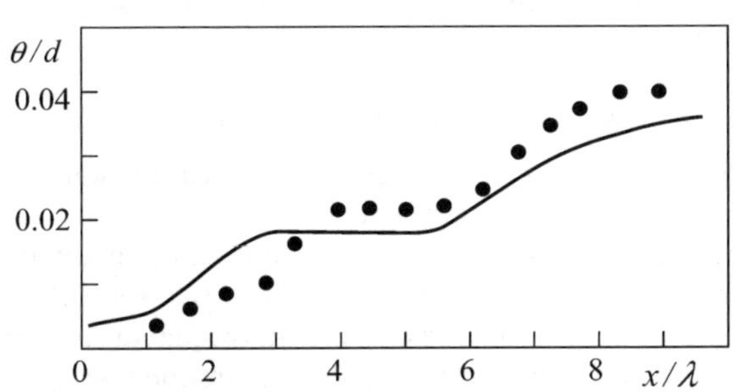

Fig. 6.17. The comparison of the calculated [6.26] and experimental values [2.48] of the momentum loss thickness, θ/d, along the mixing layer of a circular turbulent jet under high-frequency acoustical excitation ($St_\theta = 0.017$).

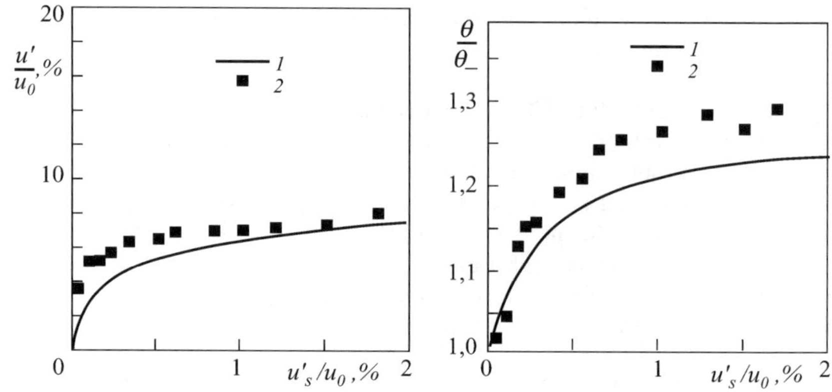

Fig. 6.18. The effect of saturation at low-frequency acoustical excitation ($St_s = 0.5$) of a turbulent jet as the acoustical excitation intensity increases in the section $x/d = 9$ ($M_0 = 0.2$). *1)* calculation, *2)* experiment.

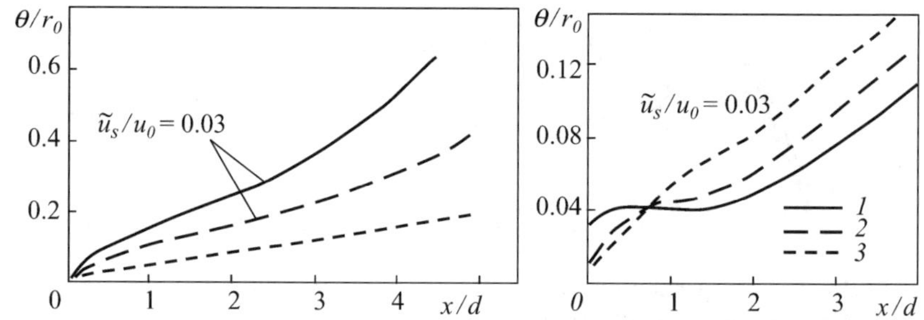

Fig. 6.19. The calculation of changes along the flow of the momentum loss thickness, θ/r_0, in the mixing layer of a circular turbulent jet under dual- and single-frequency periodical excitation with the level $\tilde{u}/u_0 = 0.03$, as well as for an excitation-free jet ($\tilde{u}/u_0 = 0$). (*a*) low-frequency excitation: *1)* $St_s = 0.4$ and 0.8, *2)* $St_s = 0.8$, *3)* an excitation-free jet; (*b*) high-frequency excitation: *1)* $St_s = 4.8$ and 2.4, *2)* $St_s = 4.8$, *3)* an excitation-free jet.

The curves $u'/u_0 = F_1(u'_s/u_0)$ and $\theta/\theta\_ = F_2(u'_s/u_0)$ in Fig. 6.18 illustrate the effect of saturation of the jet low-frequency acoustical excitation ($St_s = 0.5$) as the acoustical excitation intensity increases [6.32].

It is shown in [6.24] that the simultaneous periodical excitation of a jet at frequencies corresponding to Strouhal numbers $St_s = 0.8$ and 0.4 causes intensification of the vortex pairing and, as a result, the mixing layer thickening with respect to single-frequency excitation. Conversely, high-frequency excitation at frequencies corresponding to Strouhal numbers $St_s = 4.8$ and 2.4 causes the mixing layer to decrease with respect to single-frequency excitation (Fig. 6.19). The efficiency of such dual-frequency excitation depends on the phase ratio between the signals of the main tone and the subharmonic.

Similar results are obtained for three-frequency periodical excitation of a jet ($St_1 = 0.2$, $St_2 = 0.4$, and $St_3 = 0.8$) at the zero mode ($n = 0$, axisymmetric perturbations) with the phase shift β_{12} between the first and second periodical perturbations and with the phase shift β_{23} between the second and third periodical perturbations [6.27]. The study comprises the influence of the frequency of periodical perturbations on their interaction, the increase in the turbulence energy and in the effective thickness of the mixing layer on the interval of size 9 calibers. It is shown that for $\beta_{12} = 270°$ and $\beta_{23} = 180°$ the mixing layer thickening along the stream is maximum (Fig. 6.20).

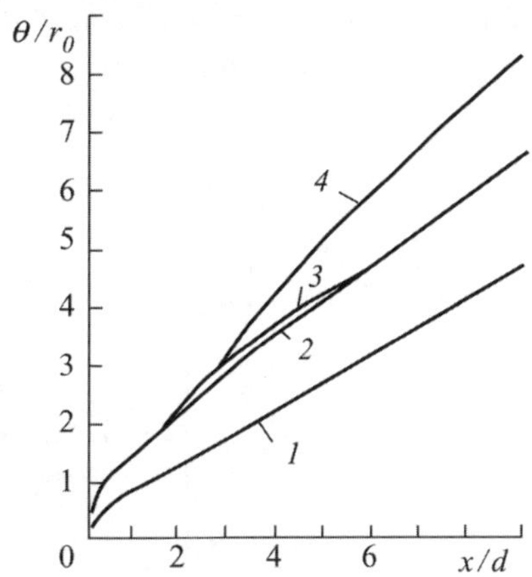

Fig. 6.20. The influence of the initial phase shift (β_{12} and β_{23}) on the momentum loss thickness, θ/r_0, in the mixing layer of a circular turbulent jet under three-frequency excitation ($St_1 = 0.2$, $St_2 = 0.4$, $St_3 = 0.8$). *1)* without excitation; *2)* $\beta_{12} = 90°$, $\beta_{23} = 180°$; *3* – $\beta_{12} = 90°$, $\beta_{23} = 0$; *4)* $\beta_{12} = 270°$, $\beta_{23} = 180°$.

The improvement of this method [6.25] allows describing a number of important effects. The stream parameters are supposed to consist of four components: the averaged stream, small-scale turbulence, main and subharmonic periodical components

$$g(x,t) = \langle g(x) \rangle + g'(x,t) + \tilde{g}_f(x,t) + \tilde{g}_{f_n}(x,t),$$
$$f_n = f/2^n, n = 1,2,... \tag{6.7}$$

The equations for the momentum of the average stream, $\langle g(x) \rangle$, the main, $\tilde{g}_f$, and subharmonic, $\tilde{g}_{f_n}$, components of motion, as well as of the small-scale turbulence, g', are integrated across the stream resulting in a set of integral relations taking into consideration interaction of all the stream components. As a result, the changes along the stream in the momentum loss thickness, amplitude components of the small-scale turbulence energy, periodical components of the main and subharmonic frequencies, as well as the number of vortex pairings are evaluated for the given initial values of these quantities, the phase shift between oscillations at the main and subharmonic frequencies, and the Strouhal number, $\mathrm{St} = fd / u_0$, determined from the periodical excitation frequency, f.

The above cited paper comprises the calculation results for the change in the momentum loss thickness along the stream for the initial levels of periodical excitation $\tilde{u} / u_0 = 0 - 10\%$ and Strouhal numbers St = 0.2 – 4.8 at the fixed low initial level of small-scale turbulence ε_0 = 0.1% in a circular jet. It is shown that for St < 0.5 the thickness, θ, for various x/d increases monotonically with the periodical excitation level (Fig. 6.21) starting from $\tilde{u} / u_0 = 0.5\%$. The vortex pairing is not observed, i.e. the subharmonic does not play any role. The mixing layer thickening along the stream for St = 0.6 – 1.0 becomes nonmonotone as the excitation level, $\tilde{u} / u_0$, increases and one vortex pairing occurs.

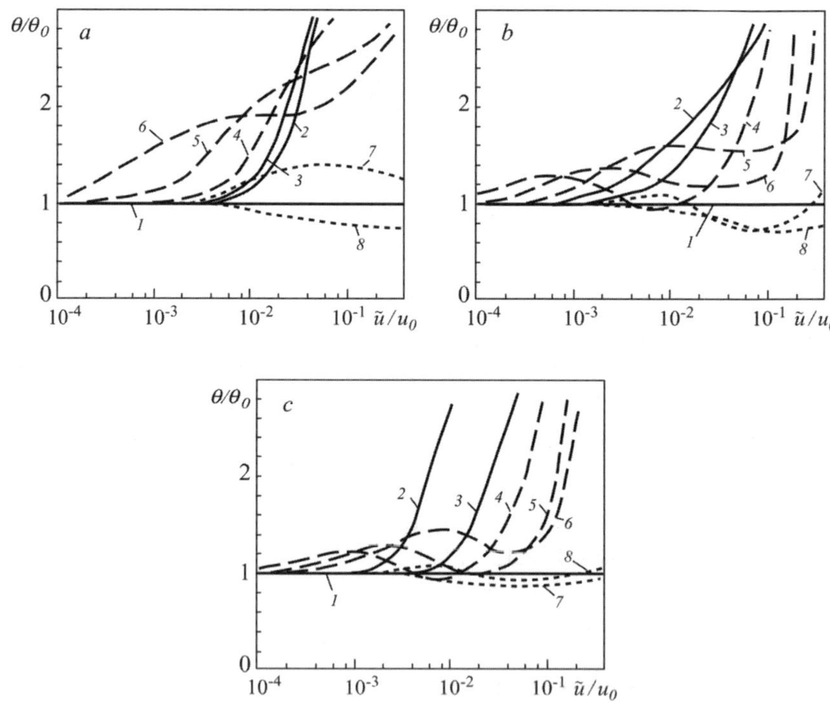

Fig. 6.21. The influence of the frequency (the Strouhal number) and level ($\tilde{u} / u_0$) of single-frequency periodical excitation on the momentum loss thickness in the mixing layer of a circular turbulent jet. (*a*) x/d = 1, (*b*) x/d = 3, (*c*) x/d = 5. *1)* St = 0, *2)* St = 0.2, *3)* St = 0.4, *4)* St = 0.6, *5)* St = 0.8, *6)* St = 1.0, *7)* St = 2.4, *8)* St = 4.8

For St = 2.4 – 4.8 and x/d = 3 – 5 the excitation of the level $\tilde{u} / u_0$ > 1% causes the mixing layer to thin and four vortex pairings. It is important to note that for large Strouhal numbers (St > 2.4) and very large levels of periodical excitation

($\tilde{u}/u_0 \geq 10\%$) the mixing layer thickens instead of thinning, i.e. the change of the action sign revealed in experiments is confirmed.

It is therefore concluded that the simulation method presented in Section 6.4 is highly suitable for the analysis of periodical excitation effects on the aerodynamic characteristics of jets.

6.5. An Interesting Analogy

This section outlines another approach to the description of the above-mentioned effects. The approach is based on the hypothesis [6.18] that turbulence in unclosed flows, in particular, in jets is determined by the presence of weak random perturbations (noise) that are abruptly amplified due to instability and move the system into a qualitatively new state. This transition is called the noise-induced phase transition. This phenomenon, similar to other complex oscillation phenomena, could be studied with an example of the simplest model that may contain such a transition [6.20, 6.19, 6.34]. It is essential that the equations of the model could fundamentally differ from that of the studied system. The reason is universality of the oscillation theory laws. A pendulum with a randomly oscillating suspension axis is an example. The effects of frequency and amplitude of harmonic oscillations on the variance of the pendulum angular displacement could be studied by the addition of a periodical component to the random vibration of the pendulum suspension axis. The results are very similar to the corresponding results for a jet under periodical excitation.

The equation for the pendulum oscillations may be written in the form

$$\ddot{\varphi} + 2\beta(1 + 2\alpha\dot{\varphi}^2)\dot{\varphi} + \omega_0^2[1 + \xi_1(t) + A\cos\omega t]\sin\varphi = \xi_2(t) \tag{6.8}$$

where φ is the pendulum angular displacement with respect to the equilibrium state, $2\beta(1+2\alpha\dot{\varphi}^2)\dot{\varphi}$ is the quantity proportional to the friction moment, ω_0 is the frequency of the small natural oscillations of the pendulum, and β and α are the coefficients of lincar and nonlinear friction respectively. Futher, $\xi_1(t) + A\cos\omega t$ is the quantity proportional to acceleration of the pendulum suspension axis and defined as the sum of the relatively broadband random process, $\xi_1(t)$ with the non-zero spectral density at the frequency $2\omega_0$ (it will be defined as κ and called as the random vibration intensity) and the harmonic component, $A\cos\omega t$ with the amplitude A and frequency ω, and $\xi_2(t)$ is the additive component of random vibration of the pendulum axis arising in the case where the vibration direction is different from the vertical line. The analytical and numerical solutions of equation (6.8) for $A = 0$ shows that the pendulum, for small intensity of the suspension axis vibration, executes small oscillations determined by the additive component $\xi_2(t)$ near the equilibrium state (if the additive component is absent, the pendulum is at rest). When the random oscillation intensity of the suspension axis exceeds some critical value, κ_{cr}, the pendulum oscillation amplitude abruptly increases resulting in the nonzero variance of the angular displacement even for $\xi_2(t) = 0$. It is found that near the threshold of such excitation of the pendulum oscillations (for $\xi_2(t) = 0$) a specific property of alternation manifests itself when the pendulum oscillates over a long period of time near its equilibrium state and rather strong bursts (the so-called laminar and turbulent phases) occur only occasionally. The laminar phase duration reduces with distance from the excitation threshold, and the turbulent one increases resulting in the disappearance of the laminar phases. The oscillation variance $\langle \varphi^2 \rangle$ increases monotonically. With the availability of the additive component, $\xi_2(t)$, alternation occurs below the threshold and its nature is less pronounced.

It is stated that additional weak harmonic oscillations of the suspension axis strongly influence on the pendulum excitation processes and low-frequency oscillations increase the intensity of the noise-induced oscillations and reduce their excitation threshold, but high-frequency oscillations, conversely, suppress the noise-induced oscillations and increase their excitation threshold. As in the case of the high-frequency acoustical excitation of a jet, the high-frequency component of the pendulum suspension axis vibration of low amplitudes do not practically influence on the existing oscillations. The intensity of the pendulum noise-induced oscillations decreases initially to a certain minimum value as the amplitude of the high-frequency vibration component increases. Next the intensity begins to increase and the converse effect occurs starting from a certain value: the intensity of the pendulum oscillations in the presence of the high-frequency harmonic component of vibrations becomes higher than without this component. The analogy with the effects of high-frequency acoustical excitation of a jet manifests itself here in full measure.

The dependencies illustrating the effect of frequency and amplitude of the harmonic component of the suspension axis vibration on the variance of the pendulum angular displacement are obtained in the framework of the presented approach. These dependencies remarkably resemble the corresponding dependencies for low- and high-frequency excitation of turbulent jets and mixing layers of different frequencies and amplitudes.

Below, dependencies are calculated for noise-induced oscillations of a pendulum with random oscillations of the suspension axis. The functions in Fig. 6.22 describe the noted effect of saturation as the amplitude increases for low-frequency excitation. These functions are similar to that for low-frequency acoustical excitation of a jet (cf. Fig. 6.18).

Figs. 6.23 and 6.24 illustrate influence of the oscillation amplitude under high-frequency excitation. Here the velocity pulsations initially decrease as the excitation level increases, attain the minimum, and then begin to increase with a tendency for the reverse effect when the rise of the high-frequency excitation level

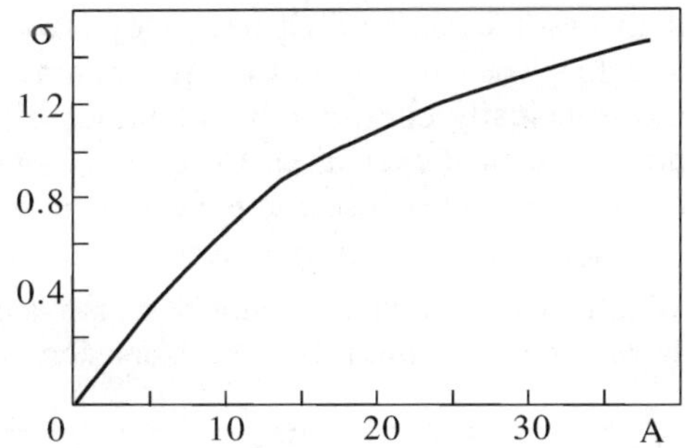

Fig. 6.22. The dependency of the parameter $\sigma = \left(\left\langle \varphi^2 \right\rangle\right)^{1/2}$ of a pendulum on the amplitude A of low-frequency (ω = 1) harmonic oscillations; $\kappa / \kappa_{cr} = 2.23$.

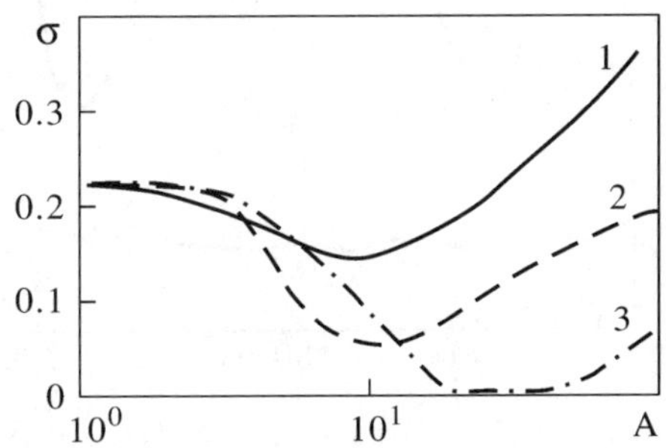

Fig. 6.23. The dependency of σ on the pendulum oscillation amplitude, A, at the frequencies (*1*) ω = 3.5, (*2*) ω = 6, (*3*) ω = 11; $\kappa / \kappa_{cr} = 5.6$.

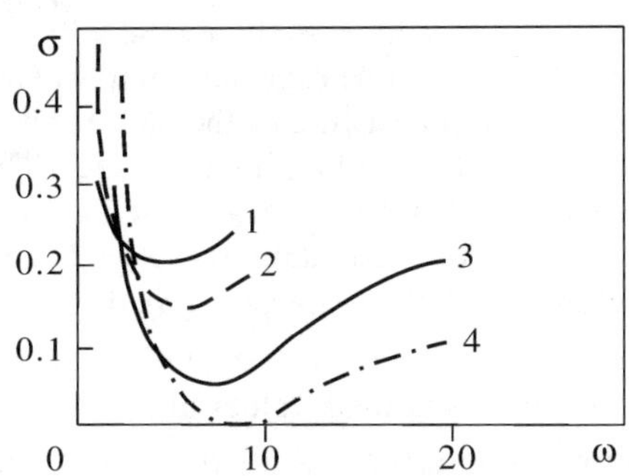

Fig. 6.24. The functions $\sigma = F(\omega)$ for different values of the pendulum oscillation amplitude: (*1*) A = 2.5, (*2*) A = 5, (*3*) A = 10, (*4*) A = 20; $\kappa / \kappa_{cr} = 5.6$.

causes the velocity pulsation increase instead of the decrease (cf. Fig. 2.12). This tendency is observed also for a jet under high-frequency excitation of the turbulent mixing layer (cf. Fig. 2.15). The calculation for a pendulum with high amplitudes of vibration excitation principally confirms the amplitude effect observed in experiments on jets under acoustical excitation. However, a sharp distinction is revealed: high levels of the amplitude cause complete suppression of the pulsations. This distinction is eliminated when taken into account that the function $\xi_2(t)$ in the right-hand side of Eq. (6.8) is nonzero, which is equivalent to the action of, aside from vertical, horizontal oscillations on the suspension axis.

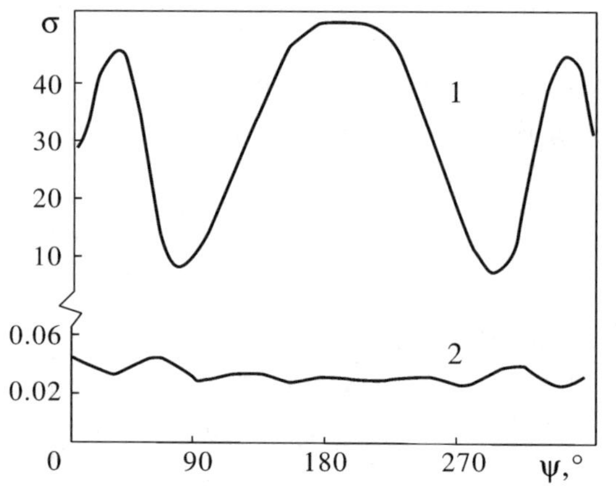

Fig. 6.25. The functions $\sigma = F(\omega)$ for a pendulum: *1)* low frequencies (ω = 0.3 and 0.15), *2)* high frequencies (ω = 11 and 5.5); κ / κ_{cr} = 5.6.

The dependencies (Fig. 6.25) illustrating the effect of pendulum dual-frequency oscillations at the main frequency and its subharmonic with the fixed phase shift, ψ, for low and high frequencies should be especially noted. The comparison of Figs. 6.25 and 2.61 demonstrate a similar nature of the phase shift effect on acoustically excited jets and pendulums for low and high frequencies. The results for a pendulum (Fig. 6.25) correspond to identical values of the harmonic component amplitudes at the main frequency and its subharmonic, A = 5. To purify the effect, the calculations for pendulum oscillations are performed for $\xi_2(t) = 0$ and different values of intensity of random vertical component of the suspension axis vibration. The other parameters are fixed and their values are $\omega_0 = 1$, β = 0.1, and α = 100. The presented correlation supports the validity of the described analogy.

Let us consider in summary the oscillograms, $\varphi(t)$, for high-frequency (ω = 19.76) harmonic excitation of the pendulum suspension axis with different amplitudes [6.19]. It follows from Fig. 6.26 that the intensity of the noise-induced oscillations decreases and the laminar phase duration increases as the amplitude increases. As the calculation implies, the oscillations are completely suppressed for

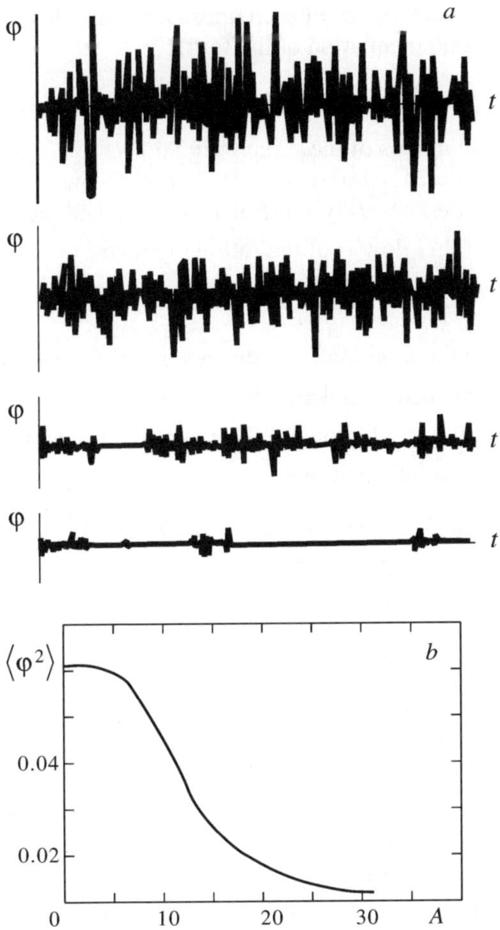

Fig. 6.26. (*a*) The functions $\varphi(t)$ at the frequency ω = 17.76 and amplitudes A = 5, 15, 30, and 40. (*b*) The values of $\langle \varphi^2 \rangle$ as a function of the amplitude, A.

the amplitude A = 42. This phenomenon is not observed in the presence of additive noise, $\langle \xi_2^2 \rangle = 0.05 \langle \xi_1^2 \rangle$ [6.19].

Eq. (6.8) discussed above is qualified in [6.34] as an 'analogy', not related to the hydrodynamics equations describing turbulent jet flow or, in any case, not resulting from them. However, not calling the reader's attention to the specific view of the equation, we may demonstrate a possible source of its structure. Actually, it is well known that the Euler infinite-dimensional representation of the hydromechanics equations could be reduced by certain procedures to a set (generally, an infinite one) of finite-dimensional relations. Such procedures are, for instance, the transition to the Lagrange description, spectral extensions, discrete-vortex modeling, etc. The flow characteristics in such approaches are described, as a rule, by a set of ordinary differential equation of the form

$$\dot{\mathbf{x}} = \mathbf{F}(\mathbf{x}, t)$$

for finite-dimensional modes of these representations.

To be clear, consider, for instance, 2-D flow with the Euler field of velocity, $u(x, y, t)$, $v(x, y, t)$, described by the Euler or Navier-Stokes equations. The Lagrange description of this flow is of the following form:

$$\dot{X} = u(X, Y, t), \dot{Y} = v(X, Y, t), \tag{6.9}$$

where $X(t), Y(t)$ are the Lagrange coordinates of a fluid particle.

Consider the perturbed problem, $X = X_0 + X_1 + \ldots$, $Y = Y_0 + Y_1 + \ldots$, where $X_1(t)$, $Y_1(t)$ are small perturbations of the Lagrange trajectory. Then Eq. (6.9) is decomposed into a chain of equations

$$\begin{gathered} \dot{X}_0 = u(X_0, Y_0, t) \equiv u^0, \dot{Y}_0 = v(X_0, Y_0, t) \equiv v^0, \\ \dot{X}_1 = a_{11} X_1 + a_{12} Y_1, \\ \dot{Y}_1 = a_{21} X_1 + a_{22} Y_1, \end{gathered} \tag{6.10}$$

where $a_{11} = u_x(X_0, Y_0, t) \equiv u_x^0$, $a_{12} = u_y^0$, $a_{21} = v_y^0$, $a_{22} = v_y^0 \equiv -a_{11}$.

The elimination of Y_1 from Eq. (6.10) results in the equation

$$\begin{gathered} \ddot{X}_1 - \beta \dot{X}_1 - \sigma X_1 = 0, \\ \beta = (\ln a_{12})^{\bullet}, \end{gathered} \tag{6.11}$$

where $\sigma = \ddot{a}_{11} + a_{11}(a_{11} - \beta) + a_{12} a_{21}$.

Hence, we get Eq. (6.11) that may describe an oscillator at some values of the coefficients β and σ. Note that the presence of arbitrary outer perturbations (determined or random) in Eq. (6.9) is not crucial and does not change the structure of Eq. (6.11). However, it is obvious that the nature of the dynamic set (6.10) is determined by the coefficients a_{ij} that characterize the main hydrodynamic field. Therefore, while analyzing the hydrodynamic field, one may estimate the probability of ability of Eq. (6.11) to describe, say, oscillating (stable or unstable) regimes.

When the main field describes flow rotations or oscillations, this probability may turn out to be high. An example is provided by the asymptotically exact solution for the dynamics of two initially discrete vortices in incompressible viscous fluid [6.2] when the perturbed motion equation for $X_1(t)$ is the Mathieu equation (oscillations of a pendulum with the suspension of the harmonically changed length). It is well known that this equation has unstable oscillation regimes

$$\ddot{X}_1 + (a + 2q\cos 2\tau)X_1 = 0, \tag{6.12}$$

where $\tau = \dfrac{4\Gamma}{\pi \nu l^2} t$, l is the initial distance between the vortices, a and q are certain functions that slowly change with the period $\cos 2\tau$, Γ is the vortex intensity, ν is the fluid viscosity, and the dots over X_1 correspond to differentiation with respect to τ.

The structure of Eq. (6.12) coincides with that of Eq. (6.8). Most likely, this fact should not be treated as a coincidence as the physical content of this analogy is rather clear – the important role in shear flows of the large vortex pairing, which is recognized as one of the governing factor of turbulent mixing in shear flows [6.3].

The situation around the coefficient β is not so clear. The point is that this coefficient in examples with pronounced vortex structures where $a_{11} \sim \Omega \sim t^{-1}\exp(-ct^{-1})$ cannot have the oscillation nature:

$$\beta \sim \frac{d}{dt}(\ln \Omega) \sim -t^{-1} + ct^{-2} \tag{6.13}$$

and its role vanishes for sufficiently large time moments.

Therefore, the presented ideas of qualitative illustrative nature demonstrate some formal and physical grounds to consider such heuristic models as a sufficiently rational approach.

REFERENCES

6.1. Acton E (1980) A modelling of large eddies in an axisymmetric jet. J Fluid Mech 98, # 1: 1 – 31
6.2. Airapetov AB (1990) Motion of an initially discrete vortex in viscous fluid (in Russian). Appl Math & Mech 54, # 3: 430 – 434
6.3. Airapetov AB (2003) Possible fundamentals for some simple mechanical analogies in hydromechanics. J Eng Phys and Thermophys 77, # 2
6.4. Andronov PR (2002) Periodical excitement of submerged turbulent jets (in Russian). Aerodynamics and Gas Dynamics # 1: 69 – 78
6.5. Belotserkovsky SM, Ginevsky AS (1995) Simulation of turbulent jets and wakes using the method of discrete vortices (in Russian). Nauka, Moscow
6.6. Belotserkovsky SM, Ginevsky AS, Khlapov NV (1995) Simulation of a round turbulent jet using the method of discrete vortices (in Russian). Phisycs-Doklady 40, # 6: 1327 – 1332
6.7. Chan YY (1977) Wavelike eddies in a turbulent jet. AIAA J 15, # 7: 999 – 1001

6.8. Danaila I, Boersma BJ (2000) Direct numerical simulation of bifurcating jets. Physics Fluids 12, # 5: 1255 – 1257
6.9. Fedorchenko AT (1988) Action of small-scale turbulence on the development of coherent structures in a mixing layer. Soviet Physics Doklady, 302, # 6: 1327 – 1332
6.10. Grinstein FF, Oran ES, Boris JP, Hussain AKMF (1987) Numerical simulation of the transition region of an axisymmetric free jet. AIAA Pap 52
6.11. Grinstein FF, Oran ES, Boris JP (1987) Direct numerical simulation of axisymmetric jets. AIAA J 25, # 4: 92 – 98
6.12. Grinstein FF, Hussain F, Oran ES (1990) Vortex-ring dynamics in a transitional subsonic free jet. A numerical study. Eur J Mech, B/Fluids 9, # 6: 499 – 525
6.13. Hilgers A, Boersma BJ (2001) Optimization of turbulent jet mixing. Fluid Dynamics Research 29: 345 – 368
6.14. Husain ZD, Hussain AKMF (1983) Natural instability of free shear layers. AIAA J 21, # 11: 1512 – 1518
6.15. Hussain AKMF, Reynolds WC (1970) The mechanics of an organized wave in turbulent shear flow. J Fluid Mech 41: 241 – 258
6.16. Hussain AKMF, Husain HS, Zaman KBMQ, Tso J, Hayakawa M, Takaki R, Hasan MAZ (1986) Free shear flows: Organized structures and effects of excitation. AIAA Pap 235
6.17. Kibens V (1980) Discrete noise spectrum generated by an acoustically excited jet. AIAA J 48, # 4: 434 – 441
6.18. Landa PS (1996) Turbulence in nonclosed fluid flows as a noise-induced phase transition. Europhys Lett 36, # 6: 401 – 406
6.19. Landa PS (1998) Turbulence generation in nonclosed fluid flows as a nonequilibrium noise-inducted phase transition of the second kind (in Russian). Tech Phys J 68, # 1: 31 – 39
6.20. Landa PS, Zaikin AA, Ginevsky AS, Vlasov YeV (1999) Turbulence and coherent structures in submerged jets. Control of the turbulence. Int J Bifurcation and Chaos 9, # 2: 397 – 414
6.21. Liu JTC (1989) Coherent structures in transitional and turbulent free shear flows. Ann Rev Fluid Mech 21: 285 – 315
6.22. Liu JTC, Merkins L (1976) On the interactions between large-scale structure and fine-grained turbulence in a free shear flow. 1. The development of temporal interactions in the mean. Proc Roy Soc London A 352: 213 – 247
6.23. Lushchik VG, Pavel'ev AA, Yakubenko AYe (1978) Three-parameter model of shear turbulence. Fluid Dynamics 13, #3: 350 – 359
6.24. Mankbadi RR (1985) The effect of bi-modal excitation on the spreading rate of a circular jet. AIAA Pap 572
6.25. Mankbadi RR (1985) The mechanism of mixing enhancement and suppression in a curcular jet under excitation conditions. Phys Fluids 28, # 7: 2062 – 2074
6.26. Mankbadi RR (1986) The effect of phase-difference on the spreading rate of a jet. AIAA J 24, # 12: 1941 – 1948

6.27. Mankbadi RR (1990) Multiwave nonlinear interactions in the axisymmetric mode. AIAA Pap 365
6.28. Morris PJ (1980) A model for broadband jet noise amplification. AIAA Pap 1004
6.29. Morris PJ, Baltas C (1981) Turbulence in sound excited jets: Measurements and theory. AIAA Pap 58
6.30. Morris PJ, Baltas C (1981) Measurements and prediction of turbulence in sound excited jets. AIAA Pap 2006
6.31. Nallasamy M, Hussain AKMF (1985) Numerical study of the phenomenon of turbulence suppression in a plane shear layer. In: Turbulent Shear Flows 4, Sptinger, Berlin, pp 8.6 – 8.11
6.32. Raman G, Rice EJ, Mankbadi RR (1988) Saturation and the limit of jet mixing enhancement by single frequency plane wave excitation: Experiment and theory. In: AIAA/ASME/SIAM/APS 1st Nat Fluid Dyn Congr, Collect Techn Pap, Pt 2, Cincinnati, Ohio, pp 1000 - 1007
6.33. Smirnykh YeA (1986) Simulation of periodical excitation of a plane turbulent jet using the method of discrete vortices. In: Thermal and Physicochemical Processes in Power Plants. Minsk, pp 92 – 96
6.34. Vlasov YeV, Ginevsky AS, Landa PS, Samodurov DL (2002) On analogy between the response on periodical excitation of a turbulent jet and the pendulum with the randomly vibrating suspension axis. J Eng Phys and Thermophys 75, # 4: 90 – 93

CHAPTER 7 SUPERSONIC NONISOBARIC TURBULENT JETS. CONTROL OF AERODYNAMIC AND ACOUSTICAL CHARACTERISTICS

7.1. Aerodynamic Characteristics of Supersonic Turbulent Jets

A supersonic anisobaric turbulent jet is a complex gas-dynamic object and is characterized by strong spatial heterogeneity of the velocity and pressure fields that are dictated by the presence of the shock system and shear layers with large velocity gradients. The strong spatial heterogeneity is favourable to the instability development leading to intense velocity and pressure pulsations. The feedback generation causes development of the self-excited oscillations resulting in intense discrete components of the pulsation spectra.

The flow in a supersonic submerged jet is usually characterized by the following similarity criteria: the static pressure ratio, $n = p_a / p_\infty$, the Mach number at the nozzle cut-off, $M_0 = u_0 / a_0$, and the inclination angle of the nozzle contour at the outlet section, θ. Here p_a and p_∞ are the static pressure at the nozzle cut-off and in the surroundings respectively, u_0 is the outflow speed, and a_0 is the flow speed. Three regimes are differentiated: $n = 1$ corresponds to the rated regime, $n < 1$ to the overexpansion regime, and $n > 1$ to the underexpansion regime.

A supersonic submerged jet at unrated outflow regimes is characterized by the shock system at its gas-dynamic region. The presence of the almost periodical shock system at the gas-dynamic region of supersonic unrated jets causes the impact pressure wave-like change along the jet axis. The initial gas-dynamic region is followed by the transition region and further the main region with anisobaric flow and the velocity maximum at the jet axis as in the case of a regular subsonic isobaric turbulent jet. By way of example Fig. 7.1 presents experimental values of the pressure along the axis of a circular submerged jet at $M_0 = 2.0$, $\theta = 10°$, and $n = 0.6, 1.0, 2.0$ characterizing the anisobarity degeneration process [7.6]. The measurements of the impact pressure along the supersonic jet axis were performed by means of the Pitot tube. Here the normal shock is generated in front of the

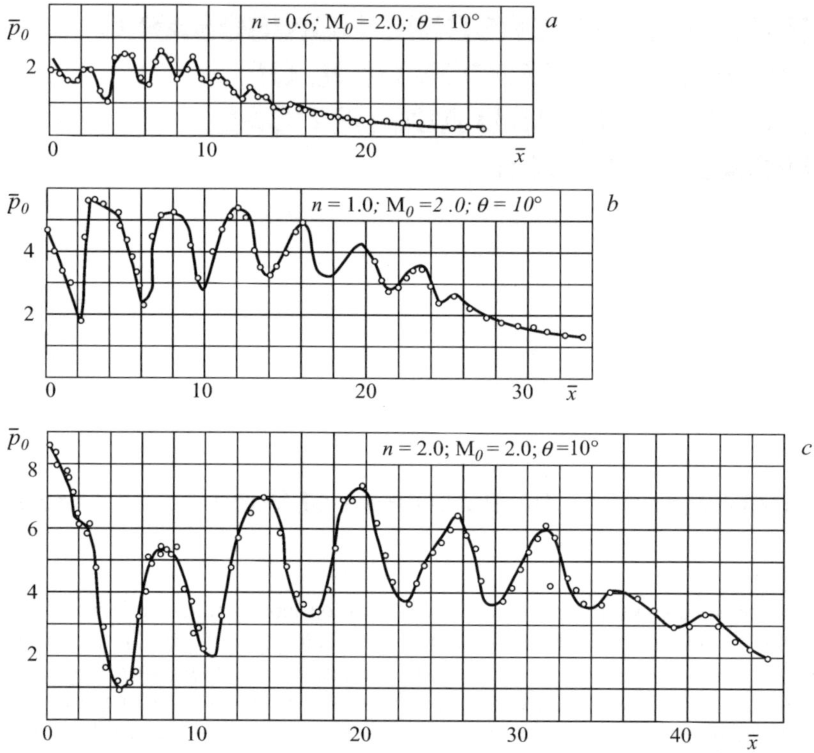

Fig. 7.1. The impact pressure behind the normal shock along the axis of a circular submerged turbulent jet for $M_0 = 2.0$, $\theta = 10°$, and $n = 0.6$, 1.0, 2.0.

measurement head, so the impact pressure values plotted as ordinates are reduced by the loss in the normal shock.

It should be noted that a different periodical structure, much as for subsonic jets, is realized in the initial region mixing layer for the case of supersonic jets issuing from the Laval nozzle at the rated regime where the shock system is absent. For example, a weak longitudinal acoustical irradiation at frequencies corresponding to $St_s = 0.2 - 0.8$ of a supersonic isobaric jet ($M_0 = 1.4$, $Re = 1.6 \cdot 10^6$) together with the flow visualization was used for experimental evaluation of the large-scale coherent structure parameters [7.15]. The changes in the mean velocity and the longitudinal velocity pulsations along the jet axis at $x/d = 0 - 20$ for the excitation-free and acoustically excited jets were practically identical. The analysis of the shadow photos for first six calibers has revealed periodical structures with the frequency corresponding to the Strouhal number $St = 0.4$ and with the phase velocity $u_c / u_0 = 0.7$ for this value of the Strouhal number.

The aerodynamic problems for supersonic anisobaric jets are considered in detail in [7.3, 7.4] where the corresponding numerical methods can be also found. The modeling of turbulent supersonic isobaric and anisobaric jets using the turbulence algebraic or differential model [7.1, 7.2] allows calculation of the mean velocity and pressure fields in such jets, as well as the velocity and pressure pulsation fields.

7.2. Mechanism of Noise Generation. Broadband Noise and Discrete Components

Noise of supersonic jets contains a number of components determined by sources of different kinds [7.5, 7.9, 7.11]: the mixing noise radiated by turbulent vortices in the mixing layer; the radiation generated by turbulent vortices carried away with supersonic speed relative to the environment; the broadband component of the shock noise generated as a result of interaction between the turbulence and the shock waves; and finally the discrete component resulting from the jet instability at some flow regimes. Another noise source of a supersonic jet is related to unstable oscillations of the velocity tangential discontinuity surface at the border of the jet supersonic part. As a result of the mathematical analysis of the noise generated by such a source, Sedelnikov [7.19] has arrived at a good correlation between the frequency of the spectrum maximum in the far acoustical field of a supersonic jet and the available experimental data.

The overall sound level of supersonic jets depends on the Mach number, M_0, the unrated degree, n, and the anisothermic degree that is equal to the ratio between the jet temperature and the environmental temperature. These parameters have a variable impact on the noise of different components creating complications for the experimental data analysis. When evaluating the mixing noise for an unrated outflow regime, the outflow speed of the equivalent jet having the same gas rate as the real jet but at the static pressure at the nozzle cut-off equal to the environmental pressure is used in place of the outflow speed of the real jet.

The most intensive sources of broadband noise are located in the jet transition region. The angle, φ, of the radiation maximum in the far field increases with the outflows speed and the jet temperature. For instance, the maximum radiation angle $\varphi \approx 25^\circ$ for a cold jet (M_0 = 1.5, T_{00} = 288 K) and $\varphi \approx 65^\circ$ for a high-temperature jet (M_0 = 3.5, T_{00} = 3000 K) [7.5]. It should be noted that the maximum acoustical radiation of the mixing noise corresponds to the angle range $\varphi \approx 30 - 60^\circ$ but the broadband noise resulting from the shock does not have any pronounced directivity. This noise intensity is determined by the shock intensity in the jet and is practically independent of the stagnation temperature at the nozzle cut-off. At the same time the mixing noise increases with the stagnation temperature. Therefore, the noise resulting from the shocks manifests itself to a greater extent in cold jets and at large angles □ when the mixing noise is relatively small.

The overall noise power, W, of a jet issuing from a supersonic nozzle at sufficiently high outflow speeds is proportional to the third power of the speed: $W \approx \rho_0 u_0^3 d^2$ and not to the eighth power as for subsonic jets [7.8, 7.11].

At unrated outflow regimes, besides the broadband noise, a discrete component could be generated with the frequency and intensity depending on the Mach number, M_0, the unrated degree, n, the anisothermic degree, the presence of the accompanying flow and other factors. In some instances the discrete component level is considerably higher than that of the broadband component and determines the overall noise level radiated by the jet. The discrete component was first discovered when studying jet outflow from a tapered nozzle at unrated outflow regimes [7.18]. A great many papers have been published on evaluating the component frequency dependency on the jet gas-dynamics parameters since then [7.5, 7.8].

It follows from the generalization of experimental results [7.14] that the discrete component in the noise spectrum of a supersonic jet is realized in a certain range of the unrated degree depending on the Mach number. The discrete component frequency monotonically decreases as the accompanying flow speed increases.

Discrete components with characteristics depending on the relative distance between the jets, $\Delta z / d$ (where Δz is the distance between adjacent jets), could emerge in the noise spectrum for a system of several parallel jets. The single jet structure is collapsed for sufficiently close arrangement of the nozzles ($\Delta z / d = 2$). As a result, the jet system does not need to radiate a discrete component. As indicated in [7.5], the discrete tone does not emerge for multi-nozzle systems with uniform arrangement of the nozzles at $\Delta z / d = 1.4 - 2.6$.

This fact should be kept in mind when studying Chap. 8 where acoustical characteristics of a system of supersonic jets are considered where the central jet is surrounded by peripheral jets with diameters an order of magnitude less than the central jet diameter. The discrete component in such a system for $n < 1$ and $n > 1$ is suppressed or shifted in the high-frequency region.

7.3. Acoustical Excitation of Supersonic Jets. Active Control

We consider some results of the experimental study on effects of an outer source of acoustical disturbances on supersonic anisobaric jets. A jet issuing from a conical nozzle and excited by a gas-jet radiator ($M_0 = 2.0$, the unrated degree $n = 0.5 - 2.0$, $f = 5 - 11$ kHz, $d = 0.02$ m) was studied by means of shadow photography [7.16]. The radiators were located at different distances from the jet axis. The sound pressure level at the nozzle cut-off $L = 156$ dB. Basically, the jet border nearest to the radiator was irradiated. Fig. 7.2 presents the semiopening angle, α_0, of the jet near border as a function of the sound relative intensity that is equal to the ratio between the sound intensity in the falling wave at the nozzle edge without the jet and the total pressure in the jet at the nozzle cut-off, $\langle p_s'^2 \rangle^{1/2} / p_0$. The

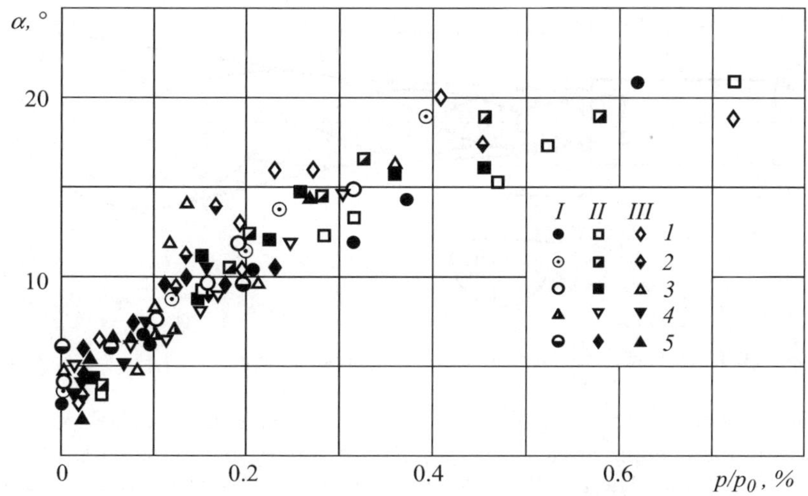

Fig. 7.2. The semiopening angle of the jet near border as a function of the transversal acoustical irradiation intensity for various *f* and *n*.

numbers *I*, *II*, and *III* correspond to f_s = 6.5, 8.5 and 11.8 kHz and the numbers *1* – *5* to the unrated degree n = 0.5, 0.7, 1.0, 1.5 and 2.0. It is significant that the influence of the outer excitation frequency on the supersonic jet expansion in the above experiments was not discovered.

The analysis of the interaction dynamics by the film has revealed the absence of the jet oscillations. The interaction between the outer source sound and the supersonic jet is accompanied by the intensive jet sound propagating at 20 – 30° to the jet axis with the outer source frequency. The fact that only the jet near border was noticeably irradiated in the above experiment is possibly related to the insufficient sound level acting on the far border.

It is significant that the above experiment has not revealed any noticeable sound effects on the existing disturbances. These disturbances emerge at the supersonic jet border near the nozzle outlet section under the action of an outer sound source.

To clarify the possibilities for control of the noise structure and spectrum in supersonic jets, the changes in the jets under high-intensity acoustical excitation at different frequencies were studied [7.21]. The jets were issued from axisymmetric and plane nozzles at M_0 = 1.2 – 2.5. Their kinetic energy varied in the range W_0 = 1.74 – 47 kW. Two gas-jet radiators of power W_s = 140 and 320 W that overlapped the frequency range from 10 to 19.5 kHz generated the noise. The noise sources were arranged at the focus of the elliptic concentrator O_1 with the eccentricity 0.5. The second focus was directed at the jet axis point O_2 (Fig. 7.3) offset by the distance l from the nozzle cut-off. The sound pressure at the point O_2 was equal to 170 – 176 dB.

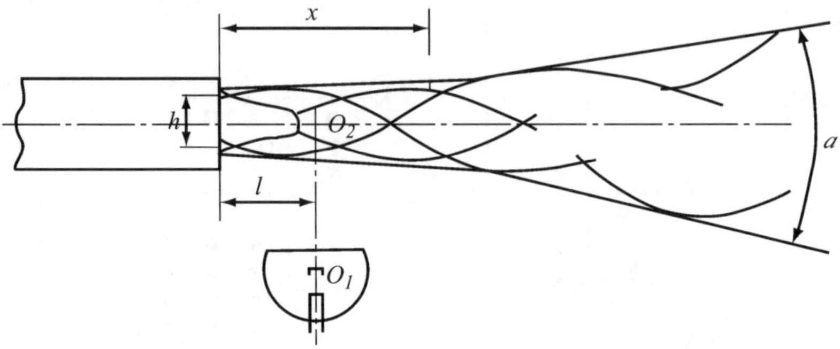

Fig. 7.3. The scheme of acoustical action on a supersonic jet.

The acoustical effect was evaluated by the changes in the jet noise spectrum, as well as in the opening angle of the jet main region, α, and in the length, x, of the supersonic region. The results show that the nature of the transversal acoustical excitation of a supersonic jet considerably depends on the ratio between the radiator acoustical power and the jet kinetic energy, W_s / W_0, as well as on the relative length, l/h, where h is the width or diameter of the nozzle. It turns out that for $W_s / W_0 > 1.5\%$ and $l/h = 0.6$ the frequency capture, i.e. the transversal oscillations of a plane supersonic jet with the frequency equal to that of the outer action, happens in the whole range under study: 10 – 20 kHz. Therefore, there is only the component with the outer radiation frequency in the noise spectrum.

In fig. 7.4, *a* depicts the noise spectrum of a free supersonic jet issuing from a plane nozzle at $M_0 = 2.0$. The spectrum contains the discrete frequency $f = 8$ kHz and multiple harmonics. The noise spectrum of the jet under the outer noise action at the frequency $f = 11.5$ kHz (Fig. 7.4, *b*) and $f = 18.7$ kHz (Fig. 7.4, *c*) for the ratios $W_s / W_0 = 5\%$ and $l/h = 0.6$ contains the component with that frequency.

Fig 7.5 presents the opening angle of the main region of an axisymmetric jet as a function of the prechamber pressure, p_0, for different action frequencies, f (Fig. 7.5, *a*), and the relative length of the supersonic region for different action frequencies (Fig. 7.5, *b*).

The efficiency of the outer radiation effect on supersonic jets decreases as l/h increases. This fact is illustrated by the functions in Fig. 7.6 for a plane jet ($p_0 = 3.4$ atm, $f = 18.7$ kHz). This conclusion conforms with the data in [7.17]. According to that paper, the effect of transversal acoustical excitation of a supersonic jet is most pronounced for acoustical irradiation of the nozzle. The supersonic jet shock wave structure could be collapsed under saw-tooth sound waves resulting in considerable changes of the sound irradiation. In this case (M = 2, n = 0, $f_s = 8.5$ kHz and $f_s = 11.8$ kHz) it is shown that the increase of the wide-band noise by 10 dB occurs in the direction of the maximum irradiation in the frequency range near the maximum of the irradiated acoustical power.

In summary, consider the experimental data on acoustical characteristics of the far field in supersonic turbulent jets issuing from a conical and Laval nozzles, i.e. the jets are underexpanded and rated [7.12]. $M_0 = 1.22$ and 1.19, $T_0 = 300$ K for

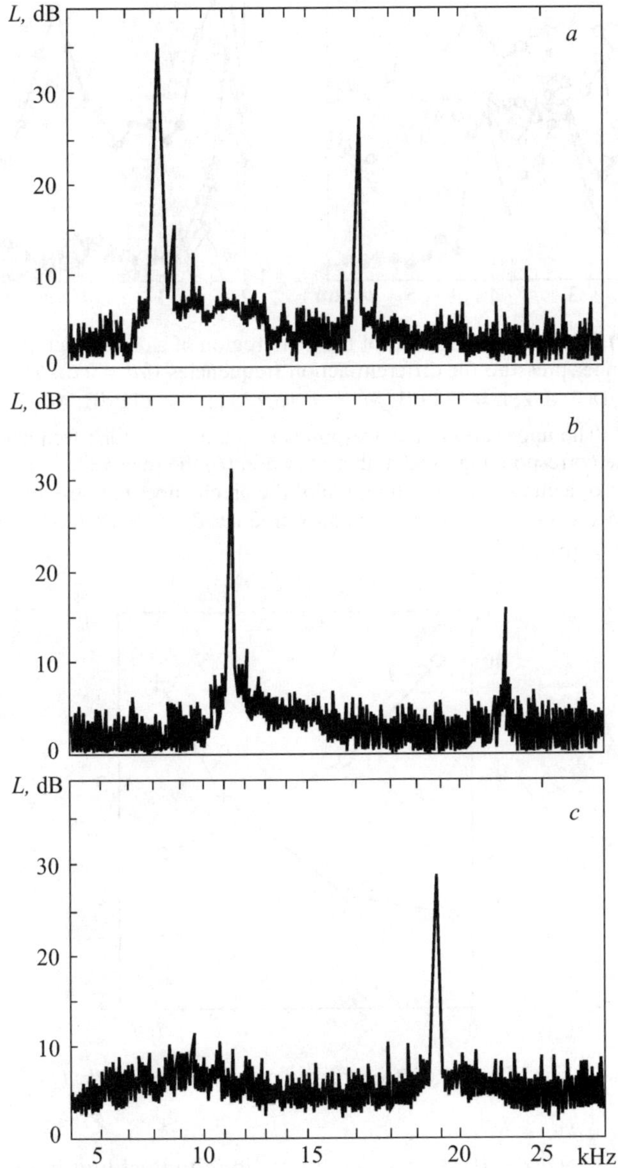

Fig. 7.4. The noise spectra of a supersonic jet.

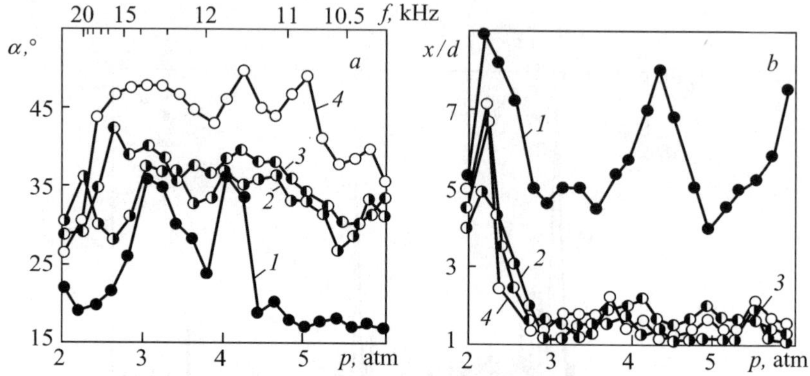

Fig. 7.5. (*a*) The opening angle, α, of the main region of a circular jet as a function of the prechamber pressure for different action frequencies (l/d = 0.6). *1)* an unexcited jet; *2)* f_s = 18.7 kHz, L = 174 dB; *3)* f_s = 16.7 kHz, L = 171 dB; *4)* f_s = 11.5 kHz, L = 168 dB. The upper abscissa corresponds to the discrete tone frequency of a free jet under the corresponding prechamber pressure. (*b*) the relative length of the supersonic region of a circular jet as a function of the prechamber pressure for different action frequencies (l/d = 0.6). *1)* an excitation-free jet; *2)* f_s = 18.7 kHz; *3)* f_s = 16.7 kHz; *4* – f_s = 105 kHz.

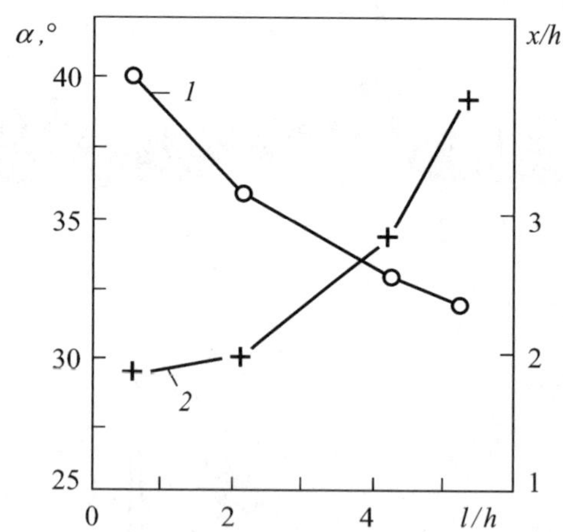

Fig. 7.6. The values of α and x/h for a plane jet as functions of l/h. *1)* $\alpha = F_1(l/h)$, *2)* $x/h = F_2(l/h)$.

both cases. The jets (d = 0.043 m) were under longitudinal low-frequency acoustical excitation with Strouhal number St_s = 0.47. Fig. 7.7 depicts the noise one-third octave spectra in the far field of the jets for φ = 30 – 120°. Two facts are noteworthy. First, as for the case of subsonic jets, one can observe here the broadband noise amplification that is most pronounced for St > St_s. Second, the excitation frequency does not manifest itself in the jet noise spectra.

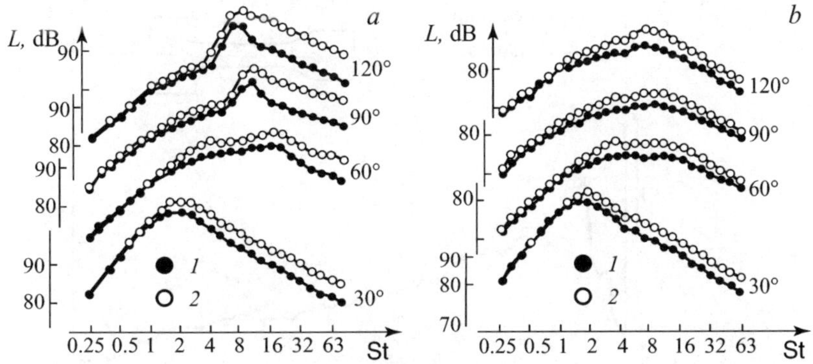

Fig. 7.7. (*a*) The noise one-third octave spectra of an underexpanded jet ($M_0 = 1.22$, $T_{00} = 300$ K) under low-frequency longitudinal acoustical excitation for $\varphi = 30°$-$120°$; *1)* an excitation-free jet; *2* – $St_s = 0.47$. (*b*) The same data for a rated jet ($M_0 = 1.19$, $T_{00} = 300$ K) for $\varphi = 30°$-$120°$; *1)* an excitation-free jet; *2)* $St_s = 0.46$.

7.4. Control of Jet Parameters Using Jet Noise Screening. Passive Control

The results of the previous section demonstrate that the acoustical excitation of a supersonic jet could cause a considerable change of the jet characteristics. The efficient acoustical action on a supersonic jet can be achieved also without outer sound sources using the sound radiated by the jet. If there are reflecting objects (the nozzle edge of finite thickness could take the part of such an object), the sound reflected from them acts on the jet causing the change in its dynamic characteristics. The action efficiency increases when using semispherical reflectors resulting in the concentration of the reflected sound at the jet.

The alternative approach to the jet acoustical excitation could be realized when placing the resonance cavity in the area of the jet output section. The jet issuing from the nozzle excites in the cavity resonance oscillations close to the pure tone that act on the jet and change its aeroacoustical characteristics. The detailed experimental studies on the effect of acoustical excitation generated by the cavity on the parameters of a supersonic rated jet issuing from the Laval nozzle with the speed corresponding to M = 2 are described in [7.22]. The variation of the cavity size permits changing the acoustical action frequency.

Fig. 7.8 presents the comparison of the pressure pulsation spectra in the jet near field with and without an exciting cavity (the cavity arrangement scheme is depicted in the figure insertion). The resulting acoustical excitation of the jet could cause considerable thickening of the mixing layer (Fig. 7.9).

The study of supersonic underexpanded jets [7.10] demonstrated that the regular increase in the dynamic pressure at a fixed point at the jet axis (in the subsonic

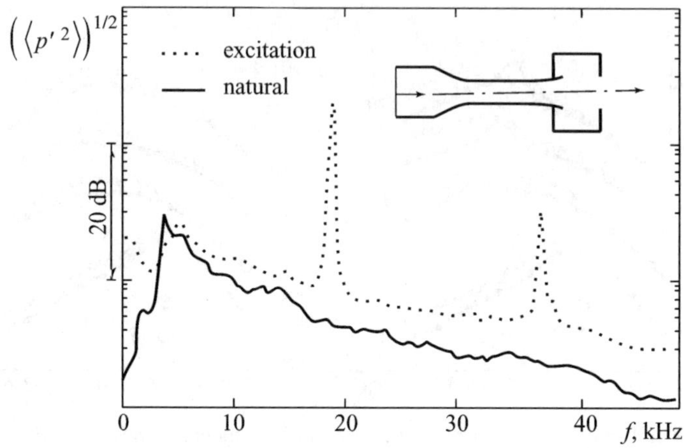

Fig. 7.8. Excitation of a supersonic jet using a resonance cavity. Deformation of the pressure pulsation spectra.

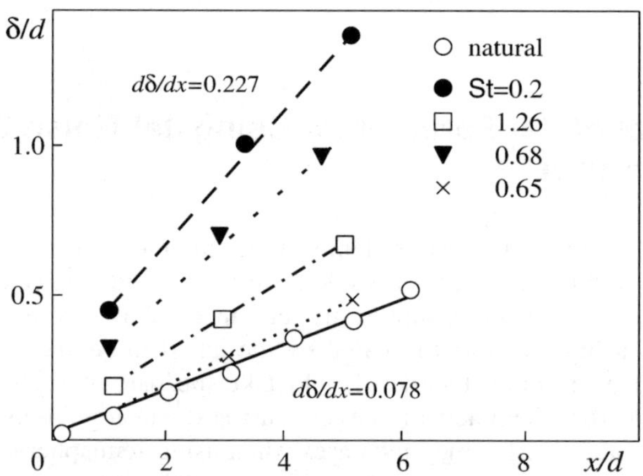

Fig. 7.9. Thickening of the mixing layer of a supersonic jet under resonance excitation for different values of St.

turbulent flow region, $x/d = 34$) with the receiver impact pressure breaks down in some cases (Fig. 7.10). When a flap inclined 45° to the jet axis is arranged near the nozzle, these violations (the abrupt decrease of the impact pressure) are amplified and the jet emits a piercing sound of high tone. Hence, the relation between the changes in the aerodynamic and acoustical jet characteristics testifies that it could be determined by the reflected sound waves. This is supported by the following experiment. The above abnormality disappeared upon cladding the reflecting flap with sound-absorbing lagging and the jet squeal was not heard.

The system 'nozzle-receiver' in the above experiments had different surfaces playing the role of acoustical reflectors. Fig. 7.11 presents the dynamic pressure at the jet axis (for $x/d = 20$) as a function of the nozzle impact pressure for different

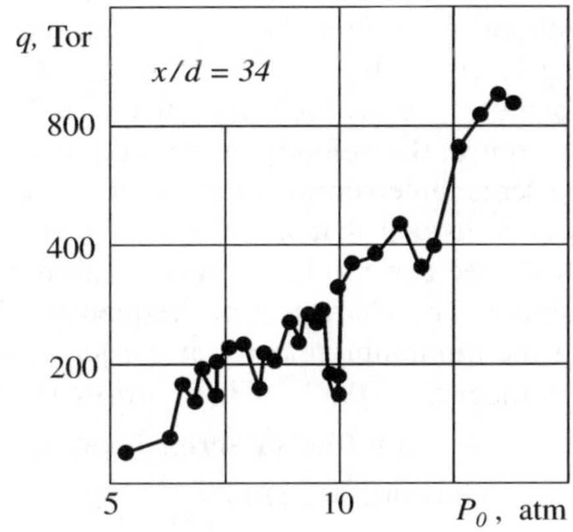

Fig. 7.10. The dynamic pressure at the jet axis in the main region (x/d = 34) as a function of the receiver impact pressure.

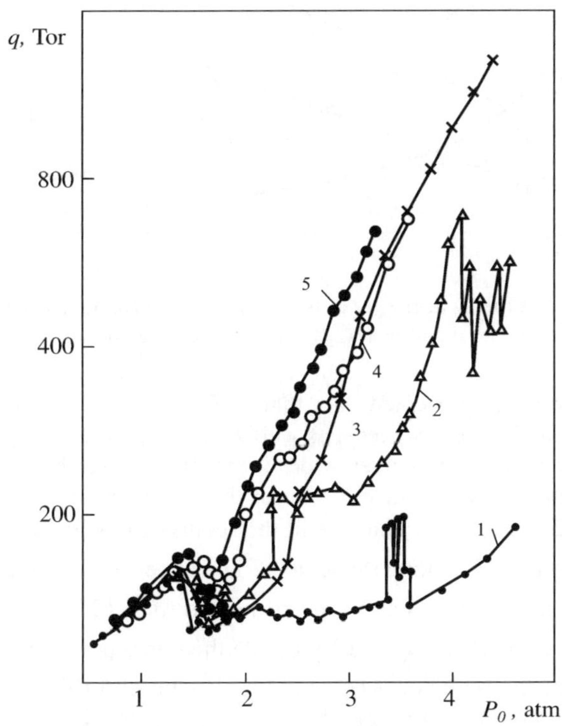

Fig. 7.11. The dynamic pressure at the jet axis (x/d = 20) as a function of the receiver impact pressure (screens without sound-absorbing lagging, with sound-absorbing lagging, and with a hole for the jet).

kinds of screen: *1)* without absorption; *2)* the receiver and the nozzle are coated with a sound-absorbing layer; *3)* the screen with sound-absorbing lagging placed in the nozzle cut-off plane, $x = 0$; *4)* the screen with a hole for the jet at $x/d = 2$; *5)* the same at $x/d = 4$. Hence, the velocity at the jet axis depends on the position of the sound-absorbing screen interrupting or attenuating the acoustical feedback.

The shadow shots demonstrate that the effect of acoustical feedback on the shock structure manifests itself in the faster degeneration of the cell structure of the jet gas-dynamic region. The characteristic frequency of the discrete tone favourable for obtaining the minimum velocity at the jet axis is $f = 10$ kHz corresponding to Strouhal number $\mathrm{St}_s \approx 0.32$ for the outflow speed $u_{0i} = 443$ m/s and the nozzle diameter $d_i = 14.1$ mm (the subscript i corresponds to the equivalent nozzle parameters for the rated outflow regime).

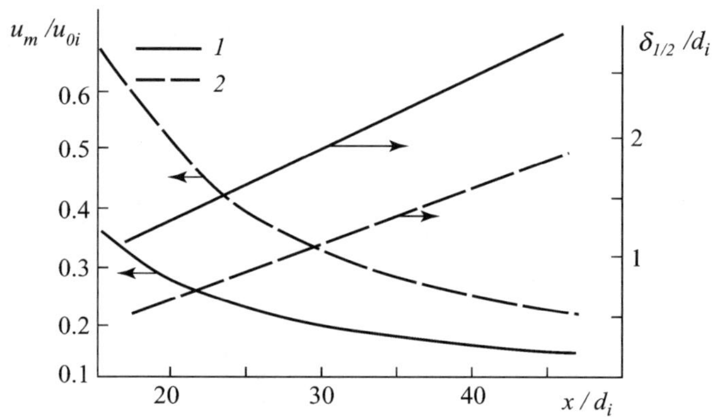

Fig. 7.12. The velocity at the jet axis and in the jet effective thickness as functions of x/d_i. *1)* with the reflector, *2)* without the reflector.

It is clear from the presented data that different kinds of screens and sound-absorbing lagging for these screens, as well as their position along the jet axis can change the jet characteristics over wide limits. Below are the results of tests for the simplest plane ring reflector placed near the nozzle cut-off. For this type of reflector Fig. 7.12 presents the changes in the relative velocity, u_m/u_{0i}, and the jet effective radius, $\delta_{1/2}/d_i$, along the jet axis. The experiments are performed for an underexpanded jet issuing from a tapered nozzle ($d = 12.7$ mm, $d_i = 14.1$ mm, $u_{0i} = 443$ m/s, $\mathrm{Re} \approx 10^6$). Fig. 7.12 presents the corresponding dependencies for two cases: where the reflector is covered with sound-absorbing lagging and without it. The main conclusion is that the acoustical feedback in supersonic unrated jets under certain conditions could decrease the velocity at the jet axis in the main region by 50 – 80% for $x/d_i = 20 - 50$ and correspondingly increase the jet effective thickness by a factor of 2 – 3.

Since a standing wave appears in the outer acoustical field between the source of the main tone radiation and the ring reflector placed near the nozzle cut-off, the reflector displacement from the nozzle cut-off upstream influences the structure of the standing wave field and on the intensity of the discrete tone radiation. Simultaneously the jet aerodynamic characteristics change also. This phenomenon is illustrated in Fig. 7.13 by presenting dependencies of the total pressure p_{00} / p_∞ at the fixed point ($x/d = 9$) at the axis of an overexpanded jet ($M_0 = 2$, $n = 0.64$) and the sound pressure level, L, near the nozzle cut-off ($x = 0$, $y/d = 1.25$) on the upstream displacement of the ring reflector ($x/d < 0$) [7.5]. The experiments demonstrate that the length of the jet supersonic region decreases when the crest of the discrete component standing wave falls on the jet base and increases when the node falls there. As the discrete component level increases, the length of the jet supersonic region linearly decreases (the maximum difference between the maximum and the minimum supersonic region lengths could be as much as 3.5).

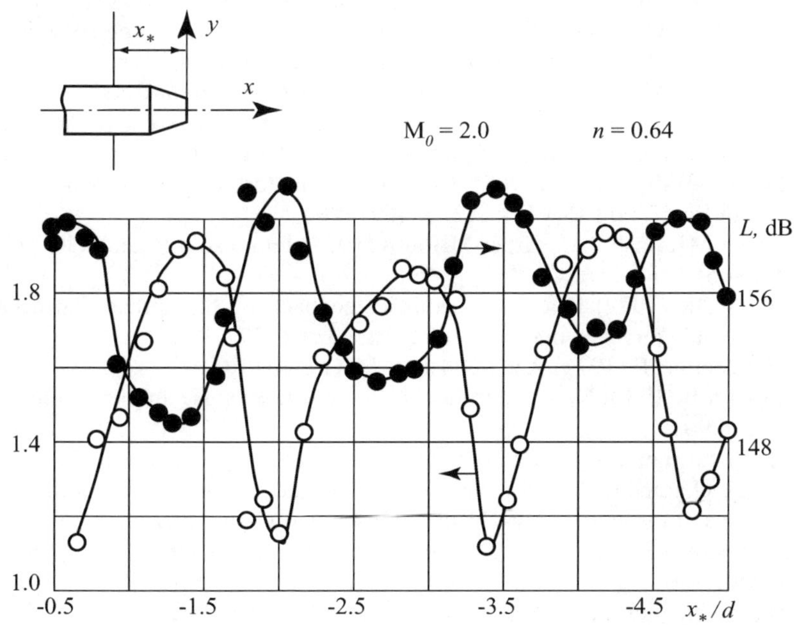

Fig. 7.13. The total pressure p_0 / p_∞ at the axis ($x/d = 9$) of an overexpanded jet . ($M_0 = 2$, $n = 0.64$) and the sound pressure level, L, near the nozzle cut-off as functions of x_* / d

Other studies of control of supersonic jets using different kinds of a screen are described in [7.13, 7.14, 7.20].

REFERENCES

7.1. Abalakin IV, Antonov AN, Antonov MA, Chetverushkin BN (2000) Numerical simulation of supersonic jets using kinetical-consistent finite difference schemes (in Russian). Math Modeling, 12, # 1: 25 – 37

7.2. Abalakin IV, Antonov AN, Antonov MA, Kozubskaya TK, Chetverushkin BN (2001) Application of kinetically consistent finite difference schemes for turbulent supersonic jet noise prediction (in Russian). Math Modeling, 13, # 10: 56 – 76

7.3. Abramovich GN, Girshovich TA, Krasheninnikov SYu, Sekundov AN, Smirnova IP (1984) Theory of turbulent jets (in Russian). Nauka, Moscow

7.4. Avduevsky VS, Ashratov EA, Ivanov AV, Pirumov UG (1989) Gas dynamics of supersonic anisobaric jets (in Russian). Mashinostroyenie, Moscow

7.5. Antonov AN, Kuptsov VM, Komarov VV (1990) Pressure pulsations in separated and jet flows (in Russian). Mashinostroyenie, Moscow

7.6. Antsupov AV, Blagosklonov VI (1976) On structure of a supersonic jet issuing into a submerged space (in Russian). Proc TsAGI, # 1781

7.7. Antsupov AV, Pimstein VG (1975) On radiation of a discrete tone by a supersonic jet issuing from a conic nozzle (in Russian). Izvestiya Academy of Sciences of the USSR, Fluid and Gas Mech, 1: 153 – 156

7.8. Ffowcs Williams JE (1963) The noise from turbulence convected at high speed. Phyl Trans Roy Soc A225: 469 – 503

7.9. Fisher MJ, Lash PA, Harper MBourn (1973) Jet noise. J Sound Vibr, 28, # 3: 563 – 587

7.10. Glass DR (1972) Effects of acoustic feedback on the spread and decay of supersonic jets. J Sound and Vibration, 23, # 1: 77 – 99

7.11. Goldstein ME (1976) Aeroacoustics. McGraw-Hill

7.12. Jubelin B (1980) New experimental studies on jet noise amplification. AIAA Pap, # 961

7.13. Kapustin YeA, Shlik OE, Yevchenko VN (1980) Acoustical effect on dynamic characteristics of a supersonic noncalculated jet (in Russian). In: Mechanics of turbulent flows. Nauka, Moscow, pp 259 – 269

7.14. Kapustin YeA, Lentsov IA, Neshcheret PA, Shlik OE (1981) Acoustical feedback of supersonic jets for different flow regimes (in Russian). In: Gas aeromechanics and elasticity theory, #28. Dnepropetrovsk, pp 23 – 27

7.15. Lepicovsky J, Ahuja KK, Brown VH, Burrin RH (1985) Coherent large-scale structures in high Reynolds number supersonic jet of mach number 1.4. AIAA Pap, # 1941

7.16. Pimshtein VG (1989) Effect of sound on a supersonic jet. Fluid Dynamics, 24, # 6: 969 – 971

7.17. Pimshtein VG (1999) Aeroacoustical interactions in turbulent jets. PhD Thesis, TsAGI

7.18. Powell A (1953) On the mechanism of choked jet noise. J Acoust Soc Am, 25, # 3: 385 – 389

7.19. Sedelnikov TH (1967) On frequency spectra of supersonic jets (in Russian). In: Physics of aerodynamic noises. Nauka, Moscow, pp 83 – 87
7.20. Shlik OE, Neshcheret PA, Ryabukhin AV, Serebriakov VA, Kort TG (1992) Study of the acoustical feedback mechanism in supersonic jets (in Russian). In: Scientific Basis for Turbulent Phenomena. Nauka, Moscow, pp 62 – 67
7.21. Yeriomin GI, Kondratiev VI (1980) Acoustical effects on the flow and spectrum of supersonic jets. In: Aeroacoustics. Nauka, Moscow, pp 119 – 123
7.22. Yu KH, Schadow KC (1997) Role of large coherent structures in turbulent compressible mixing. Experimental Thermal and Fluid Sci, 14: 75 – 84

CHAPTER 8 REDUCTION OF TURBOJET ENGINE NOISE

The urgency of the problem of jet noise reduction has been rising again at the present stage of aircraft technology development. The main reasons are the following. The studies of attenuation of turbo-machine noise have demonstrated that the jet is one of the main noise sources even in engines with high by-pass ratios. Moreover, up-to-date passenger airplanes often use engines with low by-pass ratios where the jet noise contributes predominantly to the total noise of the power engine.

Hence, studies on development of efficient approaches to reduction of reactive jet noise are of great interest. For this purpose, the so-called passive and active methods of noise reduction are used. The former methods center around reduction of noise on its way from the source. The possibilities for passive methods implementation are rather limited due to engineer difficulties associated with the sound-absorbing unit mounting near the reactive jet. The attempts of jet noise reduction using screening capabilities of aircraft wing and fuselage or ejector noise silencers with the ejector sound-absorbing lining should also be noted.

The known active methods of reactive jet noise reduction are based on the change in aerodynamic characteristics of the mixing layer within the jet initial region, for which purpose, for instance, a coaxial jet with the high-speed central jet and the lower-speed annular jet is formed causing the shear stresses to decrease. The currently developed method of reactive jet noise reduction [8.9] based on the forming of a coaxial jet with the "reverse" velocity profile of the turbojet engine exhaust where the speed in the outer contour is larger than that in the inner one is very promising. The jet noise reduction by thc change in the jet aerodynamic characteristics within the jet initial region is achieved in specific cases by means of thin transverse jet injections into the main jet near the nozzle outlet section [8.9]. These thin jets generate the flow circumferential nonuniformity ultimately attenuating the coherent structures that are the important source of jet noise.

The use of these methods provides in a number of cases the appropriate noise reduction of the reactive jet. Howcver, one has to introduce important alterations in the operating scheme of the power plant gas generator or in the engine exhaust system structure. It should be also noted that the use of multi-nozzle and lobed noise silencers causes substantial thrust loss comprising, on average, 1% for 1 dB of noise reduction.

8.1. The Acoustical Silencer of Turbojet Noise

Studies performed in recent years [8.5, 8.14] directed radically new acoustical ways to turbulent jet noise reduction. Among other things it was established that noise excitation could considerably influence aerodynamic and acoustical characteristics of turbulent jets. For instance, low-frequency noise excitation of a jet with Strouhal number $\mathrm{St}_s = f_s d / u_0 = 0.2 - 0.5$ (here d is the nozzle outlet section diameter, u_0 is the jet issue speed, and f_s is the exciting noise frequency) causes the intensification of the turbulent mixing in the initial region, and an increase in the characteristic turbulence scale in the mixing layer, resulting in an increase in the noise generation. The high-frequency noise excitation with $\mathrm{St}_s = 2 - 5$ causes the reverse effect, i.e. attenuation of the turbulent mixing in the initial region, and a decrease in the characteristic turbulence scale in the mixing layer, resulting in a decrease in the jet broad-band noise.

Thus, the idea of the employment of jet sound excitation to decrease the jet noise is appealing. However, some difficulties emerge related to the necessity of mounting a high-frequency noise radiator on the engine.

The present chapter contains studies on acoustical characteristics of model and nature reactive jets under noise excitation generated by several parallel thin jets situated around the main jet. The nozzle diameters of the thin jets are one order of magnitude less than the main jet nozzle diameter and the outflow speed for the thin jets is equal to that for the main jet. Such a jet system could be realized for both the main jet and supplementary peripheral jets from the same receiver (Fig. 8.1, *a*).

The possibility of noise reduction in such a system with respect to the initial single jet noise is based on the following considerations. It is well known [8.5, 8.9, 8.14] that the maximum of the turbulent jet noise corresponds to Strouhal numbers $\mathrm{St} = fd / u_0 = 0.2 - 0.5$. This relation is valid equally for the main jet (subscript 1) and peripheral jets (subscript 2), i.e.

$$\frac{f_1 d_1}{u_{01}} = \frac{f_2 d_2}{u_{02}} = 0.2 - 0.5. \tag{8.1}$$

We have $u_{01} = u_{02}$ for identical issue speeds of the main and peripheral jets, therefore $f_2 = f_1 d_1 / d_2$.

If the peripheral jet diameter is 10 times less than that of the main jet ($d_2 = d_1 / 10$) then $f_2 = 10 f_1$. Therefore, the Strouhal number determined from the diameter and speed of the main jet and from the exciting sound frequency f_2 equals

$$\mathrm{St}_s = \frac{f_2 d_1}{u_0} = 10 \frac{f_1 d_1}{u_0} = 2 - 5\,, \tag{8.2}$$

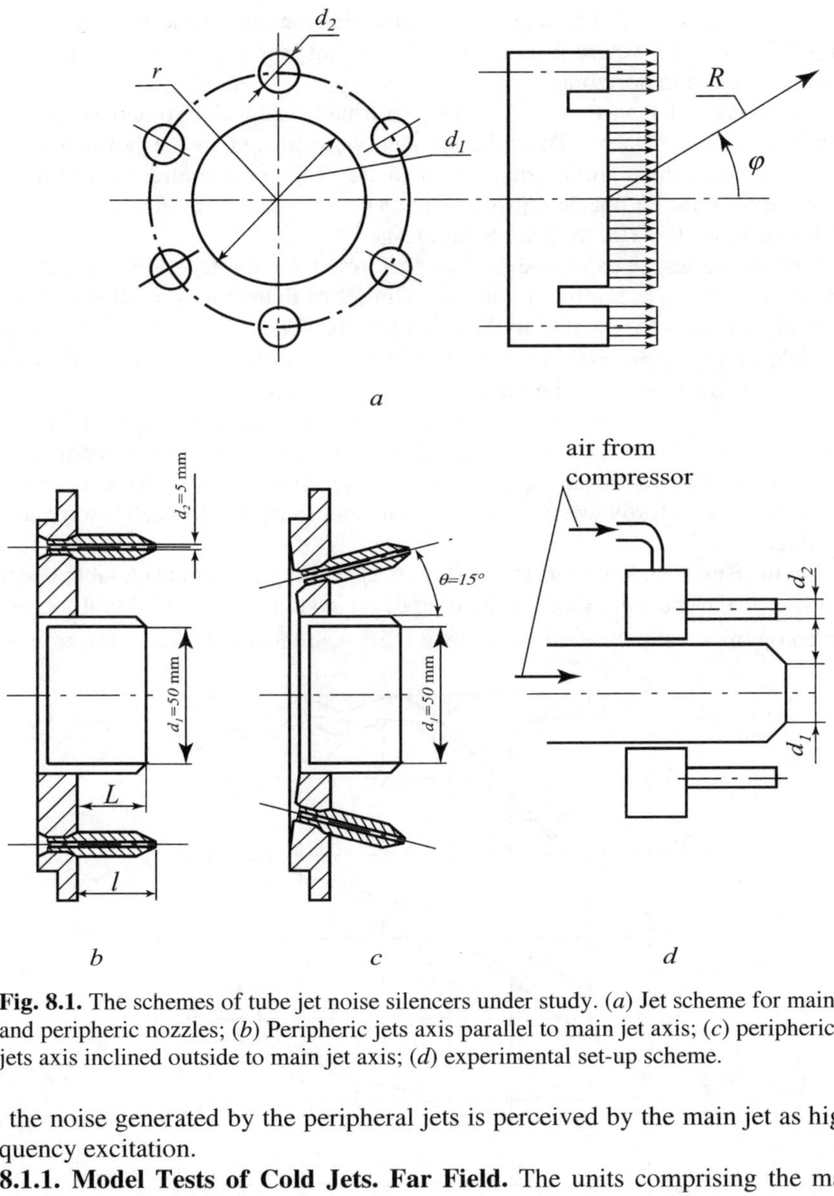

Fig. 8.1. The schemes of tube jet noise silencers under study. (*a*) Jet scheme for main and peripheric nozzles; (*b*) Peripheric jets axis parallel to main jet axis; (*c*) peripheric jets axis inclined outside to main jet axis; (*d*) experimental set-up scheme.

i.e. the noise generated by the peripheral jets is perceived by the main jet as high-frequency excitation.

8.1.1. Model Tests of Cold Jets. Far Field. The units comprising the main nozzle and several peripheral jets were arranged along a circumference around the main nozzle are studied (Fig. 8.1, *b*). The acoustical efficiency of such units was evaluated in anechoic chambers for a small-scale model with d_1 = 40 and 50 mm for cold jets with the ratios between the impact pressure at the nozzle cut-off and the static pressure $\pi_c = 1.2 - 2.4$, for a large-scale model (d_1 = 80 mm) with hot jets ($T_{00} \leq$ 475 K) and $\pi_c = 1.4 - 2.1$, as well as for a real engine mounted on an open acoustical stand. The engine operating regimes for these tests were varied from $0.5N$ to $1.2N$ where N is the thrust with respect to a nominal regime. The main jet axes were horizontal.

The acoustical measurements were performed using electro-acoustical equipment "Brüel and Kjaer". The microphones were placed in the horizontal plane passing through the main nozzle axis on an arc of the circumference with the center at the axis in the nozzle cut-off plane. The arc radius was $R = 1 - 2$ m at the model tests and $R = 100$ m at the nature tests.

The first series of experiments was performed for models with six peripheral jets of diameter $d_2 = 5$ mm around the main jet of diameter $d_1 = 50$ mm. The peripheral jet axes were parallel to the main jet axis (Fig. 8.1, *b*) or were arranged at an angle of 15° to the main jet axis (Fig. 8.1, *c*) in order to minimize the aerodynamic interaction between the main and peripheral jets.

The plane of the peripheral nozzle cut-offs could shift with respect to the plane of the main nozzle cut-off retaining parallelism of the main and peripheral jet axes (Fig. 8.1, *d*). The air for all nozzles was supplied from a single receiver, with the result that the outflow speeds for the main and peripheral nozzles were almost identical.

The measurements demonstrated (Fig. 8.2) that the most considerable decrease in noise of the main jet (and of the overall jet system) with peripheral jets is observed on the rays inclined at an angle $\varphi = 30^\circ$ to the jet axis, i.e. in the area of the

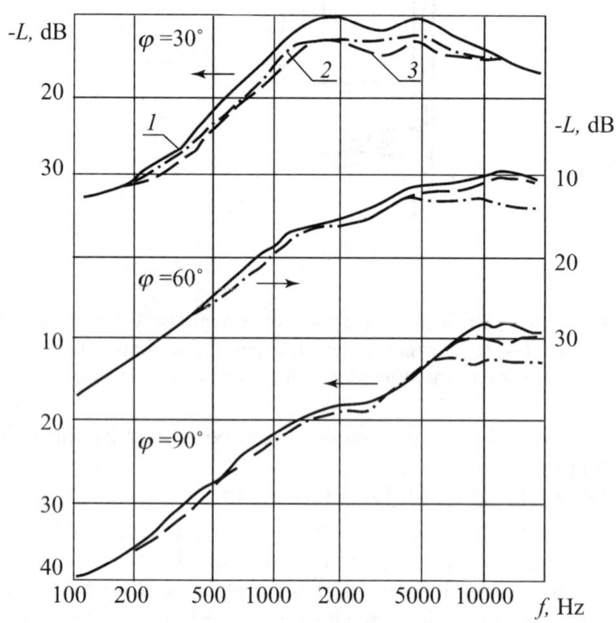

Fig. 8.2. The noise spectra in the jet far field. *1)* the initial jet; *2)* 6 peripheral nozzles, $\theta = 0$; *3)* 6 peripheral nozzles, $\theta = 15^\circ$.

jet noise maximum radiation. The maximum noise reduction is observed at higher frequencies as φ increases.

The noise reduction performance for parallel ($\theta = 0$) and inclined ($\theta = 15°$) arrangement of the peripheral tubes (nozzles) and for different length of the tubes is changed moderately lending support to the hypothesis on moderate influence of pure aerodynamic interaction between the main and peripheral jets on the jet noise.

The measurements performed for three peripheral jets (the other three were plugged) demonstrate that the noise reduction effect is conserved for this case also without any influence of the azimuthal arrangement of running jets on the noise reduction magnitude. For example, the positioning of the peripheral jets on the side of the microphone mounting gives almost the same effect as their positioning on the opposite side (Fig. 8.3). This testifies that the observed noise reduction could not be related to the noise screening and/or scattering by peripheral jets.

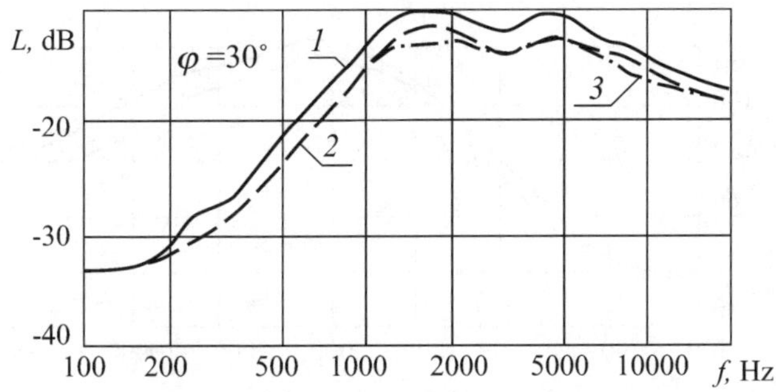

Fig. 8.3. The noise spectra in the jet far field ($\varphi = 30°$) for different azimuthal arrangement of peripheral jets ($\theta = 15°$, $\pi_c = 2.4$). *1)* the initial jet; *2)* 3 peripheral jets on the side opposite to the microphone; *3)* 3 peripheral jets on the side of the microphone mounting.

At the same time the increase in the number of peripheral jets down to one or two is accompanied by abrupt increase in acoustical performance of the unit. It seems likely that realization of the aeroacoustical interaction needs a minimum level of the sound pressure generated by peripheral jets acting on the main jet [8.5, 8.14].

The measurements of the flow mean velocity performed using the Pitot tube placed at the main jet axis at the distance from the nozzle cut-off $x/d_1 = 8$ show in full accordance with the paper [8.5] that high-frequency sound excitation of a jet causes some increase in the mean velocity (approximately by 3%).

Hence, the experiments have demonstrated that the proposed unit based on the principle of aeroacoustical interaction provides some decrease in the turbulent jet noise. The decrease in the area of maximum radiation ($\varphi = 30°$) is as much as 2 – 3 dB in a wide range of radiation frequencies.

The models available in the above experiments gave no way of performing parametric studies aimed at the search of optimal arrangement of peripheral jets with respect to the main jet. To this end the second series of experiments was performed for the model having from two to six peripheral nozzles of diameter $d_2 = 2-6$ mm arranged around the main nozzle of diameter $d_1 = 40$ mm. The distance between the axes of the main and peripheral nozzles was r = 40, 50 and 60 mm. For the most part measurements were performed for the peripheral nozzle cut-offs placed in the plane of the main nozzle cut-off; some measurements were performed for the case when the plane of the peripheral nozzle cut-offs was shifted downstream by 40 mm with respect to the plane of the main nozzle cut-off.

The analysis of the pressure pulsation spectra in the far field shows that in this case also the presence of peripheral jets causes the decrease in the sound pressure levels in the whole range of frequencies with the maximum decrease in the area of high frequencies (Fig. 8.4). The measurements for different r have shown that the maximum decrease in noise for r = 40 – 65 mm corresponds to the minimum value of r and to the maximum value from the diameter range $d\ = 2-6$ mm.

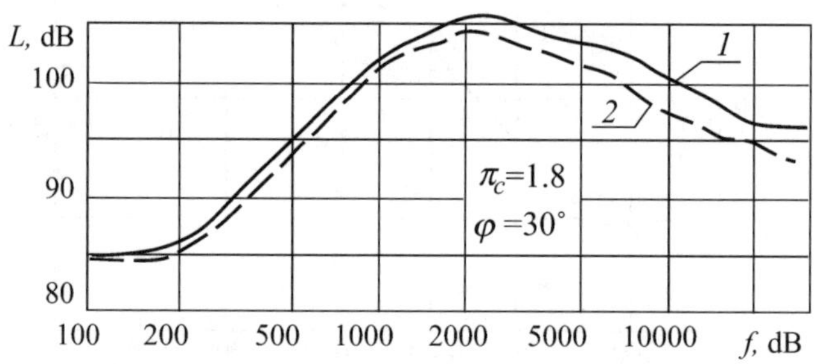

Fig. 8.4. The noise spectra in the jet far field ($\varphi = 30°$, $\pi_c = 1.8$, $n = 6$, $d_1 = 40$ mm, $d_2 = 6$ mm). *1)* the initial jet; *2)* the jet with 6 peripheral nozzles.

8.1.2. Model Tests of Hot Jets. Far Field. The measurements for hot jets were performed at models with the main jet issuing from a nozzle of diameter $d_1 = 80$ mm and with 4 – 12 peripheral jets of diameter $d_2 = 6-10$ mm. The distance, Δx between the planes of the main and peripheral nozzles was varied in the range –30 mm $\le \Delta x \le$ 30 mm (see Fig. 8.1, *d*). The distance, r, between the axes of the main and peripheral nozzles could be varied also.

The measurements have demonstrated that for hot jets (T_{00} = 470 K) the noise reduction effect is retained approximately the same as for cold jets (Fig. 8.5). The increase in the peripheral jet diameter amplifies the noise reduction effect in the same manner as the increase in the jet number. The maximum noise reduction was observed for the coincident planes of the main and peripheral nozzle cut-offs. The shift of the peripheral nozzles upstream or downstream from the main nozzle cut-off caused attenuation of the effect.

Hence, the model tests of the proposed unit lead one to conclude the following. The presence of peripheral jets causes the noise reduction in the jet system by 2 –3 dB with respect to the initial main jet noise. The noise reduction is observed in the

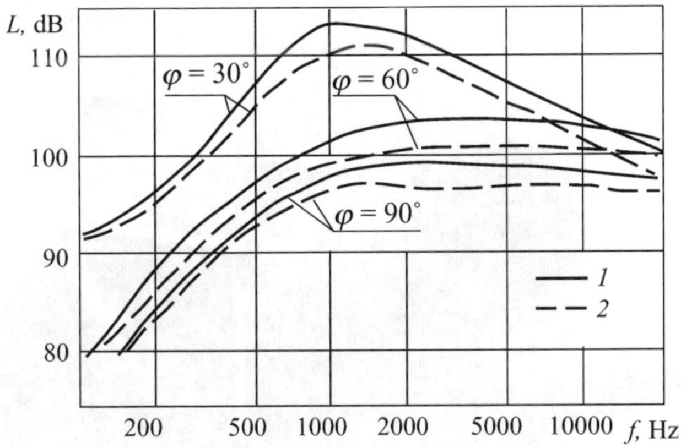

Fig. 8.5. The noise spectra in the hot jet far field (T_{00} = 470 K) with the acoustical silencer (d_1 = 80 mm, d_2 = 8 mm, n = 6, R = 2 m). *1)* the initial nozzle; *2)* the multi-tube nozzle.

whole sound frequency range. The maximum noise reduction occurs at frequencies exceeding or equal to the frequency of the maximum in the initial jet noise spectrum. The noise reduction effect is observed for both cold and hot jets. Maximum noise reduction at identical speeds is achieved for the ratio of the diameters given by $d_1 / d_2 \approx 10$.

8.1.3. Full-Scale Tests of Turbojet Engines. Far and Near Fields. The 12-tube silencer for reactive jets of a real engine was produced on the basis of the analysis of the model study results. The silencer general view is presented in Fig. 8.6. The gas from the mixing chamber arrived at the both main nozzle and 12 peripheral nozzles. The overall area of the outlet section was 0.87 m^2. All the nozzle cut-offs were in one plane. The main nozzle diameter for use in the multi-tube nozzle was reduced in order to conserve the outlet section area. The tests of the engine with the described outlet unit performed outdoors demonstrated the acoustic silencer performance almost identical to that in the model experiment (Fig. 8.7).

The recalculation of the results at the takeoff/landing conditions of real aircraft in the effective perceived noise (EPN) levels accepted as standard for aircraft noise in the vicinity of airports [8.9] shows that the acoustical silencer reduces noise at the reference points by the following values: 1.4 EPN dB at the side of the runway, 1.2 EPN dB at climbing, and 0.8 EPN dB at landing approach. The overall noise reduction at three reference points in locality is 3.0 – 3.5 EPN dB. The earlier experiments [8.15] demonstrated that the presence of the coflowing stream does not radically change the coherent structure parameters governing the jet response to acoustical excitation. This gives promise that the presented noise-abatement nozzle could be equally efficient at take-off/landing regimes as under stationary conditions.

It should be noted that the use of the multi-tube nozzle also reduces the pressure pulsation levels in the near sound field of the engine (Fig. 8.8). This is

Fig. 8.6. The real turbojet engine with the multi-tube silencer at the acoustical stand.

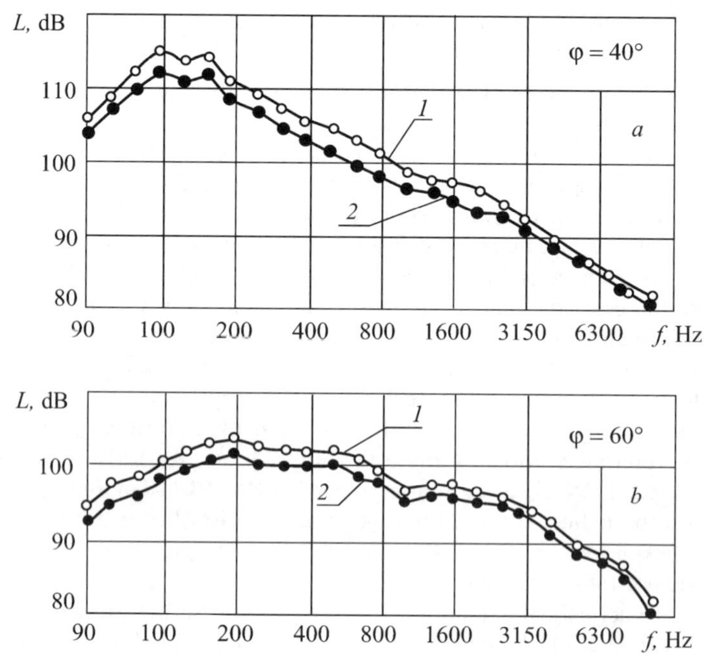

Fig. 8.7. The noise spectra in the engine far field for the nominal operating regime. *1)* the engine with the series-produced nozzle; *2)* the engine with the multi-tube nozzle ($n = 12$, $d_{1init} = 1$m, $d_1 = 0.95$ m, $d = 0.105$ m).

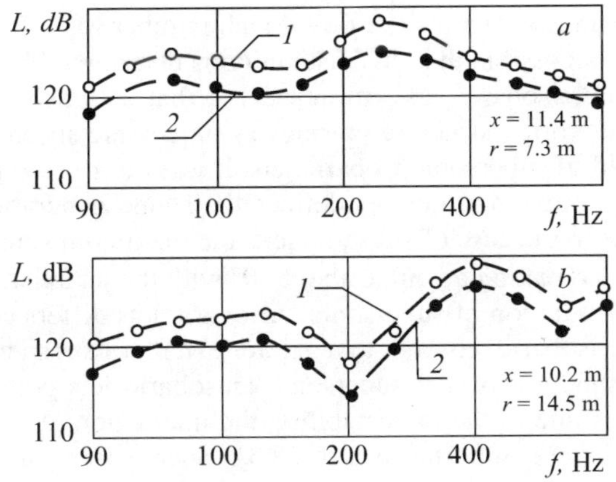

Fig. 8.8. The noise spectra in the engine near field for the nominal operating regime at the points with the coordinates (the origin is at the engine outlet section axis) (*a*) $x = 11.4$ m, $r = 7.3$ m; (*b*) $x = 10.2$ m, $r = 14.5$ m. *1)* the engine with the series-produced nozzle; *2)* the engine with the multi-tube nozzle.

possible using the presented silencer for reduction of the fuselage casing acoustical loading and, as a result, for noise reduction in the passenger cabin.

Studies on the throttle characteristics of a real engine with the multi-tube and series-produced nozzles in closed boxes and a thermal vacuum chamber show that the multi-tube silencer does not practically deteriorate the engine economical operation under stationary conditions. However, the performed calculations show that one should anticipate the increase of the thrust specific fuel consumption by 3 g/km·h in flight due to the presence of base pressure zones and some deterioration of the outer flow. Nevertheless, this increase of the thrust specific fuel consumption can be considerably lowered if the peripheral nozzles are folding.

Note, in conclusion, that the efficiency of the proposed multi-tube silencer of a reactive jet noise could evidently be raised by increase of the sound level acting on the jet. This can be obtained if the issue speed of the peripheral nozzles far exceeds that of the main nozzle.

8.2. The Jet System for Reduction of Supersonic Jet Noise. Suppression of the Discrete Component

The present section compares the noise levels for supersonic turbulent jets issuing from a single nozzle and from a nozzle system [8.18], much as it was done in Sect. 8.1 for subsonic and transonic turbulent jets.

Supersonic isobaric jets are similar in many respects to subsonic turbulent jets. For instance, parameters of large-scale coherent structures in both cases depend only slightly on the Reynolds number. At the same time the characteristic Reynolds number corresponding to coherent structures at the end of the jet initial region decreases from St = 0.5 for the flow Mach number $M_0 = 0.5$ to St = 0.15 for $M_0 = 2.5$ [8.6]. It is possible that this distinction is determined by the longer initial region of supersonic isobaric jets with respect to that of subsonic jets resulting in the increase of the vortex structure pairings at supersonic speeds. When studying the acoustical field of supersonic isobaric jets, it was discovered [8.12] that evolution of large-scale ordered structures defines the sound generation mechanism in many respects. As in the case of subsonic jets, the maximum sound levels are realized in the direction making an angle about 30° with the jet axis.

The acoustical radiation of supersonic anisobaric jets differs considerably from that of supersonic isobaric jets and, to a greater extent, of subsonic jets [8.1, 8.10]. The principal distinction is that supersonic anisobaric jets comprise shocks and new sound sources due to the jet instability; the interaction of large-scale vortices and shocks is of great importance also [8.3]. Such jets, apart from broadband noise, generate discrete components (cf. Chap. 7).

The aerodynamic noise of supersonic jets is determined by the outflow Mach number, M_0, the anisobaric degree (the unrated degree, $n = p_a / p_\infty$) and a number of other factors. For instance, the discrete tone excitation depends on the jet temperature, the presence of reflecting surfaces near the nozzle, condensed phase in the jet, the ambient humidity, etc. [8.1].

Consider another two aspects of the presented problem. First, as the peripheral jet noise at supersonic outflow speeds can act on the central jet like that is for subsonic jets, the mechanism of acoustical excitement of supersonic jets is of interest. It is shown [8.11] that there are disturbances at the irradiated side of a supersonic ($M_0 = 2$) anisobaric ($n = 0.5 - 2.0$) jet excited with the level $\left(\langle p'^2 \rangle\right)^{1/2} / p_{00} = 0.1 - 0.2\%$ of the impact pressure, p_{00}, and the supersonic jet radiates the sound at the outer excitation frequency. Any perceptible effect of the outer action frequency on the supersonic jet extension is not observed. It is important to stress that the above action of sound on a supersonic jet is observed only for irradiation of the jet near-nozzle part. Irradiation of the other parts of a supersonic jet do not influence on its structure even for very high levels of the sound pressure. Second, the sound discrete components of jets issuing from peripheral nozzles for the layout scheme with relatively large distances between adjacent peripheral nozzles [8.9, 8.17] are radiated independently [8.1].

The acoustical characteristics of a single jet and a jet system are measured in an anechoic chamber. Fig. 8.9 presents the jet unit scheme with the central nozzle (critical section diameter, d_1 = 50 mm) and six surrounding peripheral nozzles (critical section diameter, d_2 = 5 mm). The nozzles contours are almost the same as that of the Laval nozzle providing generation of a supersonic almost isobaric jet with the Mach number $M_0 = 2$ for the rated outflow regime ($n = 1$). The Reynolds number calculated from the critical section diameter, d, is equal to $Re_1 = 2.4 \cdot 10^6$. The rated degree could be changed in the limits $n = 0.5 - 1.5$ depending on the receiver pressure. The nozzle unit was mounted on the receiver with the end diameter D_1 = 250 mm.

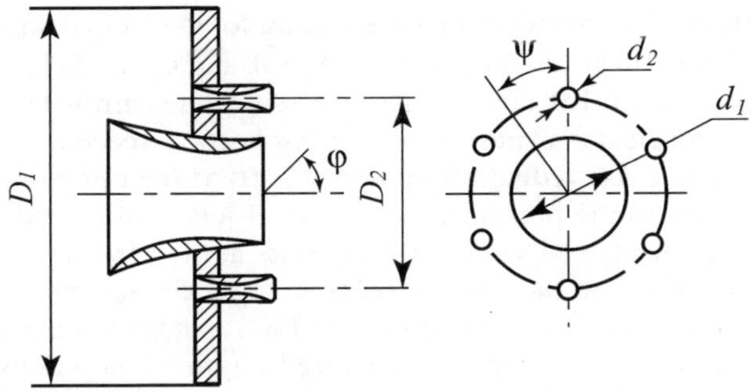

Fig. 8.9. The multi-nozzle unit scheme.

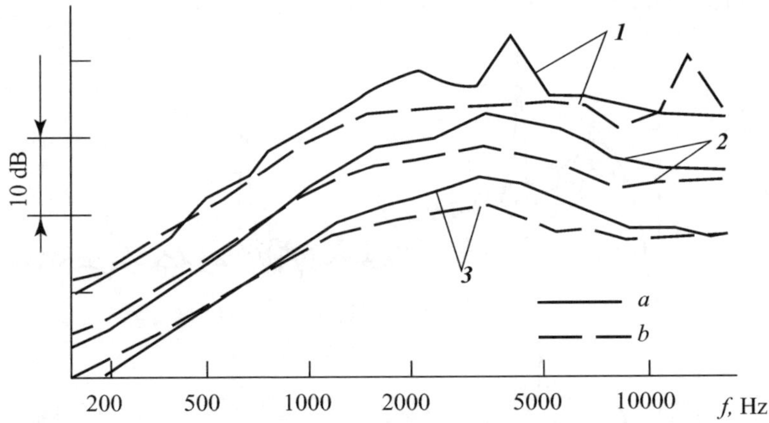

Fig. 8.10. The one-third octave noise spectra for the single jet and the jet system in the far sound field at $\varphi = 30°$ and $M_1 = 2$: (*a*) $M_2 = 0$, (*b*) $M_2 = 2$; *1)* $n = 0.75$, *2)* $n = 1.0$, *3)* $n = 1.5$.

The noise spectra in the field were measured in one-third octave and narrow ($\Delta f = 6$ and 12 Hz) frequency bands for the jet system with different numbers of peripheral jets, $N = 1$, 3 and 6. Three cases were considered: *a*) the jet without peripheral jets ($M_2 = 0$) was issued only from the central nozzle, *b*) the jets were issued from the central and N peripheral nozzles ($N = 1$, 3 and 6), and *c*) the jets were issued only from the peripheral nozzles ($M_1 = 0$).

Fig. 8.10 presents the one-third octave noise spectra for the supersonic jet far field at $\varphi = 30°$, $M_1 = 2$, $M_2 = 0$ and 2, and $n = 0.75$, 1.0 and 1.5. The comparison of the spectra for $M_1 = 2$, $M_2 = 0$ and $M_1 = 2$, $M_2 = 2$ permits evaluating the noise reduction due to peripheral jets. It may be concluded that the noise reduction due to peripheral jets at the rated outflow regime ($n = 1$) where the discrete component is not created is as much as 5 dB. The main part of the noise reduction due to peripheral jets at the unrated outflow regimes ($n \neq 1$) is determined by the discrete component suppression. However, in specific cases one could observe the discrete component shift to higher frequencies. In these cases the noise reduction may be as much as 10 dB. The narrow band noise spectra for the jet system and the central jet were measured also at $\varphi = 30°$, $M_1 = 2$, $M_2 = 0$, and $n = 0.75$-1.5. Fig. 8.11 presents them for $n = 0.75$, 1.0, and 1.5. The analysis of the narrow band spectra supports the above conclusion made for the one-third octave spectra.

A considerable noise reduction up to 5 – 7 dB at the rated outflow regimes (isobaric jets) due to the presence of peripheral jets is observed mainly in the range of low frequencies. The effect disappears at high frequencies. The influence of peripheral jets at the unrated outflow regimes (anisobaric jets) for frequencies $f < 12$ kHz decreases the broadband noise approximately by 5 dB and suppresses the discrete components exceeding the continuous noise level by 10 – 15 dB. At the same time the peripheral jet outflow is accompanied in a number of cases by generation of high-frequency discrete components exceeding the broadband noise level by 15 – 20 dB.

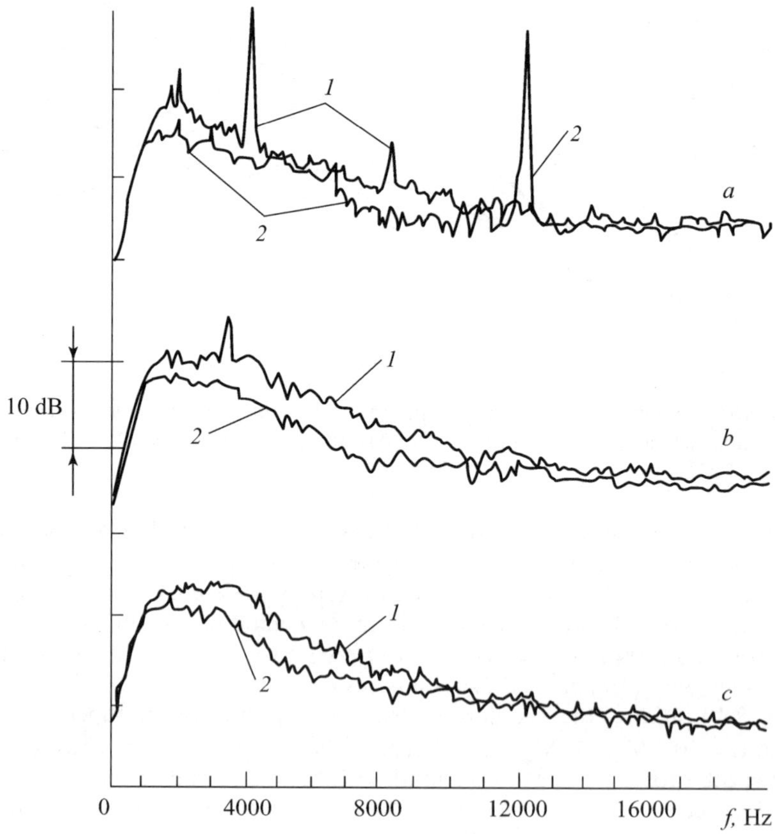

Fig. 8.11. The narrow band noise spectra for the single jet and the jet system in the far field at $\varphi = 30°$, $M_1 = 2$: (*a*) $n = 0.75$, (*b*) $n = 1.0$, (*c*) $n = 1.5$; *1)* $M_2 = 0$, *2)* $M_2 = 2$.

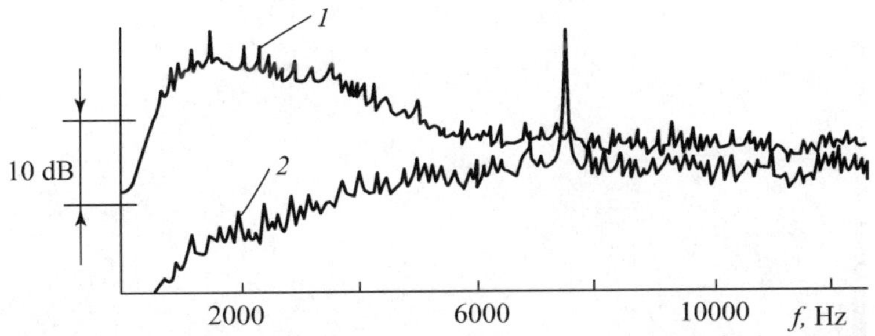

Fig. 8.12. The narrow band noise spectra in the far sound field for the jets issuing from the central nozzle (*1*) at $M_1 = 2$ and $n = 1.5$ and for the system of six peripheral jets (*2*) at $M_2 = 2$ and $n = 1.5$.

Fig. 8.12 presents characteristic noise spectra for a jet issuing from the central nozzle (*1*) and for a system comprising only six peripheral jets (*2*). The correlation between graphs in Figs. 8.11 and 8.12 makes it clear that the broadband noise of the peripheral jet system ($N = 6$) is much lower than that of the single central jet, as well as of the whole system under study. At the same time the peripheral jet system generates a discrete component at high frequencies.

It may be suggested that the peripheral jet effect reducing the noise in a supersonic jet system with respect to the noise of a single jet is determined by aerodynamic interaction of the central and peripheral jets, by their aeroacoustical interaction and/or by the screening effect of the peripheral jets. A series of experiments was performed to support this hypothesis. The flow pictures for supersonic unrated jets at $M_1 = 2$, $M_2 = 0$ and 2 and at the unrated degrees $n = 0.75$ and 1.5 were obtained by means of the flow visualization on the basis of the direct shadow method. The analysis of the photos (Fig. 8.13) demonstrates that the mixing of the central and peripheral jets (i.e. their aerodynamic interaction) starts already at distances $x / d_1 > 2 - 2.5$. This fact may change generation of low-frequency aerodynamic noise by the jet system.

Experiments for different numbers, N, of the peripheral jets (for example, $N = 3$) arranged uniformly around the central jet, as well as at one side of it, were performed also. The pressure pulsation spectra were measured at a number of angles, ψ, in a circumferential direction for the fixed angle $\varphi = 30°$ in the far sound field (cf. Fig. 8.9). It was found that the spectra differ by the broadband noise and by the discrete components only slightly. Though the noise reduction at the side of peripheral jets for specific measurements was slightly more. However, this fact gives no grounds to conclude that the screening effect of peripheral jets plays any important role. What's more, it is well known that the main source of the broadband noise in supersonic jets is located within the limits of 15 – 20 calibers (in fractions of d_1). The peripheral jets at such distances from the nozzle cut-off have time to mix with the main jet creating a complex configuration. Therefore, the screening effect cannot play any prominent role.

The question arises of whether the acoustical field of supersonic peripheral jets could have any effect on the mixing layer of the main jet near the nozzle outlet

a

b

Fig. 8.13. The photos of supersonic anisobaric jets issuing from a multi-nozzle unit. (*a*) $M_1 = M_2 = 2$, $n = 1.5$; (*b*) $M_1 = M_2 = 2$, $n = 0.75$.

section much as it is in subsonic and transonic jets. It is well known that precisely this part of the mixing layer is susceptible to the outer acoustical excitation to the greatest extent. The measurements of supersonic peripheral jet noise near the central nozzle cut-off have demonstrated that the sound pressure level for $M_2 = 2$ and $n = 0.5 - 1.5$ may be as much as $L = 145 - 160$ dB. This value corresponds to the ratio $\left(\left\langle p_s'^2 \right\rangle\right)^{1/2} / p_{00} = 0.2\%$. As previously noted, such excitation does not reduce the central jet noise independently of the frequency, at least for $n = 1$. It follows that acoustical excitation of the central jet by peripheral jets could not be the reason of the noise reduction in supersonic unrated jets.

In summary, we may conclude the following. First, the effect of the noise reduction in supersonic isobaric and anisobaric turbulent jets with a multi-tube nozzle system (the central nozzle of diameter d_1 surrounded by uniformly arranged little nozzles of diameter $d_2 \approx 0.1d_1$) in place of the single circular nozzle is discovered. It is shown that this causes reduction of the broadband noise and low-frequency discrete components. Second, the above noise reduction at supersonic speeds is mainly determined by the aerodynamic interaction effect, i.e. the mixing of these peripheral jets and the central jet. Therefore, the noise reduction mechanisms when using a multi-nozzle scheme of the jet outflow organization considerably differ from each other at subsonic and supersonic speeds. The aeroacoustical interaction effect is dominant in the first case and the aerodynamic interaction effect in the second case.

8.3. Reduction of Turbojet Excess Noise Caused by Aeroacoustical Interaction

Chap. 3 is devoted to the acoustical method of noise control in subsonic turbulent jets. Two main effects are considered: the noise broadband amplification of a jet under low-frequency tone acoustical excitation ($\mathrm{St}_s = 0.2 - 0.8$) and the noise broadband attenuation of a jet under high-frequency acoustical excitation ($\mathrm{St}_s = 1.5 - 5$). These effects are accompanied correspondingly by the broadband amplification and attenuation of turbulent pulsations in the jet (cf. Chap. 2). The effects are realized for low excitation levels exceeding a certain threshold $\left(\left\langle u_s'^2\right\rangle\right)^{1/2}/u_0 = 0.01 - 0.02$. A large body of experimental data illustrates realization of these effects in cold and hot subsonic (Chap. 3) and supersonic (Chap. 7) jets. There are considered the case where the tone forcing low-frequency excitation is missed in the far acoustical field and the case where the excitation is itself broadband. The directional pattern for noise radiated by a jet under low-frequency excitation differs little from that for a noise radiated by an excitation-free jet but the corresponding noise pressure levels at $\mathrm{St}_s = 0.2 - 0.8$ are increased for the excited jet by 6 – 8 dB.

As the broadband enhancement of the far field pulsation spectrum depends nonlinearly on the forcing excitation, the jet amplified noise intensity does not follow the eighth power law (u_0^8) that is valid for the pure jet noise at thoroughly controlled outflow initial conditions (low excitation levels) – instead of the law of u_0^8 the amplified jet noise follows the law of u_0^6. The broadband amplification of turbulence and noise is often realized in experimental studies of jets at modeling set-ups of different kinds for a broad range of Mach, Reynolds and Strouhal numbers and temperatures.

The problem of "excess" noise of turbulent jets is of practical interest, as applied to reactive jets of aircraft engines [8.2, 8.4, 8.6, 8.13]. Indeed, jet noise of some engines exceeds the rated values by 6 – 8 dB, as the inner noise sources (the compressor, combustion chamber, turbine, etc.) excite a reactive jet resulting in broadband noise increase.

To illustrate this phenomenon, consider the noise one-third octave spectra of the turbojet engine "*Olimpus 593*" in the far field at $\varphi = 70°$ and $90°$ for an unexcited jet (calculation) and for the engine as a whole (Fig. 8.14). Appreciable distinctions of the jet and engine spectra could be explained, among other factors, by the excess noise of an excited jet. Fig. 8.15 demonstrates the decrease by 2 – 6 dB in the excess noise of the engine "*Olimpus 593*" by means of suppression (using the nozzle noise-absorbing facing) of the noise propagating along the path and exciting the jet.

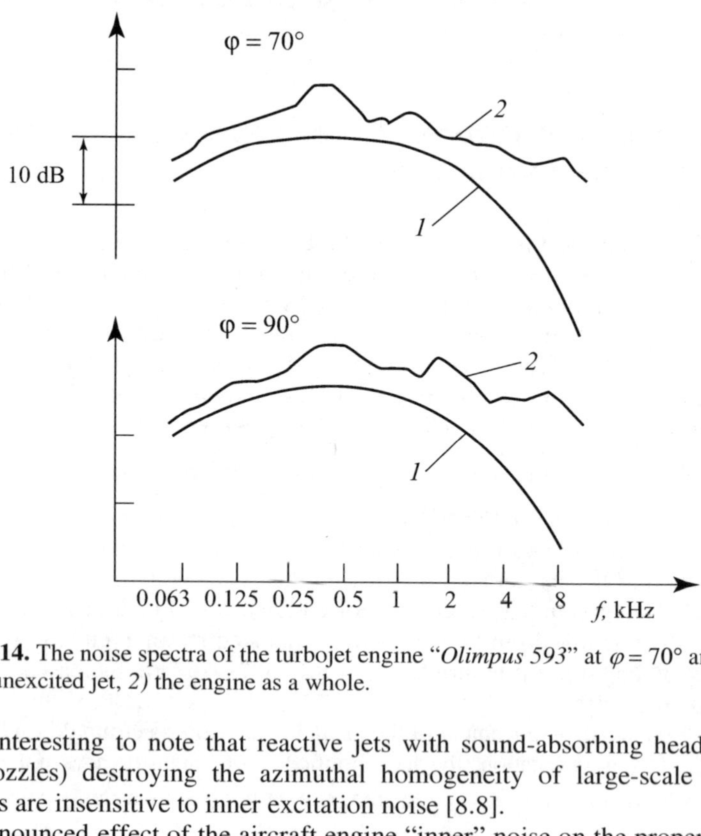

Fig. 8.14. The noise spectra of the turbojet engine "*Olimpus 593*" at $\varphi = 70°$ and $90°$: *1)* an unexcited jet, *2)* the engine as a whole.

It is interesting to note that reactive jets with sound-absorbing heads (multilobed nozzles) destroying the azimuthal homogeneity of large-scale coherent structures are insensitive to inner excitation noise [8.8].

A pronounced effect of the aircraft engine "inner" noise on the proper noise of a jet is discovered on the complex study at models of reactive nozzles and at a real turbojet engine [8.20]. It was found that the engine noise for $\pi_c > 2.6$ is

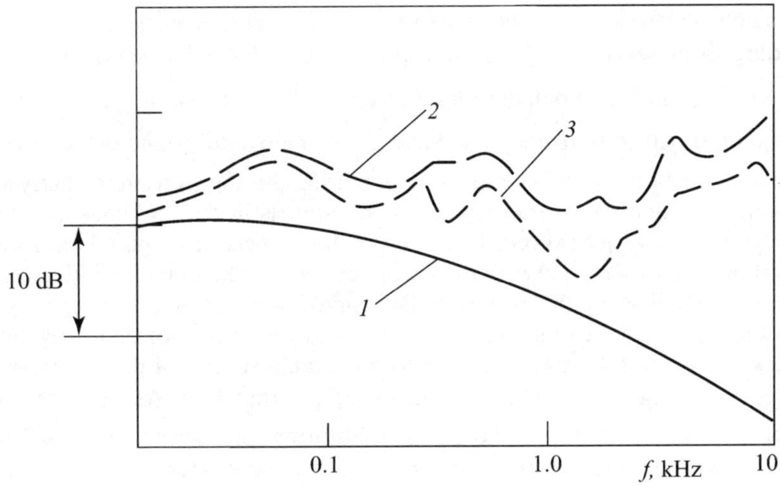

Fig. 8.15. The decrease in the broadband noise of the turbojet engine "*Olimpus 593*" by means of mounting of the noise-absorbing facing of the path: *1)* the rated noise spectrum of an unexcited jet, *2)* the noise spectrum of the engine without the facing, *3)* the noise spectrum of the engine with the facing.

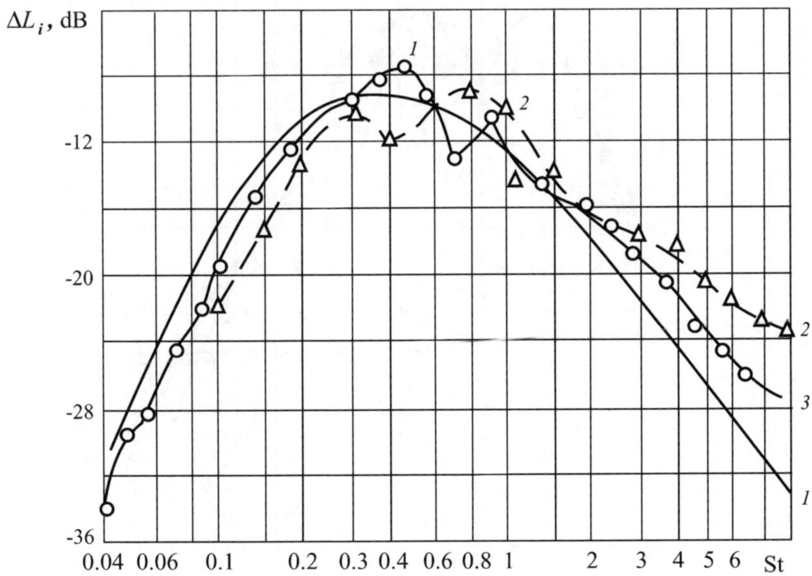

Fig. 8.16. The functions $\Delta L_i = f(\mathrm{St})$ at $\pi_c = 2.6$ and $\varphi = 135°$ for the "pure" jet, the turbojet engine and the nozzle model: *1)* the "pure" jet, *2)* the turbojet engine, *3)* the nozzle with the stabilizer.

completely determined by the reactive jet noise. The acoustical spectra corresponding to these values of π_c were presented as dependencies of the value of $\Delta L_i = L_i - L_\Sigma$ on the Strouhal number St where L_i is the sound pressure level in the one-third octave frequency band and L_Σ is the overall sound pressure level at the given direction φ to the engine axis measuring the angles from the entry side.

It turns out that the overall sound pressure levels in the far acoustical field of the engine reactive jet exceed the corresponding levels of a "pure" jet. Essential distinctions are observed also in the jet noise spectra. The initial disturbances (turbulence and noise) at the nozzle outlet approaching the corresponding disturbances at the turbojet engine outlet have been gained at the nozzle model with the flame stabilizer in its subsonic part. This moved the spectra of the model and real jets together (Fig. 8.16). The comparison of the functions $\Delta L_i = f(\mathrm{St})$ for the model and real silencers of the jet also confirms the validity of these results. Fig. 8.17 presents the functions $\Delta L_i = f(\mathrm{St})$ for the model silencer at a nozzle with a stabilizer and for the real silencer; the curves are almost identical. The same function for the model silencer at a nozzle without stabilizers does not correlate with the corresponding function for the real engine.

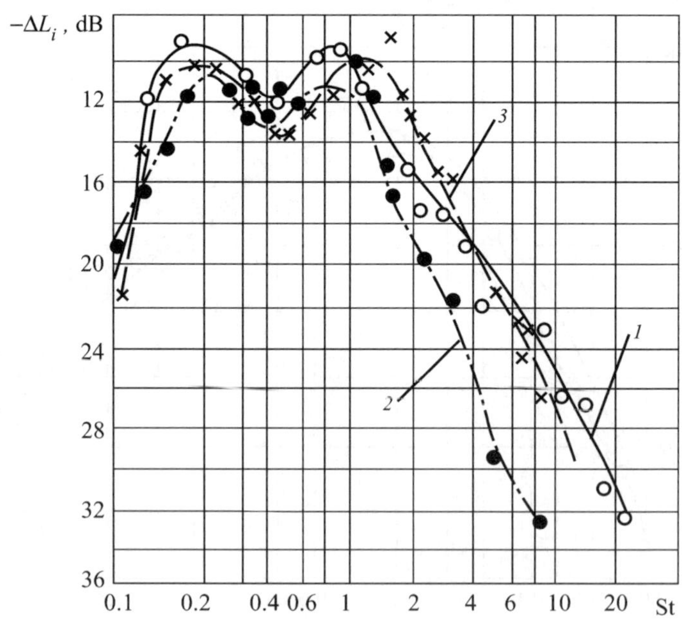

Fig. 8.17. The functions $\Delta L_i = f(\mathrm{St})$ at π_c = 2.6 and φ = 140° for the model and real silencers: *1)* the engine, *2)* the nozzle model without stabilizers, *3)* the nozzle model with the stabilizer.

REFERENCES

8.1. Antonov AN, Kuptsov VM, Komarov VV (1990) Pressure pulsations in separated and jet flows (in Russian). Mashinostroyenie, Moscow

8.2. Crighton DG (1981) Acoustics as branch of fluid mechanics. J Fluid Mech 106: 261 – 298

8.3. Davis MR (1988) Identification of vortex motion in turbulent mixing choked jets. Exp Fluids # 5: 335 – 343

8.4. Deneuville P, Jacques J (1977) Jet noise amplification: a practically important problem. AIAA Pap 1368

8.5. Ginevsky AS, Vlasov YeV, Kolesnikov AV (1978) Aeroacoustical interactions (in Russian). Mashinostroyenie, Moscow

8.6. Gliebe PR, Brauch JF, Majjigi RK, Lee R (1994) Jet noise suppression. In: Aeroacoustics of Flight Vehicles. Theory and practice. V 2: Noise control. NASA, pp 207 – 269

8.7. Lepicovsky J, Ahuja KK, Brown WH, Burrin RH (1987) Coherent large-scale structures in high Reynolds number supersonic jets. AIAA J 25, 3 11: 1419 – 1425

8.8. Moore CJ, Brierly DH (1979) Shear layer instability noise produced by carious jet nozzle configurations. In: Mech Sound Gener Flows. Joint Symp. Berlin, pp 48 – 54

8.9. Munin AG (ed) (1986) Aviation acoustics (in Russian) V 1. Mashinostroyenie, Moscow

8.10. Munin AG, Kuznetsov VM, Leontiev YeA (1981) Aerodynamic sources of noise. Mashinostroyenie, Moscow

8.11. Pimshtein VG (1989) Effect of sound on supersonic jet. Fluid Dynamics, # 6: 969 – 971

8.12. Troutt TR, Melaughlin DK (1982) Experiments on the flow and acoustic properties of a moderate-Reynolds-number supersonic jet. J Fluid Mech 116: 123 – 156

8.13. Vlasov YeV, Ginevsky AS (1980) Problem of aeroacoustical interactions. Soviet Physics Acoustics 26, # 1: 1 – 7

8.14. Vlasov YeV, Ginevsky AS (1986) Coherent structures in turbulent jets and wakes (in Russian). In: Advances in Science and Technology. Series "Mechanics of Fluids and Gases" 20, VINITI, Moscow, pp 3 – 84

8.15. Vlasov YeV, Ginevsky AS, Karavosov RK (1979) Coherent structures in coflowing turbulent jets (in Russian). In: Turbulent jet flows. Part 1. Tallinn, pp 20 – 27

8.16. Vlasov YeV, Ginevsky AS, Zaguzov IS, Karavosov RK (1991) Acoustical silencer of turbojet engine noise (in Russian). Trans TsIAM 1287: 76 – 83

8.17. Vlasov YeV, Ginevsky AS, Pimshtein VG (1992) Control of coherent structures and aero-acoustic characteristics of subsonic and supersonic turbulent jets. In: DGLR/AIAA 14th Aeroacoustic Conference. FRG, pp 672 – 678

8.18. Vlasov YeV, Ginevsky AS, Pimshtein VG (1993) Noise attenuation in supersonic turbulent jets. Acoustical Physics 39, # 5: 425 – 427

8.19. Vlasov YeV, Ginevsky AS, Karavosov RK (2000) Studies on attenuation of turbojet engine noise (in Russian). In: Polyot 2: 25 – 32
8.20. Zhvalov VP, Zhukov YeN, Sorkin LI, Tolstosheev MN (1980) Outer noise of turbojet engines. Soviet Physics Acoustics 26, # 1: 29 – 32

CHAPTER 9 ACOUSTICAL APPROACHES TO CONTROL OF SELF-SUSTAINED OSCILLATIONS IN WIND TUNNELS WITH THE OPEN TEST SECTION

9.1. The Problem Statement and Measured Parameters

The present chapter is devoted to the study of acoustical methods of suppression and generation of self-sustained oscillations in wind tunnels with the open test section based on sensitivity of the jet coherent structures to periodical excitation [9.1, 9.6 – 9.8]. The latter are attenuated under high-frequency excitation where the Strouhal number $\mathrm{St}_s = f_s d / u_0 = 2 - 5$ and are amplified under low-frequency excitation with $\mathrm{St}_s = 0.3 - 0.8$. The most sensitive jet area to periodical excitation is the thin mixing layer in the immediate vicinity of the nozzle cut-off. Vortex disturbances are generated under acoustical excitation just there determining amplification or attenuation of coherent quasi-periodical structures.

The experiments were performed in four wind tunnels with various diameters of the nozzle outlet section, d = 0.15, 0.44, 1.2, and 2.2 m. The wind tunnels had units for self-oscillation suppression that could not completely suppress the self-oscillations. There were measured the mean velocity and pressure fields, the velocity and pressure pulsations, their spectra, as well as the spatial correlations of the velocity and pressure pulsations. The resonance frequencies of the return channel were determined in the flow-free wind tunnel using a loudspeaker connecting with a generator of sinusoidal oscillations and a microphone mounted in the reverse channel.

The wind tunnels with the nozzle diameters 0.44 and 2.2 m have geometrically homothetic units for self-oscillation damping: the ring funnel in front of the diffuser entrance and three rows of holes in the diffuser walls at the entrance with the overall area 30% of the diffuser entrance area. The parameters of these units are chosen in accordance with the recommendations in [9.1]. The wind tunnel with the nozzle diameter 0.15 m has only the ring funnel to suppress self-oscillations. Finally, the wind tunnel with the nozzle diameter 1.2 m has the ring funnel around the diffuser entrance and the annular slot with the controlled width from 0 to 0.09 m at the diffuser entrance, as well as two rows of holes in the diffuser walls.

Two kinds of acoustical excitation are studied. In the first case the sound radiator is located in the wind tunnel reverse channel. The excitation frequency was adjusted to coincide with one of the resonance frequencies of the reverse channel. This provides the rise in the acoustical oscillation amplitude. The known drawback of this kind of excitation is generation of pressure pulsations in the wind tunnel test section [9.1, 9.3, 9.4, 9.8]. In the second case the acoustical disturbances are inserted into the nozzle boundary layer through a narrow slot resulting in the periodical injection/suction [9.6]. This method of excitation has two important advantages with respect to the first case. First, when controlling coherent structures in the mixing layer, only a thin boundary layer near the nozzle cut-off is excited instead of the whole volume of the wind tunnel reverse channel and the flow core in the test section. Second, as the narrow slot is a low-efficient sound radiator, one may expect not to have any pressure pulsations of perceptible magnitude.

9.2. Suppression of Self-Oscillations under High-Frequency Acoustical Excitation of the Mixing Layer

9.2.1. Return Channel Loudspeacer Location. Fig. 9.1 presents the intensity of the velocity longitudinal pulsations as a function of the flow speed at the test section axis point $x/d = 1$ of four wind tunnels with diameters $d = 0.15$, 0.44, 1.2, and 2.2 m with the electrodynamic radiator in the return channel. The Strouhal number of acoustical excitation with the frequency f_s is $St_s = 2 - 5$. From Fig. 9.1 it follows that such excitation practically causes the complete suppression of self-oscillations. More detailed data are presented in Fig. 9.2 for the wind tunnel with the nozzle diameter $d = 1.2$ m. It should be noted that the radial pulsations remain practically identical in the presence of self-oscillations and without them.

The initial boundary layer in four wind tunnels under study was laminar for the small wind tunnel (the shape parameter $H = 2.3 - 2.5$) and transition or turbulent for large wind tunnels ($H = 1.3 - 1.4$). The velocity profiles in the boundary layer of the nozzle outlet section did not practically change under acoustical excitation; the changes in the longitudinal velocity pulsation profiles were also insignificant.

The electric power consumed by the sound radiator for this kind of excitation is always less than 0.1% of the wind tunnel drive power. The radiator acoustical power is still less, as the efficiency of the used loudspeakers is 2 – 4%.

The study of statistical characteristics of the flow turbulence in the wind tunnel test section has shown that there are abrupt decreases of the velocity pulsations at self-oscillation regimes (2 – 12 times) and its slight decrease at the other regimes under high-frequency acoustical excitation. This phenomenon is accompanied by considerable increase of noise in the test section – in the jet near field (to 12 dB).

The measurements of the spatial correlation coefficients for the velocity pulsations, $R_{uu}(r)$, in the test section point $x/d = 1$ have shown that a plane hydrodynamic wave is propagating during the process of self-oscillation setting-up in the test section flow core, i.e. R_{uu} is close to 1. This correlation coefficient decreases

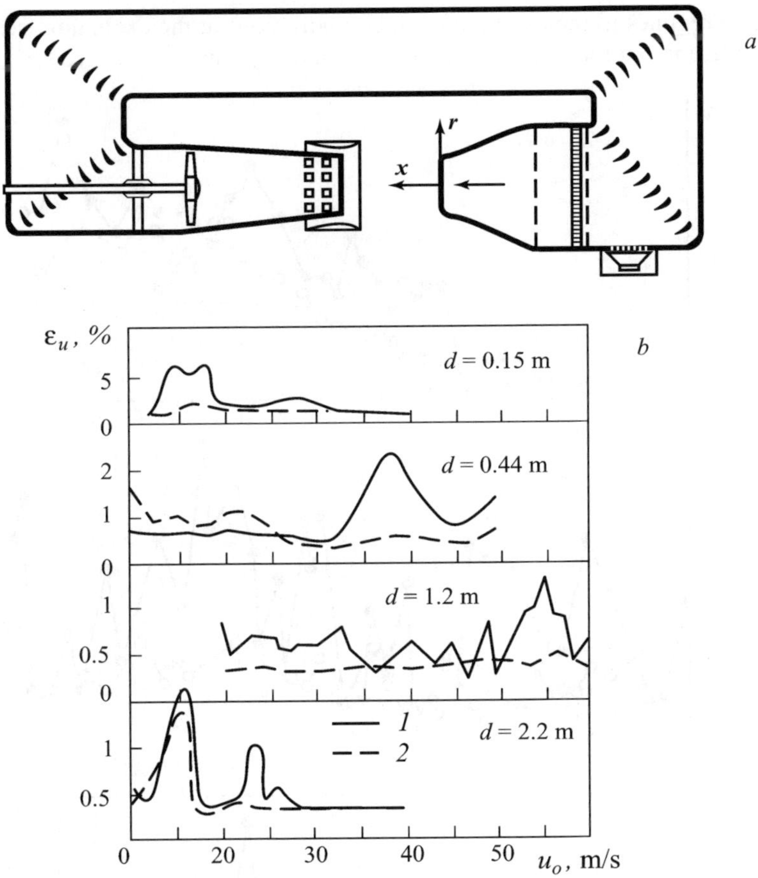

Fig. 9.1. (*a*) The wind tunnel of the closed type with the open test section. (*b*) The intensities of the velocity longitudinal pulsations, ε_u, as functions of the flow speed in wind tunnels (d = var) in the presence of acoustical excitation and without it: *1)* $\mathrm{St}_s = 0$, *2)* $\mathrm{St}_s = 2 - 5$.

at self-oscillation suppression and increases at its generation, as is shown in Fig. 9.3. This figure shows also the spatial correlation coefficients for the velocity pulsations at the wind tunnel axis in the test section and for the pressure pulsations in the return channel, $R_{up}(x/d)$ for the self-oscillation regime (at $u_0 = 16$ m/s) and for acoustical suppression of self-oscillations at the frequency $\mathrm{St}_s = 2.87$.

Fig. 9.4 presents spectra of the velocity and pressure pulsations in the wind tunnel test section with the nozzle diameter $d = 0.44$ m for the speed $u_0 = 40$ m/s at the point $x/d = 1$, $r/d = 0$ in the presence of high-frequency excitation (St_s = 3.3 and 4.26) and without it. We see how the self-oscillation suppression changes the spectra. At the same time the pressure pulsation spectrum at the excitation frequency has a considerable peak. The contribution of the flow excitation for self-excitation regimes to the excitation velocity pulsations at the excitation frequency is insignificant, but it is considerable for the other regimes.

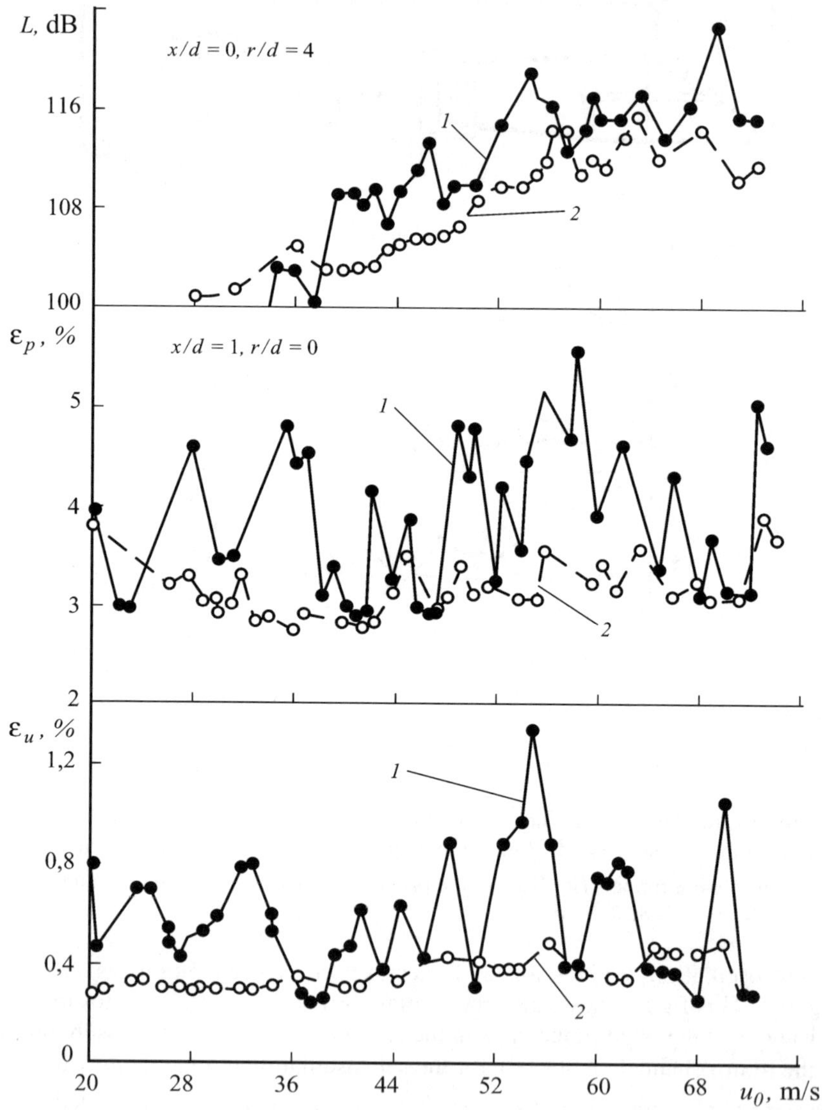

Fig. 9.2. The sound pressure level outside of the jet, the pressure and velocity pulsations at the jet axis (x/d = 1) as functions of the speed in the wind tunnel test section with the nozzle diameter d = 1.2 m under acoustical excitation with the frequency corresponding to $St_s = 2 - 5$: *1)* $St_s = 0$, *2)* $St_s = 2 - 5$.

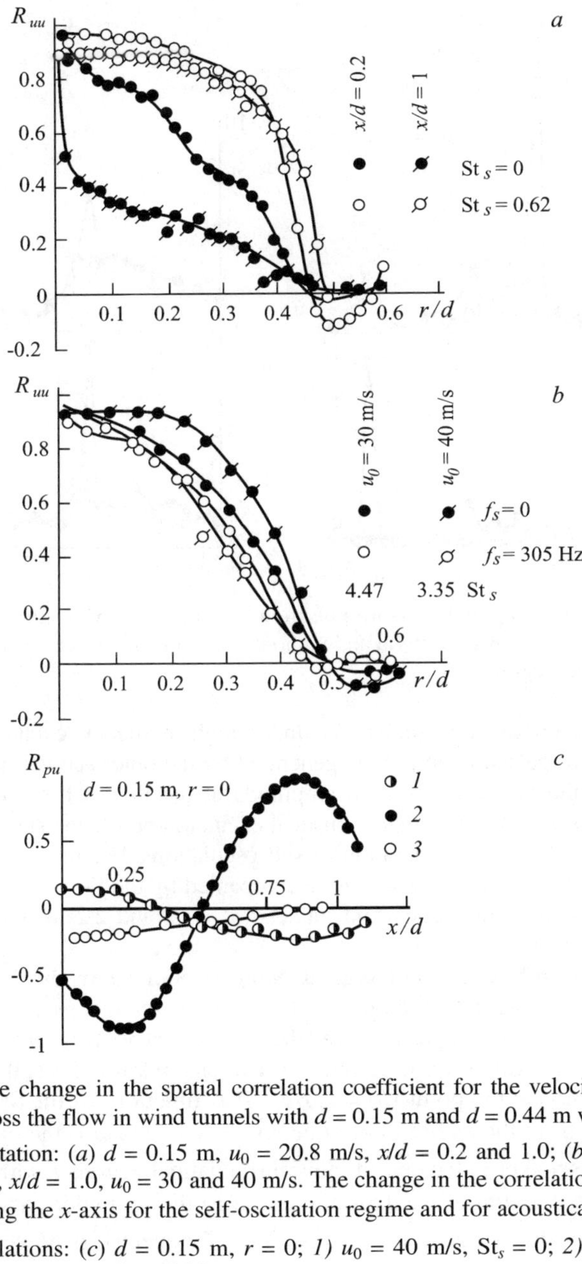

Fig. 9.3. The change in the spatial correlation coefficient for the velocity pulsations, $R_{uu}(r)$, across the flow in wind tunnels with $d = 0.15$ m and $d = 0.44$ m with and without self-excitation: (*a*) d = 0.15 m, u_0 = 20.8 m/s, x/d = 0.2 and 1.0; (*b*) d = 0.15 m, u_0 = 6.0 m/s, x/d = 1.0, u_0 = 30 and 40 m/s. The change in the correlation coefficient, $R_{up}(r)$, along the x-axis for the self-oscillation regime and for acoustical suppression of self-oscillations: (*c*) d = 0.15 m, r = 0; *1)* u_0 = 40 m/s, St_s = 0; *2)* u_0 = 16 m/s, St_s = 0; *3)* u_0 = 16 m/s, St_s = 2.87.

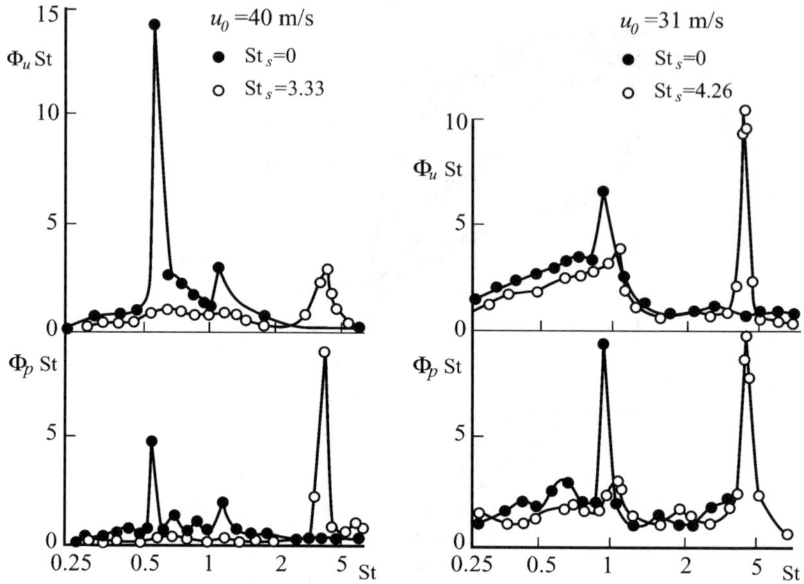

Fig. 9.4. The velocity and pressure pulsation spectra in the wind tunnel test section (d = 0.44 m, x/d = 1, r/d = 0) with and without excitation at the self-excitation regime and at the other regimes.

The suppression of self-oscillations under high-frequency excitation is realized once the pressure pulsation amplitude generated by the outer acoustical field becomes comparable to the pressure pulsation amplitude at the flow self-oscillations. Thus it makes sense to switch on the outer acoustical excitation before the steady regime coming and, better still, before generation of self-oscillations. For instance, the acoustical excitation is switched on at low speeds, accompanied by the flow speed increase. This method has been used for large wind tunnels (d = 1.2 and 2.2 m) when the radiator power is insufficient.

9.2.2. Injection/Suction Through a Narrow Slot Near the Nozzle Cut-off. Now consider the second method of jet excitation by means of the flow injection/suction through a narrow slot near the nozzle cut-off (Fig. 9.5, *a*). Fig. 9.5, *b* presents the longitudinal velocity pulsation intensities for the self-oscillation regime (u_0 = 15 m/s) at the point x/d = 1, r/d = 0 as functions of the Strouhal number St_s calculated from the excitation frequency, f_s. There is also the data for the case where the loudspeaker is located in the return channel. Curve *1* corresponds to the initial value of ε_u without excitation, curve *2* to the jet under excitation by a radiator placed in the wind tunnel prechamber (the electrical power is 5 W), and curves *3* and *4* to the periodical injection/suction through a slot in the nozzle wall (the electrical power is 50 W and 3 W correspondingly). One should pay attention to the different nature of the functions $\varepsilon_u\left(St_s\right)$ corresponding to curves *2*, *3* and *4*.

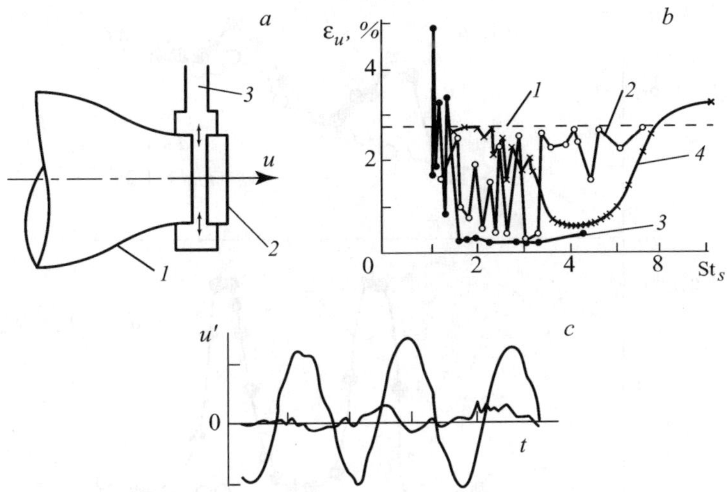

Fig. 9.5. (*a*) The slot nozzle scheme: *1)* the nozzle, *2)* the periodical injection/suction, *3)* the waveguide from the loudspeaker; (*b*) Excitation of the nozzle boundary layer by periodical injection/suction ($d = 0.15$ m). The function $\varepsilon_u(\mathrm{St}_s)$ at the axis point $x/d = 1$ for the self-oscillation regime: *1)* the self-oscillation regime ($u_0 = 15$ m/s), *2)* the loudspeaker in the reverse channel ($N = 5$ W), *3)* the periodical injection/suction through the slot ($N = 50$ W), *4)* the same for $N = 3$ W; (*c*) The velocity pulsation oscillograms.

The high-frequency excitation for curve *2* is of resonance nature and for curves *3* and *4* this is not the case: the function $\varepsilon_u(\mathrm{St}_s)$ is smooth. Fig. 9.5, *c* presents the longitudinal velocity pulsation oscillograms at the point $x/d = 1$, $r/d = 0$ in the presence of self-oscillations and without them.

The flow longitudinal velocity pulsation intensities, u', in the wind tunnel test section in the presence of high-frequency excitation ($f_s = 358$ Hz) and without it are presented in Fig. 9.6. The figure comprises also the functions $L(u_0)$ at the point $x/d = 0$, $r/d = 1.3$ outside the test section in the presence of acoustical excitation and without it.

The effect of the wind tunnel test section encumbering on efficiency of the flow sound excitation causing self-oscillation suppression was studied also. For this purpose, cylinders of different diameters and lengths were placed crosswise in the test section of two wind tunnels with the nozzle outlet section diameters $d = 0.14$ m and 0.44 m. It was found that the flow velocity corresponding to generation of self-oscillations and their intensity slightly change with the increase of the encumbering degree from 1.5% to 6.5%. At the same time, high-frequency acoustical excitation suppresses self-oscillations for the above degrees of encumbering.

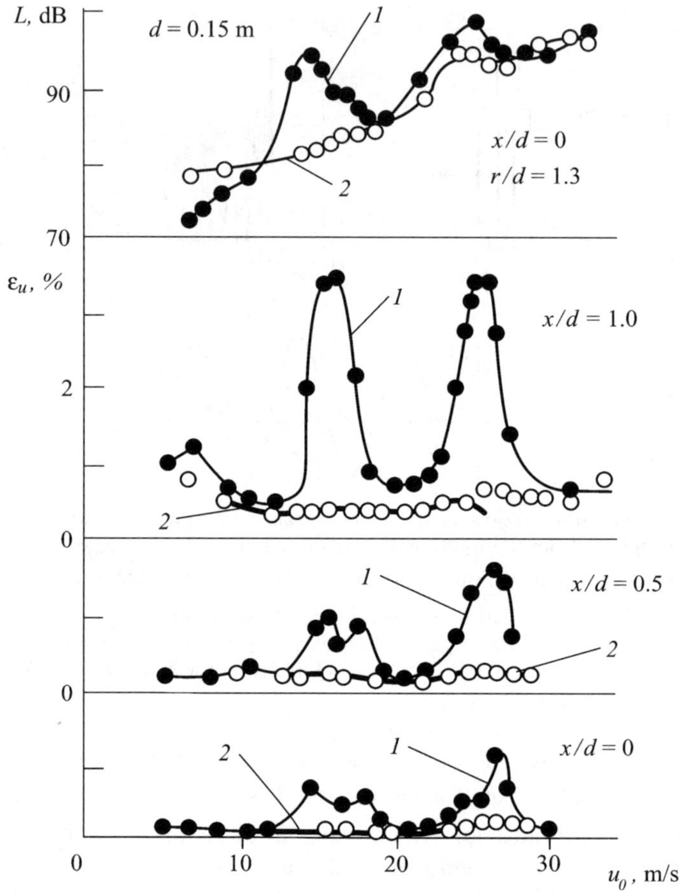

Fig. 9.6. The functions $\varepsilon_u(u_0)$ at x/d = 0, 0.5, and 1.0 and $L(u_0)$ outside the test section in the presence of periodical injection/suction through a slot: *1)* $f_s = 0$, *2)* f_s = 358 Hz (St_s = 2.15 – 3.58 at speeds u_0 = 25 – 15 m/s)

9.3. Generation of Self-Oscillations and Creation of Homogeneous Pulsing Flow in the Wind Tunnel Test Section under Low-Frequency Excitation of the Mixing Layer

Excitation in the wind tunnel test section of the velocity longitudinal pulsations sinusoidal in time could be achieved for jets under low-frequency acoustical excitation at frequencies corresponding to Strouhal numbers St_s = 0.3 – 0.8. Such an approach to self-oscillation generation is applied only to small wind tunnels with the nozzle outlet diameter $d < 0.5$ m for lack of low-frequency radiators. Fig. 9.7 presents the initial and maximum root mean square values of the longitudinal velocity pulsations excited by noise for $\mathrm{St}_s = 0.3 - 0.8$ at the wind tunnel test section axis ($x/d = 1$). Both curves *1* and *2* define the range of possible levels for the longitudinal velocity pulsations that may be achieved under self-oscillation generation. The self-oscillation generation allows the longitudinal pulsation intensity to be increased several times. The excited longitudinal velocity pulsations are almost pure sinusoidal in time and their amplitude is determined by the sound radiator power.

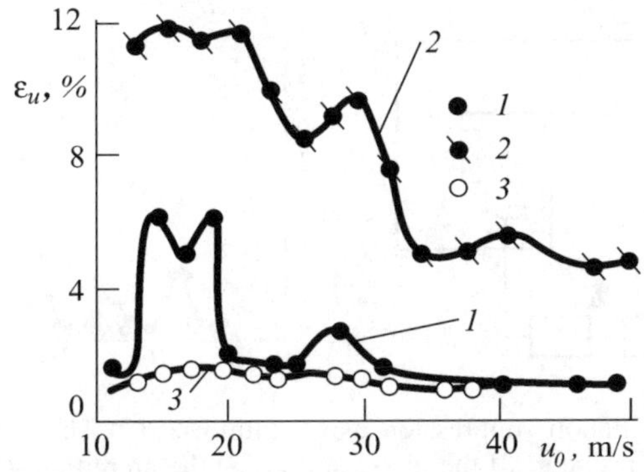

Fig. 9.7. Excitation of self-oscillations in a wind tunnel with the goal of generation of the flow sinusoidal in time ($d = 0.15$ m, $x/d = 1$, $r/d = 0$). The functions $\varepsilon_u(u_0)$ without excitation, as well as under low- and high-frequency excitation: *1)* $\mathrm{St}_s = 0$, *2)* $\mathrm{St}_s = 0.3 - 0.8$, *3* – $\mathrm{St}_s = 2 - 5$.

In conclusion special attention must be given to the effect of acoustical excitation on wind tunnel aerodynamic characteristics. In the case of low-frequency excitation ($\mathrm{St}_s = 0.3 - 0.8$) the track drag decreases and, correspondingly, the flow velocity in the wind tunnel test section increases. Conversely, in the case of high-frequency excitation ($\mathrm{St}_s = 2 - 5$) the track drag slightly increases and the flow velocity in the wind tunnel test section decreases.

9.4. Suppression of Self-Sustained Oscillations Using Antinoise

Worthy of mention is also the so-called resonance method of suppression of the acoustical field in the reverse channel by means of excitation of a compensating field in it using a supplementary sound source [9.2, 9.5]. The experiment scheme is presented in Fig. 9.8, *a*. The transducer (*1*) transforms acoustical oscillations in the wind tunnel return channel (*5*) into an electrical signal arriving through the phase shifter (*2*) and the amplifier (*3*) at the radiator (*4*) and exciting acoustical oscillations with the given phase shift in the reverse channel. The experiments were performed in the wind tunnel with the open test section having the nozzle and diffuser of the square section with the side 0.3 m. Fig. 9.8, *b* presents the velocity pulsation level at the nozzle cut-off as a function of the flow speed without self-oscillation suppression (2) and with the optimal self-oscillation suppression (*1*) that was attained through adjustment of the optimal phase shift. The authors of [9.5] note that the electrical power consumed by the radiator at the regime of complete self-oscillation suppression does not exceed 1 W for the flow speed 2 – 15 m/s in the test section.

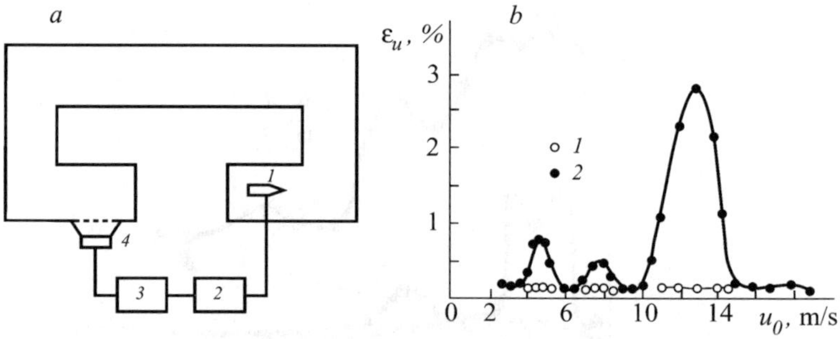

Fig. 9.8. Self-oscillation suppression using antinoise. (*a*) The experiment scheme: *1)* the pressure transducer, *2)* the phase shifter, *3)* the amplifier, *4)* the loudspeaker, *5)* the reverse channel. (*b*) The function $\varepsilon_u(u_0)$ in the wind tunnel (*2*) without self-oscillation suppression and (*1*) with suppression.

Two peculiarities of the experiment are noteworthy [9.2]. First, as the authors of the paper stated, the nozzle outlet section and the diffuser inlet section were not conformed to each other, i.e. they had the identical area and the diffuser edge was not rounded. Second, the mixing layer excitation was accomplished for this control method at the frequency of the jet large-scale structures which is essentially less than the characteristic frequency of pulsations in the thin mixing layer near the cut-off. Therefore, the presented method is valid only for small wind tunnels, as it was in the case of [9.5].

REFERENCES

9.1. Boyarchikova MYu, Vlasov YeV, Ginevsky AS, Zosimov AV (1982) Suppression of self-excited oscillations in wind tunnels with the open test section (in Russian). Trans Acad Sci USSR, Fluid and Gas Dynamics, # 1: 126 – 132

9.2. Ffowcs Williams JE (1984) Anti-sound. Proc Roy Soc London A395: 63 – 88

9.3. Ginevsky AS (1993) Self-oscillation flow control in wind tunnels. In: 1st Eur Nonlinear Oscillations Conf, Hamburg, p 43

9.4. Ginevsky AS (1997) Self-oscillation flow control in free-jet wind tunnels. In: Aerodynamics of Wind Tunnel Curcuits and their Components, AGARD, Moscow, pp 27-1 – 27-12
9.5. Gromov PR, Zobnin AB, Kiyashko SV, Sushchik MM (1988) Self-excited oscillations in wind tunnels with the open test section. Soviet Physics Acoustics 34, # 2: 206 – 207
9.6. Moskvina MYu (1990) Suppression of auto-oscillations in wind tunnels with an open test section due to periodic injection-suction into the boundary layer of a nozzle. J Eng Phys & Termophysics 59, # 1: 115 – 119
9.7. Vlasov YeV, Ginevsky AS (1986) Coherent structures in turbulent jets and wakes (in Russian). In: Advances in Science and Technology. Series "Mechanics of Fluids and Gases" 20, VINITI, Moscow, pp 3 – 84
9.8. Zosimov AV (1986) An acoustical approach to control of velocity and pressure pulsations in wind tunnels with the open test section (in Russian). Proc TsAGI, # 2292

CHAPTER 10 INTERACTION OF A MIXING LAYER WITH A CAVITY

10.1. Separated Flow over a Cavity

The flow over a cavity on a plane surface generates under certain conditions self-sustained oscillations. Their cause is generation of an acoustical wave as a result of impact of the mixing layer vortices on the cavity trailing edge. This phenomenon is illustrated by comparison of the velocity pulsations in the mixing layer for two cases: flows over a ledge facing backwards and over a cavity of rectangular section (Fig. 10.1). The pulsation spectrum in the second case contains pronounced discrete components due to acoustical feedback with excitation of self-oscillations [10.3]. A cavity could serve for some frequencies as an acoustical resonator effecting on the nature of self-oscillation excitation [10.8]. As a result, the self-oscillation characteristics are determined by the cavity geometry, the Reynolds and Mach numbers, the flow regime (laminar or turbulent) in the boundary layer in front of the cavity, and the characteristic thickness of this layer.

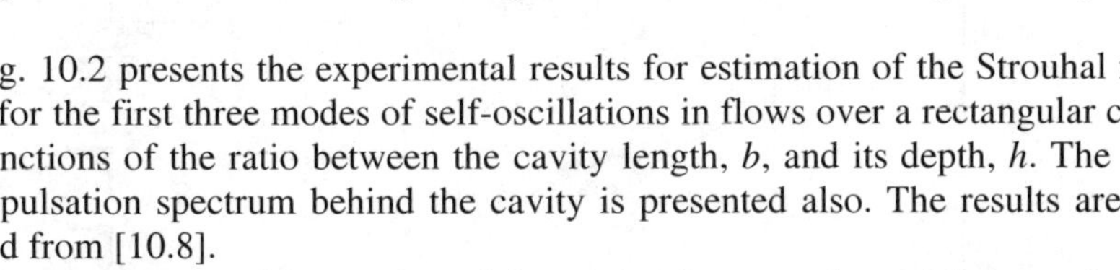

Fig. 10.1. The flows over a ledge facing backwards and over a rectangular cavity

Fig. 10.2 presents the experimental results for estimation of the Strouhal numbers for the first three modes of self-oscillations in flows over a rectangular cavity as functions of the ratio between the cavity length, b, and its depth, h. The pressure pulsation spectrum behind the cavity is presented also. The results are borrowed from [10.8].

The mechanisms of interaction of the mixing layer vortices with the cavity rectangular edge are of great interest. The mixing layer interaction (at the characteristic frequency of organized structures, β) with the edge for $\mathrm{Re}_{\theta} = u\theta_0 / \nu = 190$ generates either pulsations of one-half frequency (0.5β) for $b/\theta_0 < 90$ or pulsations of frequencies 0.4β and 0.6β for $b/\theta_0 > 90$ where θ is the momentum loss thickness in the mixing layer at the cut-out leading edge. It was found that such different effects arise from different kinds of interaction between the discrete vortex structure and the cut-out trailing edge: in some cases the vortex passes over the edge without disruption, in other cases it is subdivided into decomposition parts carrying away on either side of the edge, and in the rest the whole vortex is deflected into the cavity in front of the edge (Fig. 10.3). The presence of different interaction kinds determines its periodical changing [10.7].

A number of measures intended for the decrease of the excited self-oscillation amplitude are developed: the leading and/or trailing edges are cut off or rounded, a perforated plate or grid is placed over the cavity, etc. (cf. the review [10.7]).

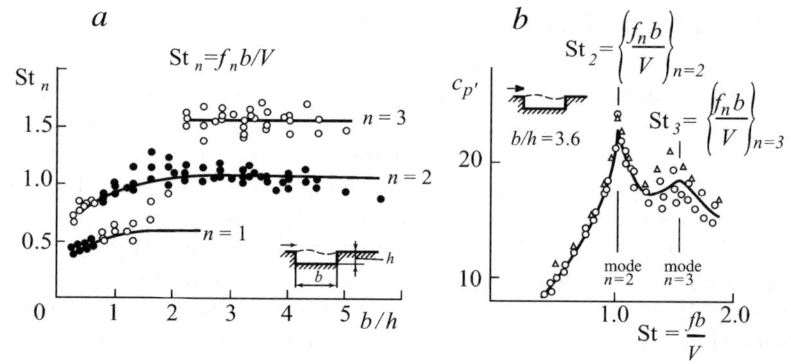

Fig. 10.2. (*a*) The Strouhal number as a function of *b*/*h*. (*b*)The pressure pulsation spectrum behind the cavity.

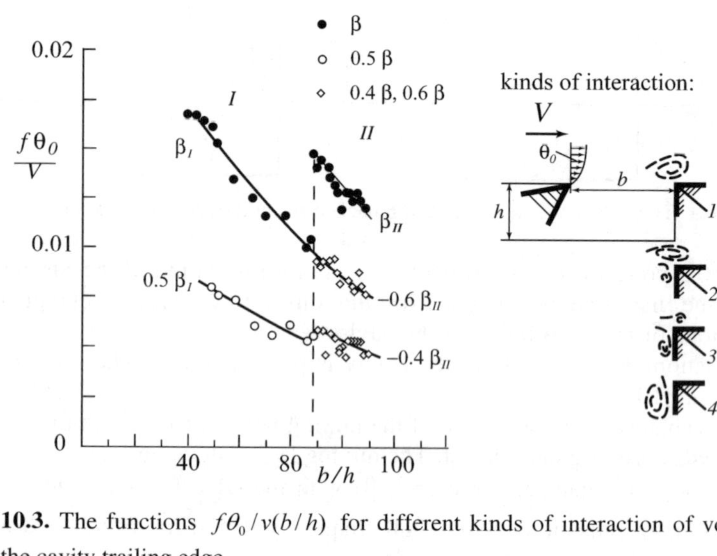

Fig. 10.3. The functions $f\theta_0/v(b/h)$ for different kinds of interaction of vortices with the cavity trailing edge.

10.2. Near-Wall Pressure Pulsations in a Cavity Flow and Approaches to Their Reduction

It is well known that separated flows over open cavities of rectangular section are accompanied by pressure pulsations. There could occur under certain conditions resonance radiations excited by airflow in the cavity. These radiations may result in the structure disruption.

The results are presented below of an experimental study on aerodynamic and acoustical characteristics of rectangular cavities on a plane surface in the presence of units intended for suppression of pressure pulsations on the cavity surface [10.6]. The studies were performed for the plate 1×1 m of thickness 0.05 m with profiled nose and stern parts and with side plates of height 0.15 m. The initial version of the rectangular cut-out had width 400 mm and cross-section 60×60 mm. The cut-out depth could vary by repositioning the removable bottom. The tests were performed in wind tunnels with the open test section of diameter 2.2 m on the plate set at the zero angle of attack. A wire turbulator of diameter 0.6 mm for turbulization of the boundary layer in front of the cut-out was fixed on the plate surface near its nose.

The boundary layer thickness in front of the cutout was $\delta = 9$ mm at the flow speed $V_\infty = 40$ m/s and the flow turbulence degree was $\varepsilon = 0.35\%$. The ratio between the cut-out width and depth is $b/h = 0.15 - 1.5$ for the used cut-out depths $h = 6 - 60$ mm. The Reynolds number is Re = $V_\infty b/\nu = 10^4$. The static pressure distribution on the cut-out surface was determined using the drain holes arranged over the central section of the cut-out on the side walls, on the bottom, and on the surface behind the cut-out. The pressure pulsations were measured at the near-cut-out section using condenser microphones flush-mounted on the surface.

Fig. 10.4 presents the versions of geometry of the studied specimens. The effect of the cavity depth on the flow nature could be traced using thc experimental results in Fig. 10.5 – 10.10 where the distributions of the averaged pressure coefficients, $c_p = (p_i - p_\infty)/\frac{1}{2}\rho V_\infty^2$, and pressure pulsations, $\varepsilon_p' = \left(\left\langle p'^2 \right\rangle\right)^{1/2} / \frac{1}{2}\rho V_\infty^2$ are shown for every version 1-6. As one would expect, the maximum pressure pulsations are observed near the cut-out trailing edge.

Consider now characteristics of units (cf. Fig. 10.4) intended for suppression of the pressure pulsations on the cut-out walls. The following cases were studied: flaps placed horizontally and overriding the cut-out leading and trailing parts (Version *1*); the same flaps deflected upwards near the cut-out leading edge and downwards near the trailing edge at the angle 15° to the horizon (Version *2*); the flaps placed near the cut-out leading edge in upright (Version *3*) and inclined position (Version *5*); and the cut-outs with the trailing edge cut off at the angle 45° (Versions *4* and *6*).

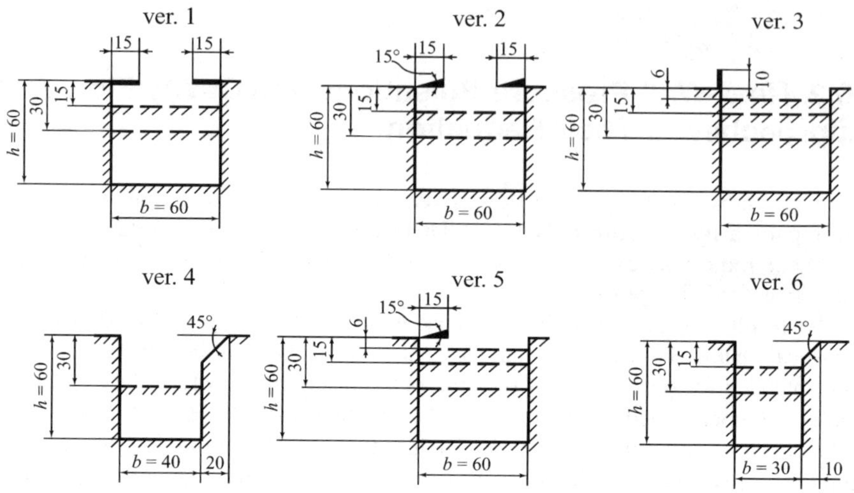

Fig. 10.4. The versions of the studied configurations of the cavity.

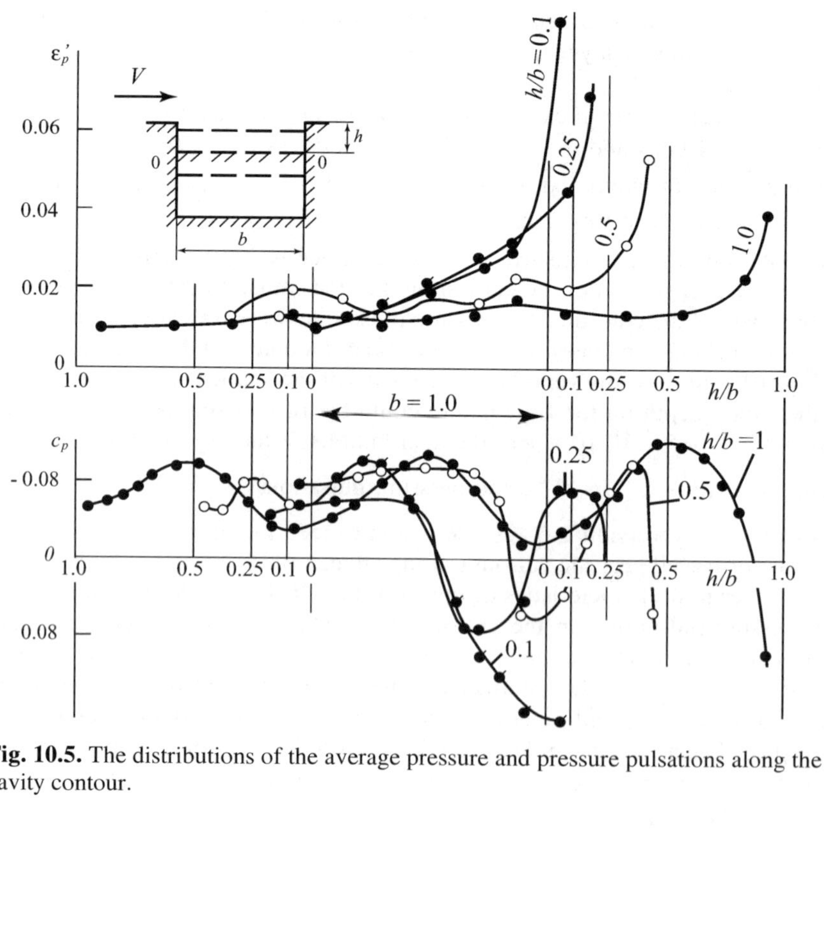

Fig. 10.5. The distributions of the average pressure and pressure pulsations along the cavity contour.

Figs. 10.6 – 10.10 present distributions of pressure pulsations and averaged pressure on the surface of cut-outs in the presence of the outer units intended for suppression of pulsations in the cut-outs. Their correlation with distributions of pressure pulsations and averaged pressure for the initial versions of cut-outs without outer units (Fig. 10.5) demonstrates that the use of the studied units could cause a considerable change in the above distributions. In parallel with the redistribution of the averaged pressure field there also have been considerable changes in the pressure pulsation distributions. The averaged pressure redistribution could occur due to the change in the flow parameters in the cut-out, in particular, as the flow connection area displaces, as well as due to the change in the resonance characteristics of the cut-out plane.

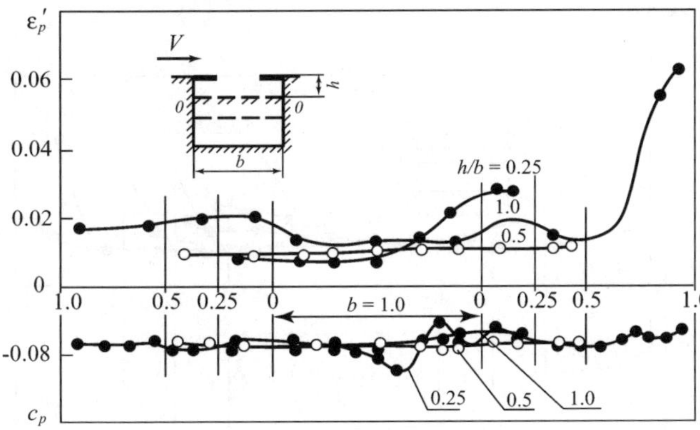

Fig. 10.6. Version 1.

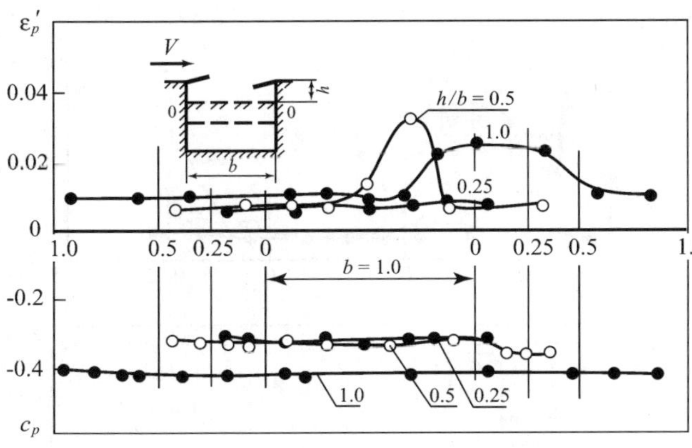

Fig. 10.7. Version 2.

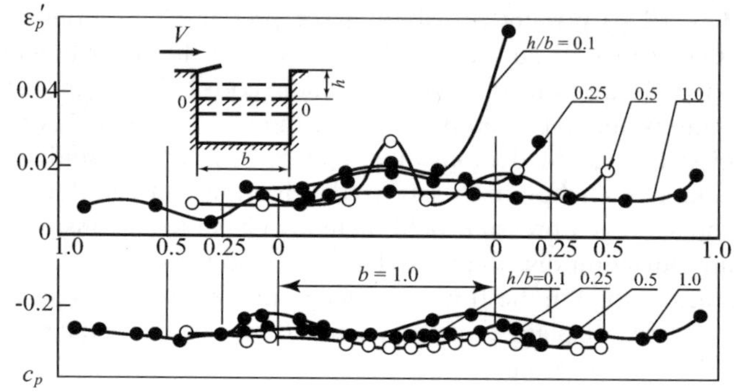

Fig. 10.8. Version 5.

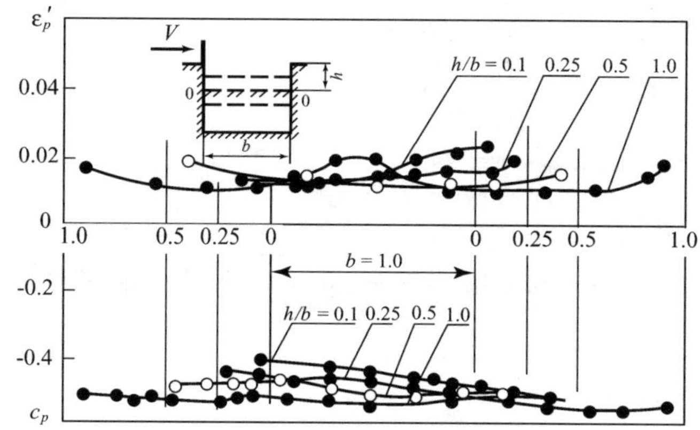

Fig. 10.9. Version 3.

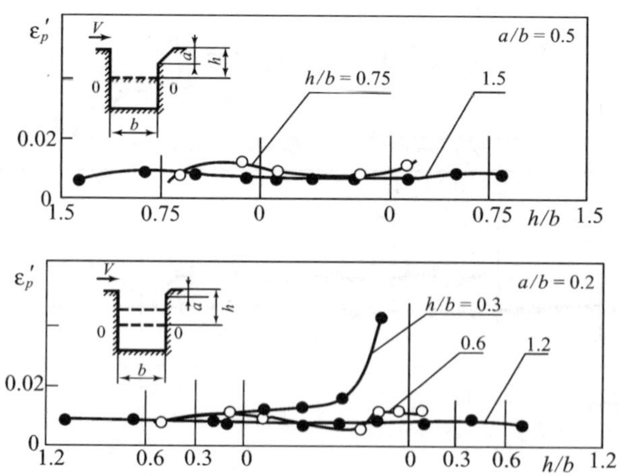

Fig. 10.10. Versions 4 and 6.

The pressure pulsation spectra inside and outside the cut-out are generally broadband: the pressure pulsations override the whole range of sound frequencies. By way of example Fig. 10.11 presents the pressure pulsation spectra measured in the flow over the initial rectangular cut-out. There occur discrete components in the pressure pulsation spectra inside the cut-out in some cases. Generally, the area with such the discrete components is not large and the discrete components in the pressure pulsation spectra on the surface outside the cut-out are not observed. However, upon installation of horizontal flaps (Fig. 10.6) on the cut-out of depth $h/b = 1$ the intensive resonance oscillations determining the overall pressure pulsation levels are observed (Fig. 10.12) everywhere in the measurement area including the area behind the cut-out. The approach to calculation of velocity, pressure and pressure pulsation fields in the flows over rectangular cavities on a plane surface is described in [10.5].

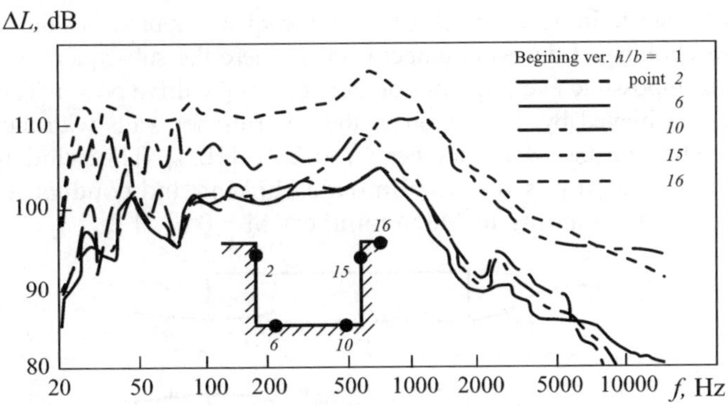

Fig. 10.11. The pressure pulsation spectra on the cavity walls.

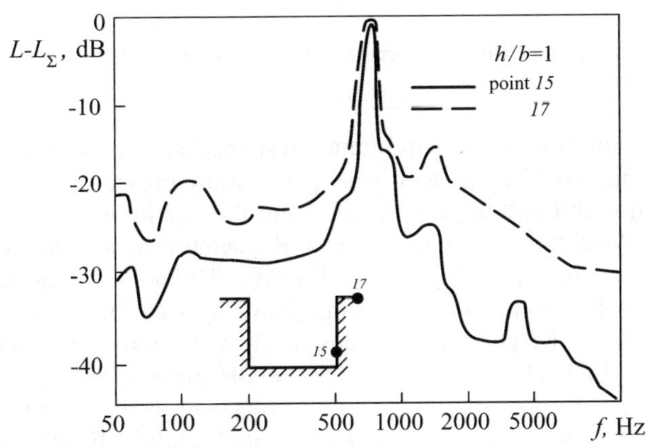

Fig. 10.12. The discrete peaks in the spectra.

The detailed experimental data on the pressure pulsations, their spectra, the overall level and discrete components in rectangular cavities at subsonic and supersonic flow speeds depending on the cavity geometrical parameters are contained in [10.1]. This book contains also results of studies on the influence of the outer flow Mach number on the characteristic Strouhal numbers determined from the discrete component frequencies and on the maximum levels of these components.

10.3. Pressure Pulsations in Transonic Wind Tunnels with the Closed Test Sections and Approaches to Their Reduction

When the flow speed in wind tunnels of transonic speeds approaches the sound speed, the so-called choking of the wind tunnel occurs where the subsequent increase in the flow speed is impossible even for considerable rise in the drive power. The clearing of this trouble is achieved by perforation of the wind tunnel walls with the perforation degree depending on the Mach number up to 20% [10.4]. The wind tunnel wall permeability makes it possible to eliminate choking and conduct tests at transonic speeds corresponding to Mach numbers M = 0.6 – 1.3.

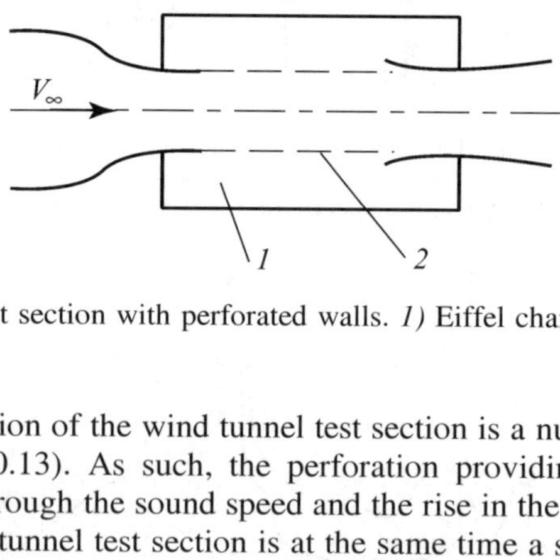

Fig. 10.13. The test section with perforated walls. *1)* Eiffel chambers, *2)* perforated walls.

The wall perforation of the wind tunnel test section is a number of rows of circular holes (Fig. 10.13). As such, the perforation providing the possibility of smooth transition through the sound speed and the rise in the velocity field homogeneity in the wind tunnel test section is at the same time a source of undesirable background noise distorting experimental results. The point is that each of the perforation holes behaves like a cavity being flown by gas. The contribution of self-oscillations induced by perforation holes in the wind tunnel test section into the overall level of the background noise is especially great at M = 0.6 – 0.8. The unsteady vortex generation at the perforation holes and excitation of resonance frequencies in the system 'test section – perforated walls – Eiffel chamber' could prove to be deciding in the pressure pulsation levels and spectra of the wind tunnel test section [10.2]. The known approaches to decrease the background noise in wind tunnels with the perforated test section are based on the disruption of the self-oscillation excitation mechanism by means of the holes. This is achieved by installation of a metal screen on the test section perforated walls [10.2] or of longitudinal lintels in each of the perforation holes.

Paper [10.2] contains descriptions of experiments with subsequent stopping holes of side and horizontal perforation of the wind tunnel test section. Elimination of the side perforation causes complete disappearance of the intense noise discrete components but retains the broadband maximum. The hole stopping in the side panels causes elimination of the broadband maximum but the intense discrete components do not change. It is shown that installation of the netted panels with simultaneous elimination of the side perforation causes elimination of the discrete components and the broadband maximum with decrease of the overall pulsation level in the whole frequency range [10.2]. This phenomenon is illustrated by the pressure pulsation spectra in the wind tunnel test section at M = 0.8 measured in narrow frequency bands (Δf = 12.5 Hz) and in one-third octave frequency bands (Fig. 10.14).

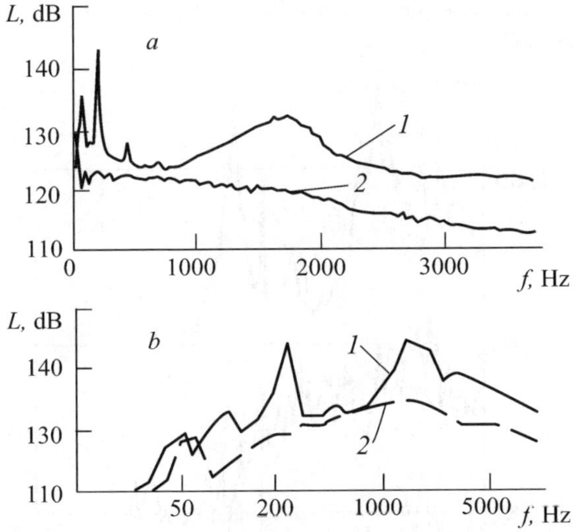

Fig. 10.14. Decrease of pressure pulsations in the test section of a transonic wind tunnel at M = 0.8. (*a*) the analysis in narrow frequency bands Δf = 12.5 Hz, (*b*) the analysis in one-third octave frequency bands, *1)* the standard test section, *2)* the updated test section (installation of netted panels and elimination of the side perforation).

10.4. Suppression of Self-Sustained Oscillations in Deadlock Branches of Gas Pipelines

The pressure pulsations of resonance nature (that are a real threat to the construction integrity) are perpetually observed while studying gas-compressor stations [10.9]. The gas flow speeds in wind tunnels of diameter 1 m may be as great as 11 – 17 m/s and the frequencies of narrow band pressure pulsations were 8 – 10 Hz in some cases and 175 – 180 Hz in other cases depending on the resonator sizes. The excitation source of these intensive oscillations was in the area of junction between the tube-header and the deadlock branch.

The flow turn in the path with a deadlock cavity creates a self-oscillation system determined by acoustical feedback. Suppression of such self-oscillations is performed decreasing the scales of periodical vortex structures causing attenuation of acoustical feedback.

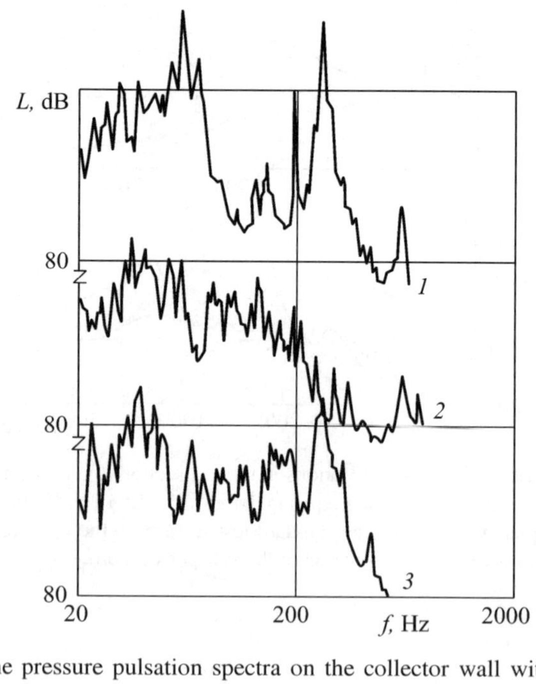

Fig. 10.15. The pressure pulsation spectra on the collector wall with the deadlock branch for $d_0 / d_k = 0.733$. *1)* $t/d_0 = 4.6$, *2)* $t/d_0 = 2.57$, *3)* $t/d_0 = 1.19$.

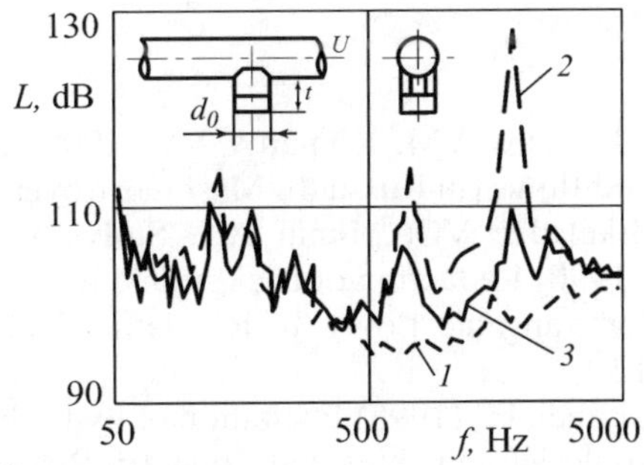

Fig. 10.16. The pressure pulsation spectra *L*(*f*) in a tube-collector (*1*) without a branch, (*2*) with the deadlock branch, and (*3*) with the deadlock branch and five longitudinal ribs.

Fig. 10.15 presents the pressure pulsation spectra on the collector wall with the deadlock branch for the ratio of their diameters $d_0 / d_k = 0.733$ and for three values of the deadlock branch depths $t / d_0 = 4.6, 2.57$, and 1.19 [10.10].

Fig. 10.16 presents the pressure pulsation spectra in a tube without a branch (*1*), in a tube with the deadlock branch (2) of depth $t / d_0 = 2$, and in a tube with the same branch and five installed longitudinal ribs (*3*) [10.9]. The described approach to the prevention of self-oscillations may be thought as applicable to real gas pipelines considering low hydraulic drag of the ribs.

10.5. Acoustical Control of Flows in Cavities

Above are considered the approaches to attenuation of acoustical feedback for flows in rectangular cross-section cavities based on the changes in the cavity geometry. It should be noted that a similar effect could be achieved by acoustical excitation of the cavity [10.8], as well as by periodical or acoustical excitation of the flow near the cavity leading edge. It is demonstrated, in particular, by studies of periodical excitation of the flow near a facing backward step of height *H* at the abrupt broadening of the plane channel [10.11].

It is shown that the mixing layer thickens at Strouhal numbers $\mathrm{St} = fH/V = 0.05 - 0.5$, causing faster attachment of the separated flow. It is believed that such an effect would be useful also for flows over cavities of rectangular cross-section. In that case, contrary to a channel with abrupt broadening, the excitation effect could be amplified due to the acoustical feedback.

REFERENCES

10.1. Antonov AN, Kuptsov VM, Komarov VV (1990) Pressure pulsations in jet and separated flows (in Russian). Mashinostroyenie, Moscow

10.2. Ereza AG, Mikeladze VG, Munin AG, Stolyarov YeP, Filippova RD, Shliagun AN (1990) Reduction of background noise in the test section of a wind tunnel for transonic flows (in Russian). Uchionyie Zapiski TsAGI 21, # 2: 10 – 12

10.3. Ginevsky AS, Landa PS (1995) Excitation of hydrodynamic and acoustical waves in subsonic jet and separated flows (in Russian). Izvestiya Vuzov. Appl Nonlin Dyn 3, # 2: 42 – 59

10.4. Grodzovsky GL, Nikolsky AA, Svishchev GP, Taganov GI (1967) Supersonic flows in perforated boundaries (in Russian). Mashinostroyenie, Moscow

10.5. Kanevsky MI, Pogrebnaya TV (2001) Modeling of turbulent flows over 2- and 3-D cavities on a surface (in Russian). Proc TsAGI 2643

10.6. Kanevsky MI, Karavosov RK, Frankfurt MO (1998) Calculating and experimental study on peculiarities of flows over 2-D cavity on a surface (in Russian). TsAGI, Preprint # 118

10.7. Rockwell D, Knisely C (1980) Vortex-edge interaction: Mechanisms for generating low frequency components. Phys Fluids 23, # 2: 239 – 240

10.8. Rockwell D, Naudasher E (1978) Self-sustaining oscillations of flow past cavities (Review). Trans ASME J Fluids Eng 100, # 2: 152 – 165

10.9. Vishnjakov VA, Zasetskii VG, Karavosov RK, Prozorov AG, Sokolinskii LI (1998) Generation and damping of flow oscillations in the region of a joint between a pipe-line and a deadlock branch piece. J Eng Phys & Thermophys 71, # 6: 1064 – 1071

10.10. Vishnjakov VA, Zasetskii VG, Karavosov RK, Prozorov AG, Sokolinskii LI (1998) Initiation of intense pressure fluctuations when a flow is turned in a duct with a dead end cavity. Fluid Dynamics 33, #2: 238 – 244

10.11. Yoshioka S, Masukawa T, Obi S, Masuda S (1999) Turbulent statistics of backward-facing step flow with periodic perturbations. Trans Jap Soc Mech Eng B 65, # 629: 179 – 184

Foundations of Engineering Mechanics

Series Editors: Vladimir I. Babitsky, Loughborough University
Jens Wittenburg, Karlsruhe University

Palmov	Vibrations of Elasto-Plastic Bodies (1998, ISBN 3-540-63724-9)
Babitsky	Theory of Vibro-Impact Systems and Applications (1998, ISBN 3-540-63723-0)
Skrzypek/ Ganczarski	Modeling of Material Damage and Failure of Structures Theory and Applications (1999, ISBN 3-540-63725-7)
Kovaleva	Optimal Control of Mechanical Oscillations (1999, ISBN 3-540-65442-9)
Kolovsky	Nonlinear Dynamics of Active and Passive Systems of Vibration Protection (1999, ISBN 3-540-65661-8)
Guz	Fundamentals of the Three-Dimensional Theory of Stability of Deformable Bodies (1999, ISBN 3-540-63721-4)
Alfutov	Stability of Elastic Structures (2000, ISBN 3-540-65700-2)
Morozov/ Petrov	Dynamics of Fracture (2000, ISBN 3-540-64274-9)
Astashev/ Babitsky/ Kolovsky	Dynamics and Control of Machines (2000, ISBN 3-540-63722-2)
Svetlitsky	Statics of Rods (2000, ISBN 3-540-67452-7)
Kolovsky/ Evgrafov/ Slousch/ Semenov	Advanced Theory of Mechanisms and Machines (2000, ISBN 3-540-67168-4)
Landa	Regular and Chaotic Oscillations (2001, ISBN 3-540-41001-5)

Foundations of Engineering Mechanics

Series Editors: Vladimir I. Babitsky, Loughborough University
Jens Wittenburg, Karlsruhe University

Muravskii	Mechanics of Non-Homogeneous and Anisotropic Foundations (2001, ISBN 3-540-41631-5)
Gorshkov/ Tarlakovsky	Transient Aerohydroelasticity of Spherical Bodies (2001, ISBN 3-540-42151-3)
Babitsky/ Krupenin	Vibration of Strongly Nonlinear Discontinuous Systems (2001, ISBN 3-540-41447-9
Manevitch/ Andrianov/ Oshmyan	Mechanics of Periodically Heterogeneous Structures (2002, ISBN 3-540-41630-7)
Lurie	Analytical Mechanics (2002, ISBN 3-540-42982-4)
Slepyan	Models and Phenomena in Fracture Mechanics (2002, ISBN 3-540-43767-3)
Nagaev	Dynamics of Synchronising Systems (2003, ISBN 3-540-44195-6)
Svetlitsky	Statistical Dynamics and Reliability Theory for Mechanical Structures (2003, ISBN 3-540-44297-9)
Neimark	Mathematical Models in Natural Science and Engineering (2003, ISBN 3-540-43680-4)
Babitsky/ Shipilov	Resonant Robotic Systems (2003, ISBN 3-540-00334-7)
Le xuan Anh	Dynamics of Mechanical Systems with Coulomb Friction (2003, ISBN 3-540-00654-0)
Perelmuter/ Slivker	Numerical Structural Analysis (2003, ISBN 3-540-00628-1)

Foundations of Engineering Mechanics

Series Editors: Vladimir I. Babitsky, Loughborough University
Jens Wittenburg, Karlsruhe University

Andrianov/ Awrejcewicz/ Manevitch	Asymptotical Mechanics of Thin-Walled Structures (2004, ISBN 3-540-40876-2)
Shorr	The Wave Finite Element Method (2004, ISBN 3-540-41638-2)
Ginevsky Vlasov / Karavosov	Acoustic Control of Turbulent Jets (2004, ISBN 3-540-20143-2)
Svetlitsky	Advanced Theory of Vibrations 1 (2004, ISBN 3-540-20658-2)

Printing: Saladruck, Berlin
Binding: Stein+Lehmann, Berlin